Complete Scattering Experiments

PHYSICS OF ATOMS AND MOLECULES

Series Editors

P. G. Burke, *The Queen's University of Belfast, Northern Ireland*
H. Kleinpoppen, *Atomic Physics Laboratory, University of Stirling, Scotland*

Editorial Advisory Board

R. B. Bernstein (*New York, U.S.A.*)
J. C. Cohen-Tannoudji (*Paris, France*)
R. W. Crompton (*Canberra, Australia*)
Y. N. Demkov (*St. Petersburg, Russia*)
C. J. Joachain (*Brussels, Belgium*)
W. E. Lamb, Jr. (*Tucson, U.S.A.*)
P.-O. Löwdin (*Gainesville, U.S.A.*)
H. O. Lutz (*Bielefeld, Germany*)
M. C. Standage (*Brisbane, Australia*)
K. Takayanagi (*Tokyo, Japan*)

Recent volumes in this series:

COINCIDENCE STUDIES OF ELECTRON AND PHOTON IMPACT IONIZATION
Edited by Colm T. Whelan and H. R. J. Walters

COMPLETE SCATTERING EXPERIMENTS
Edited by Uwe Becker and Albert Crowe

DENSITY MATRIX THEORY AND APPLICATIONS, SECOND EDITION
Karl Blum

IMPACT SPECTROPOLARIMETRIC SENSING
S. A. Kazantsev, A. G. Petrashen, and N. M. Firstova

INTRODUCTION TO THE THEORY OF X-RAY AND ELECTRONIC SPECTRA OF FREE ATOMS
Roman Karazjia

NEW DIRECTIONS IN ATOMIC PHYSICS
Edited by Colm T. Whelan, R. M. Dreizler, J. H. Macek, and H. R. J. Walters

PHOTON AND ELECTRON COLLISION WITH ATOMS AND MOLECULES
Edited by Philip G. Burke and Charles J. Joachain

POLARIZED ELECTRON/POLARIZED PHOTON PHYSICS
Edited by H. Kleinpoppen and W. R. Newell

PRACTICAL SPECTROSCOPY OF HIGH-FREQUENCY DISCHARGES
Sergei A. Kazantsev, Vyacheslav I. Khutorshchikov, Günter H. Guthöhrlein, and Laurentius Windholz

SELECTED TOPICS ON ELECTRON PHYSICS
Edited by D. Murray Campbell and Hans Kleinpoppen

VUV AND SOFT-X-RAY PHOTOIONIZATION
Edited by Uwe Becker and David A. Shirley

A Chronological Listing of Volumes in this series appears at the back of this volume.

A Continuation Order Plan is available for this series. A continuation order will bring delivery of each new volume immediately upon publication. Volumes are billed only upon actual shipment. For further information please contact the publisher.

Complete Scattering Experiments

Edited by

Uwe Becker
Fritz Haber Institute of the Max Planck Society
Berlin, Germany

and

Albert Crowe
University of Newcastle-Upon-Tyne
Newcastle-Upon-Tyne, England

Kluwer Academic / Plenum Publishers
New York, Boston, Dordrecht, London, Moscow

Library of Congress Cataloging-in-Publication Data

Complete scattering experiments/edited by Uwe Becker and Albert Crowe.
    p. cm. — (Physics of atoms and molecules)
   Includes bibliographical references and index.
   ISBN 0-306-46503-5
     1. Electrons—Scattering—Congresses. 2. Photoionization—Congresses. I. Becker, Uwe. II. Crowe, Albert. III. Kleinpoppen, H. (Hans). IV. Hans Kleinpoppen Symposium on Complete Scattering Experiments (1998: Il Ciocco, Italy) V. Series.

QC793.5 .E628 C66 2001
539.7'58—dc21

00-054582

Proceedings of the Hans Kleinpoppen Symposium on Complete Scattering Experiments, held July 12–13, 1998, in Il Ciocco, Lucca, Italy, in honor of the 70th birthday of Hans Kleinpoppen

ISBN 0-306-46503-5

©2001 Kluwer Academic/Plenum Publishers, New York
233 Spring Street, New York, N.Y. 10013

10 9 8 7 6 5 4 3 2 1

A C.I.P. record for this book is available from the Library of Congress.

All rights reserved

No part of this book may be reproduced, stored in a retrieval system, or transmitted in any form or by any means, electronic, mechanical, photocopying, microfilming, recording, or otherwise, without written permission from the Publisher

Printed in the United States of America

Hans Kleinpoppen

Preface

The Hans Kleinpoppen Symposium on "Complete Scattering Experiments" was held in honor of Hans Kleinpoppen's 70th birthday. It took place in Il Ciocco, Italy. The symposium had two purposes: to present the work that Hans Kleinpoppen has done or initiated during his remarkable scientific career, and to bring people from various fields together who perform complete scattering experiments. Hans Kleinpoppen's work included electron and photon impact experiments which were accompanied by studies of entangled states - a field of high current interest. Representatives from each of these fields gave excellent lectures on their particular subjects, and many discussions that started during the sessions were continued later in the relaxed atmosphere of the Il Ciocco resort. The breathtaking view of the beautiful landscape will be an unforgettable memory to all who participated in this extraordinary scientific event. The coherent and ideal combination of subject, people and location reflected the coherence of Hans Kleinpoppen's aims and activities in science and life.

We offer our grateful thanks to all contributers who made this volume such a worthy tribute to Hans Kleinpoppen. We also like to thank Rainer Hentges for the painstaking work to prepare this volume in its complete ready to print version. We are also grateful to the Royal Society of London and the Max–Planck–Gesellschaft who generous support of the Hans Kleinpoppen symposium made this marvelous meeting and this proceedings possible.

UWE BECKER   ALBERT CROWE

viii *COMPLETE SCATTERING EXPERIMENTS*

Participants of the symposium in honor of Hans Kleinpoppen in front of the "Il Ciocco" resort

Participants

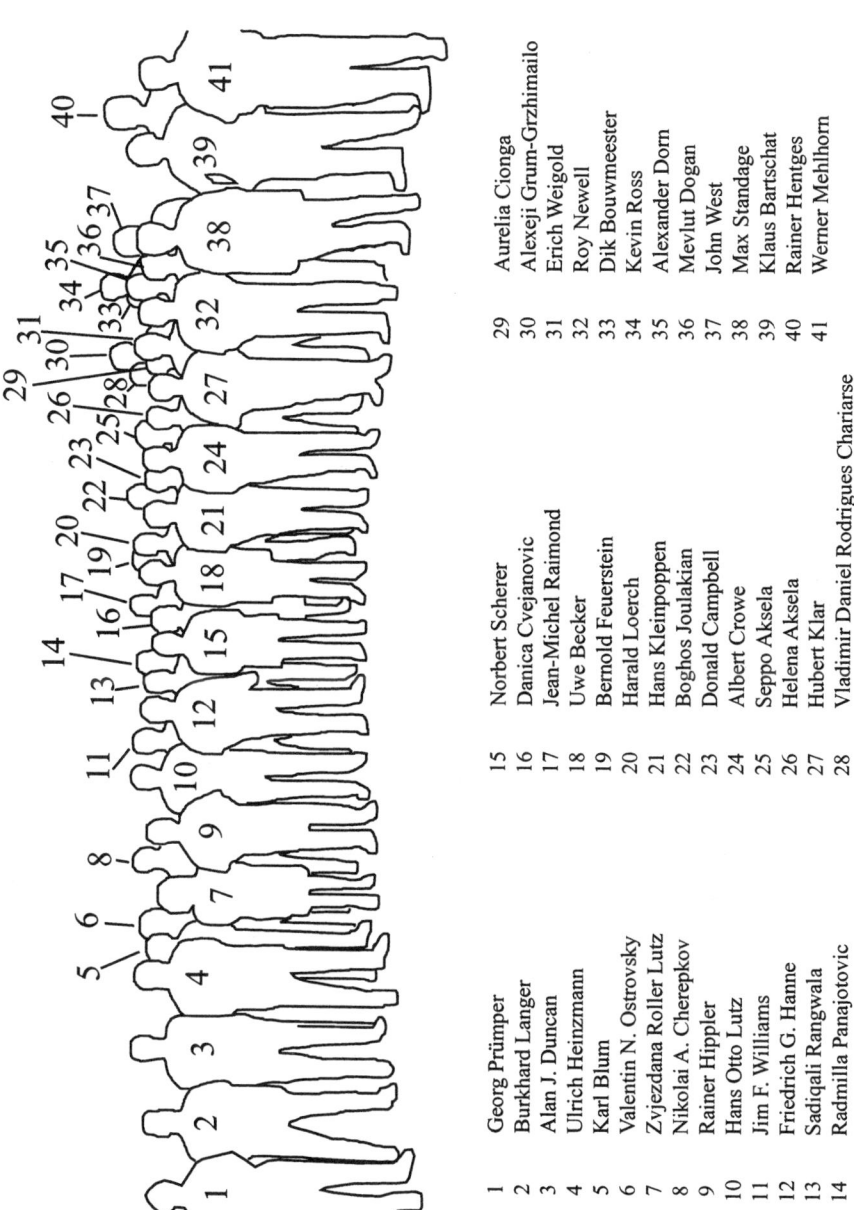

1 Georg Prümper
2 Burkhard Langer
3 Alan J. Duncan
4 Ulrich Heinzmann
5 Karl Blum
6 Valentin N. Ostrovsky
7 Zvjezdana Roller Lutz
8 Nikolai A. Cherepkov
9 Rainer Hippler
10 Hans Otto Lutz
11 Jim F. Williams
12 Friedrich G. Hanne
13 Sadiqali Rangwala
14 Radmilla Panajotovic
15 Norbert Scherer
16 Danica Cvejanovic
17 Jean-Michel Raimond
18 Uwe Becker
19 Bernold Feuerstein
20 Harald Loerch
21 Hans Kleinpoppen
22 Boghos Joulakian
23 Donald Campbell
24 Albert Crowe
25 Seppo Aksela
26 Helena Aksela
27 Hubert Klar
28 Vladimir Daniel Rodrigues Chariarse
29 Aurelia Cionga
30 Alexeji Grum-Grzhimailo
31 Erich Weigold
32 Roy Newell
33 Dik Bouwmeester
34 Kevin Ross
35 Alexander Dorn
36 Mevlut Dogan
37 John West
38 Max Standage
39 Klaus Bartschat
40 Rainer Hentges
41 Werner Mehlhorn

Contributing Authors

Alain Aspect • Institut d'Optique Théorique et Appliquée Bâtiment 503-Centre universitaire d'Orsay-BP 147 91403 ORSAY Cedex – France

Klaus Bartschat • Department of Physics and Astronomy, Drake University, Des Moines, IA 50311, U.S.A.

Uwe Becker • Fritz–Haber–Institut der Max–Planck–Gesellschaft, Faradayweg 4–6, D–14195 Berlin, Germany

Karl Blum • Institut für Theoretische Physik I, Universität Münster, Wilhelm–Klemm–Str. 9, D–48149 Münster, Germany

Dik Bouwmeester • Centre for Quantum Computation, Clarendon Laboratory, University of Oxford, Parks Road, OX1 3PU Oxford, United Kingdom

Andreas Busalla • Institut für Theoretische Physik I, Universität Münster, Wilhelm–Klemm–Str. 9, D–48149 Münster, Germany

Nikolai A. Cherepkov • State University of Aerospace Instrumentation, 190000 St.Petersburg, Russia

Alexander Dorn • Fakultät für Physik, Universität Freiburg, D–79104 Freiburg, Germany

Alan J. Duncan • Unit of Atomic and Molecular Physics, University of Stirling, Stirling, Scotland

Alexeji N. Grum-Grzhimailo • Fakultät für Physik, Universität Freiburg, D–79104 Freiburg, Germany

Rainer Hippler • Institut für Physik, Ernst–Moritz–Arndt–Universität Greifswald, Domstr. 10a, D–17487 Greifswald, Germany

Hans Kleinpoppen • Unit of Atomic and Molecular Physics, University of Stirling, Stirling, Scotland

William R. MacGillivray • The Laser Atomic Physics Laboratory, Griffith University, Brisbane, QLD 4111, Australia

Werner Mehlhorn • Fakultät für Physik, Universität Freiburg, D–79104 Freiburg, Germany

Andrew J. Murray • The Schuster Laboratory, University of Manchester, Manchester, M13 9PL, UK

Manfred Musigmann • Institut für Theoretische Physik I, Universität Münster, Wilhelm–Klemm–Str. 9, D–48149 Münster, Germany

Jian–Wei Pan • Institut für Experimentalphysik, Universität Wien, Austria

Georg Prümper • Fritz–Haber–Institut der Max–Planck–Gesellschaft, Faradayweg 4–6, D–14195 Berlin, Germany

Stephen J. Schaphorst • Fakultät für Physik, Universität Freiburg, D–79104 Freiburg, Germany • Now at: Delta Information Systems, 300 Welsh Road, Horsham PA 19044–2273

Norbert Scherer • Fakultät für Physik, Universität Freiburg, D–79104 Freiburg, Germany

Volker Schmidt • Fakultät für Physik, Universität Freiburg, D–79104 Freiburg, Germany

Zahoor A. Sheikh • Department of Physics, Bahaudin Zakariya University, Multan, Pakistan

Max Shurgalin • Harvard–Smithsonian Centre for Astrophysics, Cambridge, MA 02138, USA

Max C. Standage • The Laser Atomic Physics Laboratory, Griffith University, Brisbane, QLD 4111, Australia

David G. Thompson • Department of Applied Mathematics and Theoretical Physics, Queen's University of Belfast, Belfast BT7 1NN, UK

Harald Weinfurter • Sektion Physik, Ludwig-Maximilians-Universität München, Schellingstrasse 4/III, D–80799 München, Germany

Jim Williams • Centre of Atomic, Molecular and Surface Physics, Department of Physics, The University of Western Australia, Nedlands, Perth, Western Australia

Anton Zeilinger • Institut für Experimentalphysik, Universität Wien, Austria

Björn Zimmermann • Fritz–Haber–Institut der Max–Planck–Gesellschaft, Faradayweg 4–6, D–14195 Berlin, Germany

Contents

1
A "laudatio" for Professor Hans Kleinpoppen 1
Jim Williams, and Rainer Hippler

    1. Introduction 1
    2. Early studies of line polarization 2
        2.1 Level widths, fine and hyperfine structure 2
        2.2 Threshold excitation and resonances 5
    3. Level crossings, anticrossings and Lamb shifts 6
    4. Correlation and coherence 14
        4.1 Theory of the Measurement 14
        4.2 Excitation of two–electron atoms, 2^1P and 3^1P states of helium 16
        4.3 Hydrogen 21
        4.4 Sodium $[(Ne)3s^1]$ and Potassium $[(Ar)4s^1]$ 22
        4.5 Calcium $[(Ar)4s^2]$ and Strontium $[(Kr)5s^2]$ 23
        4.6 Mercury $[(Xe)4f^{14}5d^{10}6s^2]$ 24
    5. Photoionization (Synchrotron) Studies 25
        5.1 Calcium $[(Ar)4s]$ 25
        5.2 Atomic oxygen $[1s^2 2s^2 2p^4]$ 27
    6. Outer shell excitation in heavy particle collisions 29
    7. Inner shell ionisation 35
    8. Two–photon studies and the Einstein–Podolsky–Rosen Paradox 38
    9. Conclusion 40

PART I: ELECTRON SCATTERING

2
Complete Experiments in Electron-Atom Collisions – Benchmarks for Atomic Collision Theory 61
Klaus Bartschat

    1. Introduction 61
    2. S →P excitation in helium 62
    3. S →D excitation in helium 70
    4. S →P excitation in sodium 76
    5. S →P excitation in mercury 79
    6. S →P simultaneous ionization–excitation in helium 84
    7. Summary 87

3
Differential Cross Section and Spin Asymmetries for Collisions between Electrons and Oriented Chiral Molecules 93
A. Busalla, M. Musigmann, K. Blum, and D. G. Thompson

    1. Introduction 93
    2. Discussion of cross section 94
        2.1 General theory 94
        2.2 Numerical results 96
    3. Spin asymmetries 100
        3.1 Introduction 100
        3.2 Discussion of the asymmetry A 101
        3.3 Discussion of the asymmetry η 103
        3.4 Numerical results and discussion 104

4
Towards a Complete Experiment for Auger Decay 111
A. N. Grum–Grzhimailo, A. Dorn, and W. Mehlhorn

    1. Introduction 111
    2. Example for a possible complete experiment 113
    3. Proposal for an "almost" complete experiment 115
    4. Performance of an "almost" complete experiment 117
    5. Conditions and applicability of the proposed method 124
    6. Conclusions 125

5
The Transfer of Angular Momentum for Electron–Atom Collision Processes Involving Inelastically Excited $S{\rightarrow}P$ and $P{\rightarrow}S$ Transitions 129
M. Shurgalin, A. J. Murray, W. R. MacGillivray, and M. C. Standage

PART II: PHOTON IMPACT

6
Complete Photoionization Experiments Using Polarized Atoms   141
G. Prümper, B. Zimmermann, U. Becker, and H. Kleinpoppen

    1. Introduction   141
        1.1 Approximations   142
    2. The observables in photoionisation   143
        2.1 Experimental set-up   144
        2.2 The thallium $5d$ spectrum   147
        2.3 Results for the dynamical spin polarization   151
        2.4 Differences between the LMDAD-asymmetry and
        dynamical spin polarisation   151
        2.5 Summary   153

7
On the Contribution of Photoelectron Auger Electron Coincidence Spectrometry to Complete Photoionization Studies   155
N. Scherer, S. J. Schaphorst, and V. Schmidt

    1. Introduction   155
    2. Selected process   157
    3. Results from a particular correlation pattern   158
    4. Proposed study   160
    5. Conclusions   165

8
Complete Experiments in Molecular Photoionization   167
N. A. Cherepkov

    1. Introduction   167
    2. On the possibilities of complete experiments with atoms and molecules   168
    3. Principles of complete experiments with molecules   170
    4. Photoionization of excited aligned molecules   174
    5. Photoionization of oriented (fixed–in–space) molecules   177
        5.1 General consideration   177
        5.2 Particular results for O K–shell of CO   180
    6. Conclusions   184

PART III: EINSTEIN–PODOLSKY–ROSEN EXPERIMENTS

9
Experimental Tests of Bell's Inequalities with Correlated Photons   189
Alain Aspect

    1. Introduction   189

2. Why supplementary parameters? The Einstein–Podolsky–Rosen–Bohm Gedankenexperiment	190
2.1 Experimental scheme	190
2.2 Correlations	192
2.3 Difficulty of an image derived from the formalism of Quantum Mechanics	192
2.4 Supplementary parameters	194
3. Bells inequalities	195
3.1 Formalism	195
3.2 A (naive) example of supplementary parameters theory	195
3.3 Bell's Inequalities	197
4. Conflict with quantum mechanics	198
4.1 Evidence	198
4.2 Maximum conflict	198
5. Discussion: The locality condition	200
6. Gedankenexperiment with variable analyzers: The locality condition as a consequence of Einstein's causality	201
7. From Bell's theorem to a realistic experiment	203
7.1 Experimentally testing Bell's inequalities	203
7.2 Sensitive situations are rare	203
7.3 Production of pairs of photons in an EPR state	204
7.4 Realistic experiment	205
7.5 Timing conditions	206
8. First generation experiments	206
8.1 Experiments with one channel polarizer	206
8.2 Results	208
9. Orsay experiments (1980–1982)	208
9.1 The source	208
9.2 Detection – Coincidence counting	209
9.3 Experiment with one–channel polarizers	210
9.4 Experiment with two–channel analyzers	211
9.5 Timing experiment	213
10. Third Generation: Experiments with pairs of photons produced in parametric down conversion	216
11. Conclusion	218

10
Polarization and coherence analysis of the optical two–photon radiation from the metastable $2^2S_{1/2}$ state of atomic Hydrogen   223
A. J. Duncan, H. Kleinpoppen, and Z. A. Sheikh

1. Introduction	224
2. On the Theory of the Two–Photon Decay of the Metastable State of Atomic Hydrogen	225
3. The Stirling Two–photon Apparatus	231
4. Angular and Polarization Correlation Experiments	233
4.1 Two–Polarizers Experiments: Polarization Correlation and Einstein–Podolsky–Rosen–Tests	233

4.2 Tests of Garuccio–Selleri Enhancement Effects	239
4.3 Three–Polarizer Experiments	240
4.4 Breit–Teller Hypothesis	243
5. Coherence and Fourier Spectral Analysis	247
6. Time Correlation	252
7. Correlated Emission Spectroscopy of Metastable Hydrogen	253
8. Conclusions	257

11
Quantum–State Transmission via Quantum Teleportation   261
Dik Bouwmeester, Jian-Wei Pan, Harald Weinfurter, and Anton Zeilinger

1. Introduction	261
2. Quantum Teleportation Protocol	262
3. Experimental Quantum Teleportation	264
3.1 Polarisation Entangled Photons	264
3.2 Bell-State Analyser	266
3.3 Experimental Setup	268
3.4 Experimental Predictions	269
3.5 Experimental Results	270
3.6 Teleportation of Entanglement	272
4. Concluding Remarks and Prospects	273
Index	277

Chapter 1

A "LAUDATIO" FOR PROFESSOR HANS KLEINPOPPEN

Jim Williams
Centre of Atomic, Molecular and Surface Physics, Department of Physics, The University of Western Australia, Nedlands, Perth, Western Australia

Rainer Hippler
Institut für Physik, Ernst–Moritz–Arndt–Universität Greifswald, Domstr. 10a, 17487 Greifswald, Germany

1. INTRODUCTION

This 'laudatio' gives tribute to the contributions of Hans Kleinpoppen to atomic physics. To indicate the timeliness and historical significance of his influence, all his papers, editorships of conference proceedings and chapters in books are presented in chronological order in the bibliography [1–195]. This paper presents his work in seven categories:

(i) early studies of line polarisation

(ii) Lamb shifts, level crossing and anticrossing,

(iii) correlations and coherence

(iv) photon impact (synchrotron) studies,

(v) excitation in heavy particle collisions,

(vi) inner shells, and

(vii) two–photon decay and the EPR paradox.

Many people contributed to the work; important steps occurred at about the same time and often repeatedly in slightly different contexts. The contributions of his colleagues are implicitly acknowledged. The outcome has been a significant advancement in knowledge and in the development and understanding of the areas of atomic physics presented in this volume.

The development of the quantum description and observation of the interactions between particles and atoms during the past 40 years is reflected in the publications of Professor Hans Kleinpoppen. His first paper [1] in 1958 on *Dichroism and Luminescence Polarization of Stretched Polyvinylalcohol Films* reflects the influence of his co-author and mentor, Professor W. Hanle in Giessen, and probably established his scientific inclinations. His Ph.D. studies with Professor Krüger in Tübingen led into electron–atom collisions with measurements of the Lamb shift [2] and fine structure [6] in hydrogenic atoms and of polarization phenomena [3–14]. Subsequent studies, mostly at the University of Stirling until about 1990, concerned observations of the polarization, energy and angular distributions and correlations between photon, electron and ions emerging from various scattering processes. The emerging information led to the determination of the fundamental quantum information of scattering amplitudes, and their relative phases, for magnetic sublevels which motivates much research. The history of the subject indicates the naturally close developments of theory and experiment and the parallel advances in computational methods and technology. These features have underpinned decades of exciting developments in the understanding of the interactions between electrons, photons and atoms, that determined atomic structure and scattering phenomena. In the last decade Professor Kleinpoppen has collaborated with groups using synchrotron radiation in Daresbury and Berlin and has helped to bring to fruition and extended some of his early ideas. Some of the fundamentals are discussed at length in the first part of this paper while other major aspects, such as the development of the *theory of the measurements* which echoes many of his seminars, are discussed more briefly than their significance deserves. Details are given in the references and other papers in this volume.

2. EARLY STUDIES OF LINE POLARIZATION

2.1 LEVEL WIDTHS, FINE AND HYPERFINE STRUCTURE

The early studies [1–17] helped to establish the foundations of polarization phenomena to verify basic quantum ideas, from which numerous significant and fundamental ideas and experiments followed. Some of these concepts, although accepted long ago and taught in undergraduate physics courses, are revisited here to underpin subsequent and 'state-of-the-art' measurements and new ideas. His initial observations [3, 4, 7, 11, 12, 14] concerned line polarization, that is of the unequal populations of the magnetic sublevels, in singly charged helium ions [4,7], hydrogen [3,8,12,14], alkali [6] and mercury atoms [23,30]. It was shown how these populations could be changed by a collision process, an external field or by the presence of a level crossing or anticrossing. The change in the intensity or polarization characterized, for example, the fine and hyperfine structure, line widths, magnetic sublevel cross sections and relaxation times. In the 1960s these quantities were the main quantum observables from which information was obtained to guide the development of theories and methods of measurement. The research of Professor Kleinpoppen inspired these developments and it is instructive to trace their evolution.

Initially consider the excitation of a simple atom. The use of an electron beam [3] introduces the naturally defined directions, parallel and perpendicular to the beam. The excitation of a P–state has definite orbital angular momentum components $m_l = 1, 0, -1$ with the quantization axis as the direction of the incident beam and with different excitation probabilities for those substates. Observations of the intensities of photons, radiated from those sublevels with the same lifetime, determine the polarization. The observations are made of either the anisotropic angular distribution of photons or the ratio of the difference divided by the sum of the photon intensities measured in the above directions. With observational symmetry about the quantization axis, the observed photon intensities are independent of the sign of m_l and are proportional to the sublevel cross sections σ_{m_l} whose sum is defined as the total cross section for excitation. The application of these ideas led to four decades of research at the cutting edge of atomic physics. Kleinpoppen quickly confirmed [3, 4] that a simple description by earlier workers was applicable only to P states without fine or hyperfine structure, i.e. only to ^{1}P states of two electron atoms or to ^{2}P states whose level width is large compared to both fine and hyperfine structure splitting. Polarization measurements yielded, for P states with fine structure, only the ratio σ_1/σ_0 while for D states the polarization is a function of σ_0, σ_1 and σ_2 and the total cross section is needed to extract a partial cross section. It was also shown [6] that the polarization of the 497.2 nm photons from the $4^2S_{1/2}$–$2^2P_{1/2,3/2}$ transition of lithium was zero as expected for a spherically symmetric excited S–state. The expected dipole pattern was confirmed for excited P states in hydrogen [3] and the helium ion [4]. These concepts were setting the foundations for future studies.

To see the effects of the hyperfine structure and the level widths, two considerations of the excitation process were required. First, the natural time sequence of events for electron–atom interactions usually separates the excitation processes from the relaxation processes. The electron orbital periods and the collision times are approximately 10^{-15} sec while the times for collisional spin–orbit, atomic spin–orbit and nuclear hyperfine relaxations are usually less than the dipole radiative decay time of about 10^{-9} sec. However, there are variations of these times between atoms and the observational effects of these relaxation processes are most conveniently explored when there are large differences between these times. Kleinpoppen chose [6, 11, 12] the alkali atoms ^{6}Li,

^{7}Li and ^{23}Na, whose lowest resonance lines have hyperfine relaxation times which are approximately greater than, equal to and less than the level decay times, respectively.

These studies also took advantage of observation of the decay radiation near the excitation threshold which has become widely used as a region for attempts to differentiate between atomic interactions. For excitation from S to P states observed in the beam direction, the incoming electrons, before and after excitation, have zero orbital angular momentum and the atom cannot change its value for m_l since the selection rules prescribe that $\Delta m_l = 0$. For an excited state with fine structure the selection rules allow only m_j states composed of the possible values for m_s and $m_l = 0$ eigenstates. The combined effect of these

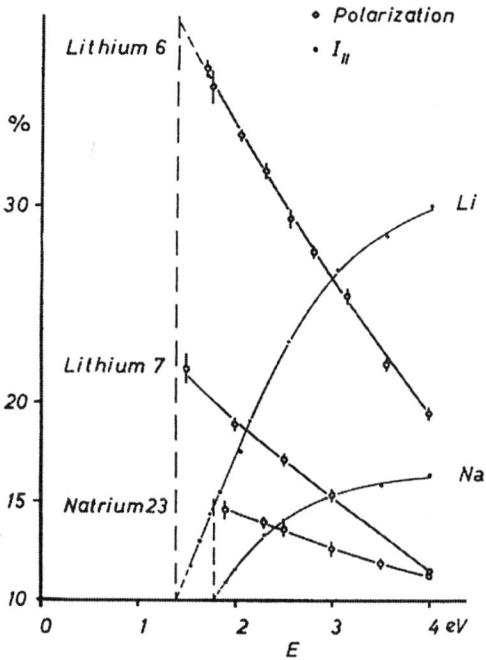

Figure 1.1 Threshold polarization for the lowest resonance lines in ^{6}Li, ^{7}Li and ^{23}Na as a function of the incident electron energy E (from Ref. [6]).

angular momenta conservation and relaxation times for threshold excitation are shown in figure 1.1 for ^{6}Li, ^{7}Li, and ^{23}Na for the polarization of the lowest resonance lines. For ^{6}Li the hyperfine structure is not resolved and so the polarisation is reduced only by the fine structure. Then the threshold polarization for S-to-P state excitation is reduced to 37.5% compared to 100% without fine structure. In a similar way with hyperfine structure splitting, only those m_F substates with $m_l = 0$ are excited at threshold and the threshold polarization is reduced further depending on the nuclear spin. For ^{23}Na the level width is not larger than the smallest splitting so the hyperfine structure of the states is resolved, hyperfine effects are large and the polarisation is reduced to 14.1%. In ^{7}Li there is partial overlapping of the hyperfine states and the polarization of 21.6% is intermediate between ^{6}Li and ^{23}Na. These data indicate the significant differences and behaviour of these threshold polarization measurements that were defining the knowledge of the field. Polarization measurements and consideration of the number of π and σ transitions at threshold for nuclear spins confirmed the Percival–Seaton model [196, 197] and showed that the electron impact threshold polarization values were equal to the resonance fluorescence optical values. This work was also significant because it motivated further thought about the time development of the excitation and decay processes, about interference phenomena as well as the nature of coherence of the excited states.

2.2   THRESHOLD EXCITATION AND RESONANCES

Later work [23, 24, 29, 30, 36, 56] built upon these constraints of the angular momentum at threshold to look at the excitation process and the formation of resonant states. Atomic hydrogen [5, 8, 9, 24], neon and argon [29], xenon [56] and mercury [23, 30, 36] each had special interest. At the time much interest centered on atomic hydrogen for which theoretical studies, particularly in Belfast, were showing significant advances with the close coupling models. Kleinpoppen confirmed [24] the absolute cross section for excitation of atomic hydrogen to the 2s state had a finite value at threshold. Other large resonances, about 0.5 eV below the n = 3 threshold but not separable with 110 meV resolution, were consistent with theoretical descriptions of a superposition of the ^{1}S and ^{3}P resonances. These features and their energy dependencies were consistent with the best close coupling theory [198], in which the complete set of target states was approximated by 23–states within 0.5 eV of threshold and 6–states up to 12 eV. This gave added confidence to the close coupling model that was later developed for wider dynamical applicability.

At this time ultraviolet (UV) polarization data were rare and were needed to broaden the general description of threshold excitation behaviour. The limits of technology were generally the Brewster angle reflecting polarizers. Kleinpoppen used [29, 56] the dipolar cosine dependence of the angular distributions of the UV photons to determine polarizations near threshold for the 5p^5 6s ^{3}P$_1$ (146.9 nm) and ^{1}P$_1$ (129.4 nm) states of xenon. Negative ion resonance structures were seen in the excitation function and in large negative polarizations which added to the general characterization of resonance phenomena by establishing the quantum numbers, series structure and bonding mechanisms. Mercury atoms were showing similar phenomena with more prospect of theoretical study and ease of photon detection so attention switched to the $6^1S_0 - 6^3P_1$ 253.7 nm intercombination line of mercury, with a threshold at 4.9 eV. Good (100 meV) electron energy resolution studies showed [23, 30, 36] the presence of three features, possibly resonant states at 4.92, 5.23 and 5.50 eV, and a large negative polarization at threshold which was modified by the resonances. An identification of the resonances was made on the basis of a theoretical indication that a $J = 1/2$ resonance would have a zero polarization and a $J = 3/2$ state a polarization of 60%. The polarization data, measured parallel and perpendicular to the quantization axis, indicated a $J = 3/2$ negative ion state at 4.92 eV. These data are seen now as laying the foundations for subsequent studies using polarized electrons and polarized atoms and better–defined quantum states so that definite threshold behaviour could be established and the studies extended to other atoms. They were at the leading edge of technology at the time.

3. LEVEL CROSSINGS, ANTICROSSINGS AND LAMB SHIFTS

Professor Kleinpoppen's second paper was the measurement of the $3^2S_{1/2} - 3^2P_{1/2}$ Lamb shift of atomic hydrogen [2] in 1961. That work developed from the background of the 1947 radio–frequency measurements of Lamb and Retherford [199–201] which separated the $2^2S_{1/2}$ and $2^2P_{1/2}$ levels of atomic hydrogen. They followed the classic work of Rabi in 1937 [202] that established how two discrete states were coupled under an external perturbation. New ways were being developed to look at the relativistic fine structure, which includes spin–orbit and spin–spin and quantum electrodynamic (QED) contributions, and at the electrostatic fine structure, which includes electron exchange, inner electron screening and core polarisation, the latter occurring in all atoms except hydrogenic atoms. Also methods were being sought to explore the strength of internal couplings of atomic states, the influence of external fields, level intervals and atomic lifetimes. Previously, knowledge of atomic structure had come mainly from conventional spectroscopy. For example, in helium, the relativistic fine structure intervals had been measured directly up to n = 9 whereas the only electrostatic intervals known from direct observation in about 1970 were the triplet and singlet 2S–2P intervals of 1 and 2 μm, respectively, and the 95.8 μm $3^1P_1-3^1D_2$ and the 216.3 μm $4^1P_1-4^1D_2$ intervals. Other intervals were determined by taking the differences of wave numbers of optical transitions which rapidly become less accurate for higher levels.

Renewed interest arose from the successes of the optical double resonance *photon plus rf field* method of Brossel and Bitter [203]. However the accuracy of the radio–frequency (rf) spectroscopic approach decreases as the atomic lifetime increases or the transition moment decreases and the states do not receive enough power to be saturated. The attractions of the level crossing methods in external fields, as well as the zero field Hanle effect [204], were known to Professor Kleinpoppen and they were pursued in a long collaboration with HJ Beyer at Stirling University [59,205]. In this section of the paper some of the basic ideas about level perturbations are described, particularly those aspects which were enhanced by the Stirling group and which led onto the subsequent studies of coherence and correlation in electron impact excitation.

For level crossings, the observed signal results from internal interactions caused by coupling between two Zeeman levels of an excited state. In contrast, anticrossing levels may arise, for example from the use of a static external field to couple levels, and a signal may result even if the coupling is otherwise absent. The name *anticrossing* describes the way in which levels repel one another in a region of closest approach as a result of the state mixing caused by the coupling. The levels that may be studied depend on the symmetry properties of the *crossing* sublevels. With an external magnetic field, there is only the rotational symmetry with respect to the field direction, the Zeeman sublevels are usually non–degenerate and may be classified according to the eigenvalues m of the angular momentum operator $J_z = L_z + S_z$. Hence Zeeman sublevels m and m' can cross when $\Delta m = m - m' \neq 0$ whereas anticrossings can be formed by sublevels with $\Delta m = 0$. In contrast, the electric field case has reflection symmetry as well.

Their work also developed an understanding of coherence since there must be coherence of the excitation process between the crossing levels which must have the same parity and belong to the same fine structure or hyperfine structure manifold of states. Consequently, interference can occur between the amplitudes describing the emission and absorption of the resonance line at the crossing of the Zeeman levels and give rise to an observed coherent signal. However, particularly if cascading is present, coherent excitation is not always possible since it requires only the components of the incident particle velocity which are perpendicular to the magnetic field. In contrast, for an anticrossing, coherence is not required; the levels are prevented from crossing by either an external field or an internal field (for example, a magnetic dipole interaction) and intensity changes are observed near the avoided crossing point. An understanding of these concepts emerged from professor Kleinpoppen's work and they emerged again in underpinning the importance of the geometry and particle dynamics in subsequent scattering studies, in particular the studies of parity coherence effects on angular correlations in hydrogen (see section 4.3). In today's language we might also say that these experiments allowed the bound electrons to be manipulated within the confines of the atomic orbitals in the presence of external fields.

Beyer and Kleinpoppen tackled the problem of how to measure the intervals between the n = 4 and 5 energy levels for one– and two–electron atoms for which there was a need for accurate observations to guide theory. The

Stirling work concentrated on the techniques using electric field induced anticrossings [19, 21, 205] (EFIACS), singlet–triplet crossings and electric field induced singlet–triplet anticrossings (EFISTACS) [53, 116]. These techniques differ in the perturbation which could be the external electric field, the internal singlet–triplet interaction due to part of the spin–orbit interaction or the combination of external and internal perturbations, respectively. The last mentioned technique had the extra advantage of allowing the interaction to be controlled so that the position and the Stark shift of a single anticrossing between a particular Zeeman branch of an n^1D and a particular branch of an n^3D level could be studied separately.

The principle of the method is indicated by a time–dependent Schrodinger equation with a Hamiltonian of three terms describing the unperturbed atom, the radiative decay and the external (electric or magnetic) or internal (spin–orbit and spin–spin etc) interactions. The excited coupled states are described by a wave function containing a linear superposition of time–dependent amplitudes. The fluorescent intensity is separated into terms that describe the non–resonant photons from the two states, the level crossing component [requiring interference (coherence) terms in the excitation and decay processes as function of the perturbation] and an anticrossing component [which does not require coherence in the excitation and detection parts]. Here, coherence implies that the optical excitation must occur from the same level of the initial state to the two intermediate states and that the detector must observe transitions from the two intermediate states to the same final state. The method requires observation, with appropriate geometry (and usually orthogonal fields), of the change in intensity or polarisation of the fluorescence light as selected sublevels are tuned to near degeneracy by an external field. The shape of the intensity I of the resonant radiation becomes apparent from an expression of the Breit form

$$I = C \sum_{\mu\mu'} (f_{\mu m} f_{m\mu'} g_{\mu'm'} g_{m'\mu})/(\Gamma + i\omega_{\mu\mu'}) \quad (1.1)$$

where C is a normalizing constant for each state, Γ is the radiative decay constant of the excited state, $\omega_{\mu\mu'}$ is the frequency interval between the excited substates and depends on the external fields and f and g are matrix elements describing the excitation and emission processes for specific polarizations of the exciting and observed photons. There is a resonant change in the intensity when $\omega_{\mu\mu'}$ passes through zero, that is whenever the levels μ and μ' cross. A typical situation is shown in figure 1.2 for the n = 4 fine structure Zeeman states of He^+. The shape of the coupling signal in figure 1.3 depends on the relative phases of the coupled states. This description of resonance behaviour is quite general. A brief comment is made on the theory and then representative data are discussed. These studies also aimed to establish precise and accurate energy level intervals and test calculated vales of energy levels, atomic

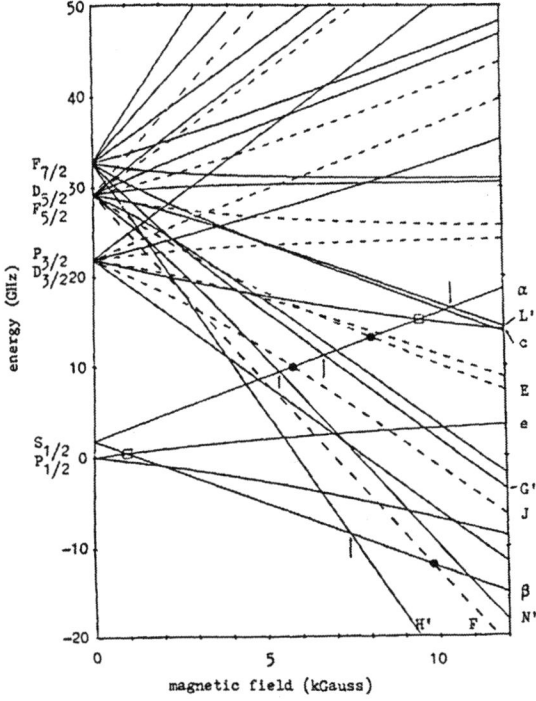

Figure 1.2 Level crossings in He$^+$ (From Ref. [59]).

wave functions and approximation methods. Progress in Atomic Spectroscopy (eds. Hanle and Kleinpoppen) [57] contains excellent reviews appropriate for the time and also placing the Stirling work in context. Here it is sufficient to indicate that the fine structure and radiative energy shifts are given by the summed QED contributions from the self energy, the vacuum polarisation, the anomalous magnetic moment of the electron, recoil corrections and nuclear size corrections, which are usually given in a series expansion as a function of two parameters, the binding parameter αZ and the ratio of the electron to the nuclear masses. Contributions from the self energy are the largest for S–states while for other states the anomalous magnetic moment of the electron becomes larger. The other contributions are always small though not negligible. High accuracy energy level measurements were required at the time when theory was making significant progress particularly for low Z atoms. This need was widely recognized and motivated much research. Professor Kleinpoppen was able to select the experimental method, which gave accurate measurements of levels, and their separations that were not readily accessible by other means. In

10   COMPLETE SCATTERING EXPERIMENTS

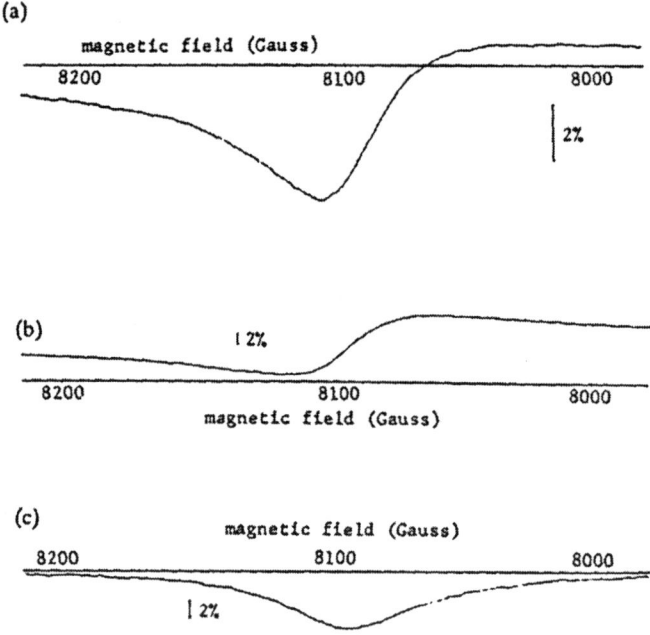

Figure 1.3 Phase shifted level crossing signals in He$^+$ (From Ref. [59]).

the long term, this interaction with theory has been underpinning much of the subsequent studies; significant advances in either theory or measurement have urged the other to keep pace. For the Stirling studies of one–electron atoms, the theory, summarized by Ermolaev [206], indicated that all of the above QED contribution were important for the accuracy achieved about 1978. However experiment produces only a single number and the techniques did not identify components that may be separately attributable to the individual QED effects. Consequently the outcome of a comparison between theory and measurement depends strongly on good experimental accuracy which was a strong point of the Stirling studies.

(a) **He$^+$ n= 4 for S–F and S–D intervals.**
The helium ion was most attractive because it is free of complicating hyperfine structure and easier to use than atomic hydrogen. While electric dipole ($\Delta l =1$) transitions between states may be induced easily by *rf* fields, much higher fields are required to induce $\Delta l > 1$ transitions. The various experimental difficulties associated with high *rf* fields were avoided by using static electric fields which are better defined, more uniform and resulted in improved accuracy. Since a homogeneous electric field can only produce

electric dipole coupling, higher order coupling is achieved through the appropriate, and still relatively near, intermediate states. Zeeman branches belonging to states with different L are tuned to near-degeneracy by a magnetic field and the necessary coupling between states provided by a uniform DC electric field. In a typical measurement, a static electric field ($\sim$150 V/cm) was applied to He$^+$ ions formed by electron impact and the readily detected n = 4 to 3 (468.6 nm) photons were observed. As a magnetic field ($\sim$ 5375 G) is swept through the S–F fine structure level crossing points, changes in the overall intensity ($\sim$ 4%) are recorded and the width of the resonant signals is determined by the natural widths of the two states and the strength of coupling that is relevant to the spacing of the two states at the crossing point. The measured intervals [19] were $4^2S_{1/2}$–$4^2F_{5/2}$ = 27435.7 $\pm$ 13.8 MHz and the $4^2S_{1/2}$–$4^2F_{7/2}$ = 31098.8 $\pm$ 10.5 MHz and consequently the $4^2F_{5/2}$–$4^2F_{7/2}$ interval = 3663.1 $\pm$24.3 MHz compared well with 3657.78 $\pm$ 0.01 MHz from theory [59]. However there was clearly scope for improvement of the technique in order to extend theory. Similar data emerged using the electric field induced anticrossing signals (EFIACS) method to determine [21] the $4^2S_{1/2}$–$4^2D_{3/2}$ and $4^2S_{1/2}$–$4^2P_{3/2}$ intervals from which the $4^2P_{3/2}$–$4^2D_{3/2}$ Lamb shift (= 34.4 $\pm$ 9.0 MHz) agreed within relatively large experimental uncertainty with 37.19 $\pm$ 1.36 MHz of theory.

(b) **Fine structure splitting of the He$^+$ 5S, 5D, 5F and 5G levels and cascade effects.**

The intervals between the fine structure S and D levels and between the S and F levels for n = 5 of He$^+$ were determined [40] in the direct decay 320.3 nm line (n = 5 to n = 3) from the anticrossing signals. The intervals were measured between the $5^2S_{1/2}$–$5^2D_{5/2}$ states at 14055.4 $\pm$ 12.6 MHz and the $5^2S_{1/2}$–$5^2F_{5/2}$ at 14058.2 $\pm$ 13.6 MHz. The measured intervals between the $5^2S_{1/2}$–$5^2F_{7/2}$ states were 31100.5 $\pm$ 1.7 MHz compared with 31103.84 $\pm$ 0.04 MHz from theory [59].

For higher levels, the transition frequencies are in the infrared or vacuum ultraviolet (VUV) and are not readily detected. However the high sensitivity of the electric field induced anticrossing signals (EFIACS) method allowed [20] the cascade effects of crossings from higher principal quantum number levels to be detected in the fluorescent intensity changes of lower n. Kleinpoppen and colleagues exploited this fact for the 5S state whose long lifetime gave rise to a narrow level crossing signal and consequently increased accuracy of the energy intervals. The anticrossing signals from the coupled 5S and 5G levels of He$^+$ were sought in the 468.6 nm photons from the n = 4 to 3 transition. The sensitivity and stability of the method allowed the nearly pure Lorentzian cascade signals of only 1 to 2% of the main intensity to be observed. The fine structure intervals were determined at 15924.4 ± 5.9 MHz for the $5^2S_{1/2}$–$5^2G_{7/2}$ and 17040.5 ± 8.0 MHz for the $5^2S_{1/2}$–$5^2G_{9/2}$ interval, and the difference interval $G_{7/2}$–$G_{9/2}$ was 1116.1 ± 14.0 MHz compared with 1123.6 MHz from theory [59].

(c) 5^1D and 5^3D intervals of helium.

Electric field induced singlet–triplet anticrossing signals (EFISTACS) were used by Kleinpoppen to measure [53, 116] the 5^1D–5^3D interval in helium at 34066.0 ± 1.7 MHz. At this time laser studies achieved an accuracy of ± 3 MHz for n ≤ 3. The microwave optical method is suitable for the high n region n ≥ 7 for nD, F and G states. The EFISTACS method gives 1.7 MHz accuracy (compare with the accuracy in (a) and (b)) for n = 5 to 7 for the region of n not covered by the accurate low–n laser or high–n microwave studies. From the electric field shift of the position of each anticrossing, the differential Stark shift ΔW of the corresponding pair of 5^1D and 5^3D Zeeman substates can be determined from ΔW = const x E^2 ± 1 % and agrees with calculated second–order perturbation theory using P and F states of the same n and hydrogenic wavefunctions. For example, for $m_j = -2$ of the 5^1D level, the quadratic Stark constants are –2.5999 ± 0.022 and theory gives –2.64 (kHz[V/cm]$^{-2}$) based on hydrogenic wave functions. The upper limit of n for which the experimental method was applicable was n = 9 when the levels are more sensitive to small stray electric fields (predominantly motional electric fields in the magnetic field required for the crossing) and the experimental uncertainty becomes too large to be useful for comparison with theory.

As there is no simple extrapolation formula [53] between states of different L, the energies of states with higher L cannot be derived from low–L results and must be calculated for each L. The n–dependence of the fine structure intervals, given empirically by $\Delta v = An^{-3} + Bn^{-5} + Cn^{-7}$ for $7 \leq n \leq 10$ and $3 \leq L \leq 7$ for helium, was found to be consistent with observations.

(d) The quadratic Stark effect of the n = 4 He$^+$ and the $2^2P_{1/2}$ state of atomic hydrogen.

In the above studies the shift of the crossing position by the external electric field had to be measured and extrapolated to zero electric field. In general, as the external electric field strength is reduced, the Stark shift eventually becomes small compared with the separation of the interacting fine structure levels [26, 90] and the linear Stark effect changes into the quadratic Stark effect. The above data were found to be almost exclusively governed by the quadratic Stark effect in agreement with theoretical results obtained from diagonalization of the complete upper state fine structure matrix using Zeeman formulae including higher–order corrections such as for the nuclear motion and the anomalous magnetic moment of the electron and allowing for the perturbation of the electric field.

For the n = 4 sublevels of He$^+$, and using electric fields from 50 to 220 V/cm to remove the anticrossing points, the quadratic Stark constants of the 3 S–F and 3 S–D signals studied at magnetic fields between 0.5 and 1 Tesla varied between (-7.68 ± 0.16) kHz(V/cm)$^{-2}$ for the anticrossing $\alpha N'$ and (-79.04 ± 0.17) kHz(V/cm)$^{-2}$ for the anticrossing αJ. Figure 1.4 is a

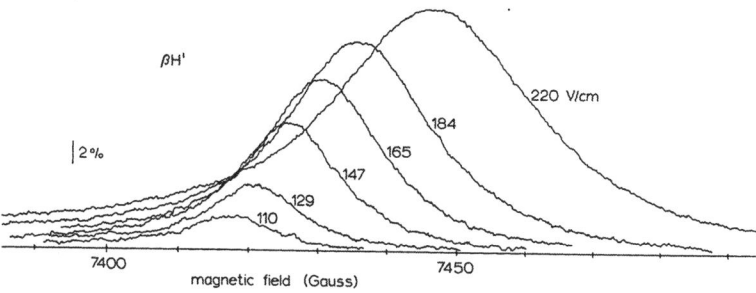

Figure 1.4 The third order anticrossing $\beta H'$ signal in 4S–4F of He$^+$ is shown as a function of magnetic field for various values of the static electric field which provides the coupling. The variations of amplitude, width and position of the anticrossing signal with field, and the quadratic Stark shift, are shown. The bar indicates a change of about 2% of the total light intensity (From Ref. [26]).

good illustration of typical anticrossing signals showing a quadratic Stark effect. All results agreed within one standard deviation with the calculation, apart from the $\alpha N'$ which is also the only signal departing noticeably from a purely quadratic Stark shift (1.7 standard deviations) and at the time allowed that aspect of quantum mechanics to be compared with observation.

About 1975 the question was whether a non–linear Stark shift existed for field strengths smaller than when the Stark shift equals the fine structure separation for n = 2 atomic hydrogen. The information was needed to determine the accuracy of Lamb shift and Rydberg constant measurements. Only a linear Stark shift had been observed when the energy splitting of states with different orbital quantum numbers was negligible compared with the Stark shift. Consequently a measurement was made [32] with $\Delta m = 2$ for the $2^2P_{1/2}$ state of H from genuine level crossings between the $2^2P_{3/2}$ ($m_j = -3/2$) and $2^2P_{1/2}$ ($m_j = 1/2$) levels. The method was to find the level crossing position for the two m_j levels by varying the magnetic field (near 0.35 T) for various (low) field strengths less than 3000 V/cm and observing the resonant Lyman–α signal. The observed non–linear perturbations were in good agreement with their quantum calculations of the second–order shift. The work is notable for the excellent experimental skills in a difficult measurement. Extensive details of their level crossing studies are given in several reviews [52, 59, 61, 108].

4. CORRELATION AND COHERENCE
4.1 THEORY OF THE MEASUREMENT

Numerous significant studies began to indicate how observations could be made to reveal fundamental quantum phenomena with fewer unobserved scattering dynamical quantities. Percival and Seaton [196, 197] had described the polarization of atomic line radiation and Seaton [207] had extended the theory to excitation and ionization. Burke and Schey [208] had described elastic scattering amplitudes and phases. The validity of the assumption that spin and orbital angular momenta may be separately conserved during a collision for light atoms was being questioned. Burke and Mitchell [209] considered spin–orbit interaction effects and related the observables of polarization and of singlet and triplet differential scattering cross sections to scattering matrix elements and scattering amplitudes. The links with more detailed observation were developing. A *perfect scattering experiment* using polarized particles in terms of amplitudes and phases had been described by Bederson [210–212] while Rubin et al. [213] measured the ratio of spin–flip to total differential cross sections from polarized atom beam recoils. In Europe, Hertel and Stoll [214–216] observed super–elastic scattering of electrons from laser–excited sodium and related the observations to the time inverse of the electron ground state excitation process and Kessler [217] had already published his authoritative book on polarised electron scattering.

Kleinpoppen, at JILA in 1970, was indicating [18] how scattering amplitudes and their relative phases could be deduced for polarized electron scattering for $^2S_{1/2}$ to $^2P_{1/2,3/2}$ excitation, as well as for elastic collisions, from one–electron atoms. The early analyses related the one–electron scattering observed quantities of differential cross sections directly to amplitudes and their relative phases. In a simple case and neglecting spin–orbit effects, only four scattering amplitudes $f^\pm$, $g^\pm$, f^0 and g^0 and three relative phases were shown to be required for a 'complete quantum description' of the scattering. The superscripts refer to the magnetic sublevels $M_L = \pm 1, 0$ and f and g refer to direct and exchange scattering. The next development analyzed the scattering amplitudes for polarized electron–polarized atom scattering for one–electron atoms and indicated how amplitudes and relative phases could be extracted. Then Hils et al. [22] scattered 3.3 eV unpolarised electrons from polarized potassium atoms to obtain direct differential elastic cross sections with a similar shape to two–state close coupling calculations but with significant differences in angular and shape variations of over three standard deviations. This experiment was significant also in showing the feasibility and limitations of measuring spin–dependent amplitudes.

The impetus had been given now for an extension of the earlier work to two–electron spin–1 atoms [28] in the first paper of a significant collaborative period with Karl Blum. Initially, spin–dependent interactions, such as spin–orbit coupling, hyperfine interactions and nuclear spin, were neglected, i.e. only polarization phenomena arising from electron exchange were considered. Expressions were derived relating the polarization vectors and tensors before and after scattering and in which the dynamics were separated from the terms describing the initial state polarisation. This feature allowed scattering amplitudes to be extracted from polarization measurements. The observables were described in terms of amplitudes which are proportional to the product of factors representing the geometry (vector coupling coefficient), Legendre Polynomials, and scattering matrix summed over unobserved spin and angular momentum quantities. The tensor components result in a characteristic angular dependence of photons. The observational challenge was how to decompose the components. The basic realization emerged from the symmetry of the transformation properties of the tensor components under various symmetry requirements which then determines the observational quantities e.g. $I(\theta, \phi) \pm I(\theta, -\phi)$ or $I(\sigma_-) \pm I(\sigma_+)$. The vector coupling coefficients relate the reference frame, or quantization axis, that is appropriate for the collision to that required for observation. Much effort was, and still is, spent in obtaining the simplest and physically most intuitive expression of the geometry and dynamics. Professor Kleinpoppen's contributions concerned particularly the S–to–P excitation of one– and two–electron atoms which led to the next significant observations.

There has always been close links between theory and measurement in atomic physics, particularly for scattering studies. Advances in either theory or measurement usually closely follow one another and new observations usually rely on theory for explanation and for full acceptance. A unifying feature of the theory and observations emerged from the work of Fano [218] in a physical picture of how scattering amplitudes and phases were related to state multipole moments and in turn to expectation values of angular momenta or equivalently alignment and orientation. The orientation vector was related to the expectation value of the angular momentum $<J>$ and the alignment tensor to quadratic combinations of components of angular momenta, such as $<J_x^2 - J_y^2>$. Macek and Jaecks [219] and Fano and Macek [220] related these quantities to the coincidence detection of the angular distribution and polarization of radiated photons and scattered electrons from excited atomic states. All of the above efforts contributed to the disentanglement of the geometric and dynamic effects of the excitation process.

4.2 EXCITATION OF TWO–ELECTRON ATOMS, 2^1P AND 3^1P STATES OF HELIUM

Eminyan, MacAdam, Slevin and Kleinpoppen [25, 31] applied these basics to measurements of the excitation of the 2^1P state of helium and extracted the magnetic sublevel amplitudes and their relative phase. Before their work, experiments concerned the measurement of scattered electrons, recoiling atoms or measuring either the intensity or polarization of the radiated photons or absolute cross section values, that is information of the amplitude squared without phase information. Now the coincidence technique identified identically prepared final states. This seminal work showed the way forward for observations and has been followed by many studies. The studies [25, 31] measured angular correlations between the 21.2 eV scattered energy loss electrons and the 58.4 nm photons radiated from the 2^1P state of helium. Figure 1.5 gives an indi-

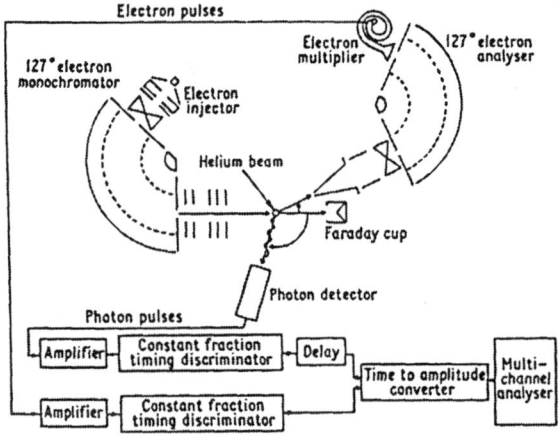

Figure 1.5 Coincidence apparatus used for helium 2^1P angular correlation measurements. (From Ref. [31])

cation of the apparatus and the coincidence timing circuit. They deduced the ratio of the differential cross sections, and hence the amplitudes, for exciting the degenerate magnetic sublevels, and their relative phases, for electron scattering angles from 16° to 40° for incident electron energies from 40 to 200 eV. The observations showed that angular momentum transfer caused the electron charge cloud to rotate and the alignment of the atomic dipole radiation departed significantly from the linear momentum direction at large scattering angles (e.g., Fig. 1.6). This information was previously hidden in the differential and total cross sections which essentially integrated over the details.

The work was extended to the 3^1P helium [37] using both the 53.7 nm (3^1P–1^1S) and the 501.6 nm (3^1P–2^1S) lines for scattering angles from 10 to 30° and incident energies from 50 to 150 eV. The observations showed the same general trends as for excitation of the 2^1P state. The main feature was the rotation of the charge cloud away from the momentum transfer axis near 80 eV at large scattering angles; this movement again corresponds to the near maximum angular momentum transfer to the atom when 100% circularly polarized photons are emitted from the atom and then a symmetry axis is not well defined.

A complete quantum description was given by Blum and Kleinpoppen [34, 35, 41] for the scattering process and, for the ideal situation, observations could be interpreted without the need to integrate over unobserved variables. Underlying these considerations is how symmetry principles determine conservation laws and their consequential simplifications of the relationships between the scattering amplitudes and their guidelines for measurements. These ideas manifest themselves such that the coincident intensity can be related to a sum of terms of different multipole rank. i.e. one term proportional to a scalar operator constant for all states, a second term representing the anisotropy of the radiation summed over all polarizations and a third term representing the anisotropy of the linear and circular polarizations. The design and interpretations of many subsequent measurements have been based on this description. For the coincident detection of the scattered energy loss electrons and the radiated photon it was shown that the same excited quantum state was sampled by each true scattering event. The observations which illustrated the strength of this approach included the excitation of the 2^1P state of helium (Eminyan et al [25, 31, 42]); the Lyman-α excitation by charge transfer [221] and orientation and alignment with heavy ion collisions with tilted carbon foils [222].

Blum and Kleinpoppen [41] then considered the formalism of electron–photon coincidence studies with the concept of the polarisation density matrix for the emitted photon as opposed to the intensity of photons emitted in a given direction in a given polarisation state. This concept thus characterizes the emission process rather than the polarisation state which is a function of the photon detectors. In particular the polarisation density matrix is given in the helicity representation so that the invariance of the helicity under rotations can be used for simplifying the formalism. This approach allowed the observations of Eminyan et al. [25, 31], for example, to be expressed simply through the Stokes parameters and state multipoles.

Then the concepts from optics of the vector polarization $P^2 = P_1^2 + P_2^2 + P_3^2$ where $0 \leq P \leq 1$ and the degree of coherence $\mu = |\mu| \exp(i\beta) = (P_2 - iP_3)/(1 - P_1^2)$ where $0 \leq \mu \leq 1$ were introduced [44, 47]. These ideas indicated that electron–photon coincidence measurements normal to the scattering plane, rather than only in the plane (where $P_3 = 0$), are necessary to determine μ and P. Only then can the equivalence of the phase between the two light vectors be made with the quantum mechanical phase between two amplitudes. Subsequent work applied and tested the ideas, particularly Standage and Kleinpoppen [44] using the 3^1P–2^1S 501.6 nm excitation in helium. Observation of the three Stokes parameters confirmed that the total vector polarization was unity and was possible only for full coherence. The measurement of right and left handedness meant that the photons were emitted from oriented atoms with orbital angular momentum equal to either +1 or −1 and that a quantum mechanical phase had a one-to-one correspondence with the photon phase and with the relative phase between amplitudes. An equivalent way of looking at electron scattering asymmetries with regard to circularly polarized light was measuring the left–right electron scattering asymmetry when the circular polarization of the coincident photons is kept fixed. Also it was shown that the angular distribution of the electron charge cloud is rotated by 90 degrees compared to the photon polarization distribution, and the alignment angle is just identical to the minimum angle of the photon intensity for an angular correlation experiment in the scattering plane. Again, in hindsight, these are fundamental classical and quantum ideas that needed observational confirmation of their significance in these studies.

The scattering amplitudes for electron impact excitation of the 3^1P state of helium at 80 eV was measured [137] for positive and negative scattering angles detecting the 501.6 nm photons in coincidence with the energy loss electrons. The sign conventions of helicity (LHC and RHC polarized light), and positive and negative scattering angles are discussed and the differences between references [62] and [44] noted. Such instances are common and indicate the consequences of various sign conventions in different fields of physics. Nevertheless, the behaviour of the Stokes parameters for the 3^1P state excitation process were shown to be similar to 2^1P excitation; i.e. P_1 is symmetric but P_2 and P_3 are asymmetric with respect to positive and negative scattering angles and P_2 and P_3 are negative for positive scattering angles. Papers [33,34,45,47,49,50,62,63,96] reviewed this progress but with slightly different emphases in succeeding papers.

Further ideas [93] on the shape and dynamics of states excited in electron–atom collision, particularly as they relate to the effects of attractive and repulsive forces effecting the orientation and alignment parameters began to emerge. Andersen, Hertel and Kleinpoppen [113] revised some aspects of the earlier work [93] and clarify the formulation in relation to the existing parameters. The subject was subsequently expounded at length in a review paper by Andersen et al. [223]. It is sufficient here to indicate two features. First, the attractive f_A and repulsive f_R amplitudes are related directly to the $f(M_L = +1)$ and $f(M_L = -1)$, respectively, in the natural frame. Secondly, the wavefunction $|\psi>$ of a P state is determined by the two parameters γ and $<L_\perp>$ in a coherence measurement whereas a correlation measurement yields only γ and $|<L_\perp>| = (1 - P_{lin}^2)^{1/2}$. A Stokes parameter analysis thus determines the P state completely while correlation analysis only reveals the shape but not the

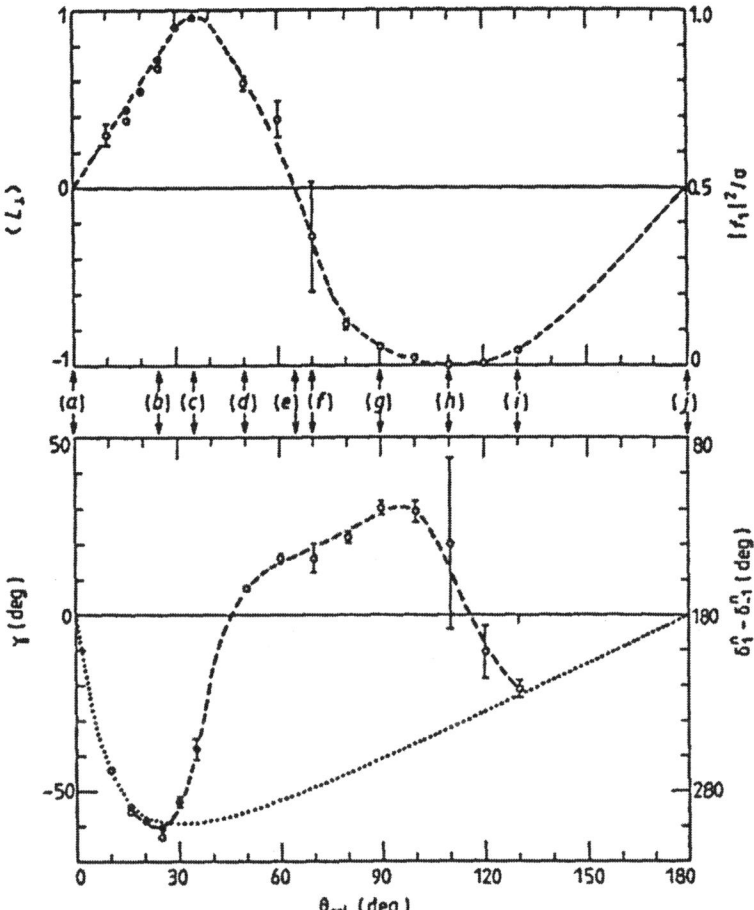

Figure 1.6 Helium 2^1P excited charge cloud parameters $<L_\perp>$ and alignment angle γ as a function of electron scattering angle. The small letters (a) to (j) indicate the angles for which the charge cloud is plotted in figure 1.7 (From Ref. [113]).

dynamics of the electron cloud. The paper illustrates their thoughts by plotting those parameters for the 80 eV excitation of the 2^1P state of helium that is reproduced here in figures 1.6 and 1.7. The paper is informative of thinking in 1984.

For decades, both theory and measurement struggled to describe the excitation of the lowest states of one– and two–electron simple targets. Now, the R–matrix with pseudo–states (RMPS), the convergent close coupling (CCC) and the distorted wave methods have obtained coherence and correlation pa-

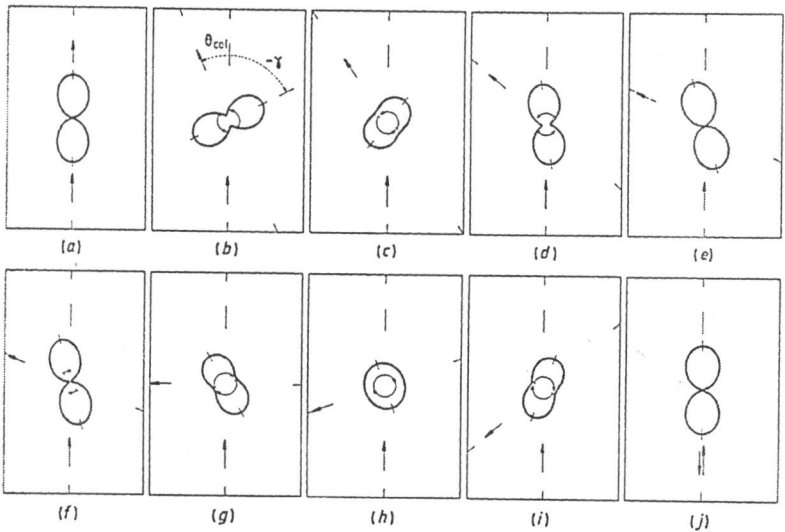

Figure 1.7 Data of figure 1.6 in charge cloud form. (From Ref. [113])

rameters, for 2^1P and 3^1D helium and 2^2-P states for hydrogen, from scattering dynamics from near threshold to about 200 eV. There is a migration of both experimental and theoretical values towards generally accepted CCC with a large number of pseudo states values. The theoretical successes owe much to their ability to include many (up to 99) pseudo states representing bound and continuum states [224], and the rapid growth in computer hardware. The RMPS and CCC descriptions of the differential cross section for excitation of hydrogen 2^2P and helium 2^1P and $3^{1,3}$D states, as well as of the angular and correlation parameters, are now in general agreement with one another and with most measured data, particularly at low energies.

It is frequently stated that one of the aims of particle scattering and structure studies is to guide theoretical studies. Basic to this approach is how to describe the target. For electron scattering this means the effects of configuration interaction concern the N electrons in the target or the N–electron correlations from CI effects caused by the interaction of the incident electron with the target or (N+1)–electron correlations. This can be done sometimes by the different energy dependencies; the (N+1)–electron correlation effect can be dominant at low energies and becomes less important at high energies, whereas the N–electron correlations can be important at all energies. The latter is seen in the allowed transitions, which at high energies is proportional to the oscillator strength that in turn depends on the wave functions used. For the low energy case of atomic hydrogen exact target states are known and only (N+1)–electron correlation effects can occur. The excitation of the helium atom is particularly interesting since the lowest and accessible levels illustrate basic symmetry and selection principles. From the ground state, the excitation of the 2^1P (21.2 eV) state is optically allowed, the 2^3P (20.96 eV) state is spin forbidden, the 2^1S state (20.61 eV) is symmetry forbidden and the 2^3S state (19.81 eV) is spin and symmetry forbidden. The consequences of these correlations remain to be explored fully.

4.3 HYDROGEN

The association of Blum and Kleinpoppen flourished with two papers [51,60] showing the strengths of the density matrix in describing even–odd parity coherence effects in electron–hydrogen scattering for n = 2 and n=3 state excitation. Earlier work concerned excitation which was axially symmetric around the incident beam. Thus this symmetry restricted coherence to states of the same z-component of total and orbital angular momentum of the excited states. Information was obtained only about the M_J diagonal density matrix elements. Here [51] for the planar symmetric geometry of coincidence experiments, the states no longer have *sharp* angular momentum or parity. The basic assumptions were that Russel–Saunders coupling was valid during the excitation process and that, after collision, the atoms relaxed into the JM_J coupling without further interaction. In this case with the same n, but different L and M and without resolution of the initial and final spins which partially destroy the coherence, an external field must be present to observe coherence quantum beats between states with different parity. Then, with and without an external electric field, using the dipole approximation in which only multipoles of rank equal to or less than two can contribute, they derived expressions for the Stokes parameters and use a first Born approximation to give sample values for the alignment and orientation parameters. The influence of fine structure in excitation and decay was considered using perturbation coefficients to describe the time modulation. The n = 2 excitation studies described detection of the Lyman–alpha photon in coincidence with the scattered energy loss electron. Coherent excitation of n = 3 hydrogen in a field free region [60] developed the concepts required when the excitation process spanned atomic levels whose energy differences are smaller than the energy spread of the exciting electron beam (for example). The theory indicated the required observations of the n = 3 → 2 Balmer photons and the effects of cascades with detection of the Lyman–alpha but the Balmer–alpha undetected. The paper clearly indicates the implications of reflection symmetry, of the observational consequences of the rank and components of the state multipole moments. The strength of these papers is bridging the gap between formal theory and observations. Generally there is still scope for further experimental work but the measurements are very difficult.

22 COMPLETE SCATTERING EXPERIMENTS

The collaborative efforts of Blum and Kleinpoppen are well represented and expounded in their reviews [62,63]. Paper [103] was based on K Blum's Habilitation thesis from Münster and indicates the considerable theoretical strengths brought to bear on the descriptions of the scattering of spin–polarized electrons in inelastic collisions.

4.4 SODIUM [(NE)3S^1] AND POTASSIUM [(AR)4S^1]

The considerable experiment skills at Stirling were shown by Hils and Kleinpoppen [55] in 1978 in significant observations of the ionisation process using polarised P_e electrons and polarised P_a potassium atoms. The potassium beam was 72 % polarised after passing through a hexapole magnet and then traveled to the interaction region in the presence of a weak magnetic field in which hyperfine coupling of the electronic and nuclear spins ($I = 3/2$) reduced the polarisation by a factor of $(2I + 1)^{-1}$ to 18 %. A beam of about 10^{-10} amps of 20% polarised electrons were produced by scattering 80 eV electrons from mercury atoms at an angle of 80°. By defining an interference cross section Q_{int} and a spin–averaged cross section Q_0, the asymmetry A of the ionization process is $A = P_e P_a (Q_{int}/Q_0)$. Since there was no theoretical value for comparison, they deduced that in reduced energy units (i.e. E/I), the A factors for Li and Na were similar and would probably be similar for potassium. This experiment was some years ahead of its time (1978), now it is seen to be an outstanding effort of technical ability with great potential to measure all of the currently interesting and needed data for a complete quantum measurement.

Hils et al. [91] later on improved the polarisation of a sodium beam to 80% by laser pumping and this enabled measurements of the ionization asymmetry from threshold to energies about seven times the ionization energy. The asymmetry A was about 0.3 near threshold at 5.12 eV and decreased to about 0.1 near 35 eV. The value near threshold varied significantly from a first Born calculation as expected since electron correlation effects were not included in the models. The data also enabled a test of the Wannier [225, 226] theory which predicted that ionization should proceed via the singlet channel with the A factor equal to unity and the ratio of triplet–to–singlet cross sections should vary as $E^{2.756}$. The range of validity of the threshold law was less than 0.5 eV above threshold while the singlet and triplet wave functions obey the same threshold law, in which case A would be energy independent near threshold and could assume any value. It was clear that further work was required. The work is also notable for the experimental skills, the assessment of false asymmetries and a time–of–flight detector of ions.

This apparatus was now working very well and more ambitious measurements were envisaged. The theory of Burke and Mitchell [209] had described the elastic scattering of electrons from spin–1/2 targets including spin–orbit interaction by six independent amplitudes and five relative phases. Khalid and Kleinpoppen [98] indicated how eleven measurements allowed these 11 quantities to be determined. Measurements by Hanne [227] obtained some data, for example, on the ratio of final to initial transverse polarisation of electrons scattered in the forward direction, but it became clear that such measurements were becoming extremely difficult and that observations of excitation processes would enable similar fundamental data to be obtained.

Electron exchange was studied in the 3s–3p impact excitation of spin–polarized Na atoms by unpolarised electrons [109]. The fluorescence light emitted in the decay of the excited state exhibits linear polarization caused by the collisionally induced 3p alignment as well as by circular polarization resulting from the spin orientation of the 3p electron at the time of the decay. Since the exchange interaction causes a spin flip, the degree of circular light polarization is a measure of the contribution of the exchange channel to the total excitation cross section. At small impact energies where exchange becomes important, the circular polarization decreased. From analysis of the hyperfine coupling between the excited electron and the polarized nucleus, and calculating the light polarization from theoretical excitation cross sections, experiment and theory showed reasonable agreement.

Stokes parameters were observed from 21% polarised sodium and potassium atoms excited [168] by unpolarised electrons to the lowest $^2P_{1/2,3/2}$ states for 589.0 / 589.6 nm for Na ($3^2P_{1/2,3/2} - {}^2S_{1/2}$) and 404.4 / 404.7 nm for potassium. Within rather large experimental uncertainties, the finite P_3 values near threshold indicate the strength of electron exchange, P_2 may be zero as expected for very weak or zero spin–orbit interaction and P_1 is about 12% as predicted from theory.

4.5   CALCIUM [(AR)4S²] AND STRONTIUM [(KR)5S²]

Excitation cross section and polarisation of the Ca II 393.3 nm line by electron impact on Ca atoms ($4^2P_{3/2}$ to $4^2S_{1/2}$) line following simultaneous ionisation and excitation from Ca ground–state atoms have been measured [148]. The absolute excitation cross section of the Ca II $^2P_{3/2}$ state has been determined to be $(0.72 \pm 0.27) \times 10^{-16}$ cm² at an electron energy of 40 eV. The linear polarisation of the 393.3 nm line of Ca II has been measured from the $4^2P_{3/2}$ threshold to 200 eV. The polarisation peaked at 17% near threshold and converged to zero at high energies.

Electron impact coherence and polarisation parameters [161] for the $5^1P–5^1S$ at $\lambda = 460.7$ nm state (5 nsec) of helium–like two electron ns² outer shell alkali–earths [Be, Mg, Ca, Sr and Ba] strontium [161] were determined at 45 eV and angles 30° to 113°. Here the lower excited n^1P states are more strongly coupled to the ground state and electron correlation effects are expected to be important in the excitation process. These studies tested the influence of the atomic configurational correlation effects on the state parameters and test the dynamical collision approximations. From calcium onwards, additional interest results from the empty (n-1)d shell below the ns^2 shell of the ground state and the influence of that fact on the scattering process could be studied by comparison of Ca and Sr excitation parameters with those of Zn and Cd which have the same ground state electron structure but with the inner d–shells filled. In the heavier atoms, the spin–orbit coupling in the excitation process might become noticeable. Since $I(\alpha) = \frac{1}{2}[1 + P_1 \cos 2\alpha + P_2 \sin 2\alpha]$ at small θ_e (approximately less than 50°) the excited charge cloud rotates away from the incoming beam up to an angle about 30°, then rotates back through the incoming beam. The P_3 parameter changes significantly around 70°. Very generally, the Stokes data show similar behaviour to 2^1P helium excitation.

4.6 MERCURY [(XE)4F^{14}5D^{10}6S^2]

Zaidi et al [73] measured electron–photon coincidence rates for 5 to 7 eV electrons exciting the 6^3P_1 - 6^3S_0 253.7 nm (4.89 eV) state of mercury and for scattering angles from 50° to 70°. Stokes parameters measured, an isotope cell in front of the photon detector absorbed radiation from odd mercury isotopes and eliminated the effects of hyperfine structure in the observed radiation. They found partially coherent excitation and analyzed the data in terms of the da Paixao et al [228] model allowing for spin–orbit interactions. Later [81], a complete set of parameters (λ, χ, δ and ϵ incorporating spin–orbit effects) was determined from linear and circular polarization coherence measurements. Exchange was the dominant excitation mechanism from threshold up to 8 eV collision energy. The alignment parameter lambda agrees well with a previous measurement using incident polarized electrons. The results were explained in terms of intermediate coupling. The work indicates that these measurements for many–electron atoms require theory to catch up and provide further interpretation.

Paper [156] reported a novel technique for the elimination of the hyperfine effect on the polarization of atomic line radiation, specifically for the 253.7 nm (6^1S_0–6^3P_1) transition from the even isotopes of mercury. Earlier work used an isotopically pure (separated) beam. The technique used an isotope cell which absorbed that part of the radiation originating from the odd isotopes in the natural isotope mixture of mercury. The even isotopes with nuclear spin zero have an abundance of 70% while the two odd isotopes ^{199}Hg and ^{201}Hg have abundances of 17% and 13% respectively have nuclear spins of 1/2 and 3/2. The cell contained only odd isotopes which absorbed the hyperfine radiation. While the idea is straightforward it had not been used in over 40 years of polarization measurements. The substantial influence of the hyperfine states was clearly shown; the resonance structure was enhanced and the threshold polarization, extrapolated from above resonance energies, indicated values of $P = -0.9$ without hfs and -0.7 with hfs in agreement with early theory.

5. PHOTOIONIZATION (SYNCHROTRON) STUDIES

The intensity and polarization of synchrotron radiation has broadened the scope for a complete quantum measurement of a collision process to photoionization, particularly for those cases where a limited number of dipole amplitudes and their relative phases are adequate. In recent years Professor Kleinpoppen has pursued the *complete quantum measurement* by collaborating with groups using the Daresbury synchrotron to study Ca atoms and the Berlin synchrotron to study atomic oxygen. As with electron impact there are a number of combinations of either angular or polarization correlations or spin polarization either before or after scattering from which partial or complete information on the scattering amplitudes and phases can be obtained. The story of measurements concerns the selection of observational method and target atom. Heinzmann et al. (Ref. [229, 230] and references therein) made a complete measurement for Xe 5p photoionization by observing the angular distribution and the spin polarization of photoelectrons. For non-resonant inner shell photoionization, measurement of the angular distribution of the photoelectrons and Auger electrons allows the angular distribution asymmetry parameter β and the alignment parameter A_{20} to be deduced. Similarly, the polarization of the fluorescence radiation instead of the angular distributions of the Auger electrons [231] has been observed. When this information is combined with a model, using valid *LS* coupling such as for light atoms, the partial photoionization cross section and the angular distribution asymmetry parameter β, then a *complete quantum description* follows.

5.1 CALCIUM [(AR)4S²]

The Daresbury group found the 3p–3d Ca (and 3p–4d Sr) resonance regions [155, 157] attractive because of large (10^{-15} cm^2) excitation cross sections and the significant configuration interaction *on and off* resonance. Polarization measurements of the fluorescence radiation $4p^2P_{3/2}$ to $4s^2S_{1/2}$ ground state of Ca$^+$ (and similarly from the $5p^2P_{3/2}$ level of Sr$^+$) determined the alignment tensor and the angular distribution parameters. The resonance $3p^64s\ ^1S_0 \to 3p^54s^23d\ ^1P_1$ enhanced the excitation to the $4p^2P_{3/2}$ state by a factor greater than 100. The applicability of *LS* coupling with only two possible outgoing photoelectron channels restricts the model to a ratio of two amplitudes and their relative phase.

Their next work [167] identified the cascade radiation from Rydberg levels following autoionizing decay after 3p–3d excitation, using coincidence detection of the fluorescent photon 4p–4s transition with the photoejected electrons leaving the Ca$^+$ ion in the 4p^{2}P$_{3/2}$ level. Basic to the geometry of this work, as identified in the earlier studies [17, 23], the linear polarization and momentum vectors of the incident synchrotron photons defines the reference frame. They deduced the ratio of the dipole amplitudes D_s and D_d for s– and d–wave scattering to be $D_s/D_d = 1.4$ with a relative phase of about 44°.

Then [174], for reversed helicities σ_+ and σ_- of the incident photons, the difference between the intensities of the ejected photoelectrons, gives the circular dichroism. For a given ejected electron angle θ_e, the circular dichroism CD is given by $CD = I(\theta_e, \sigma_+) - I(\theta_e, \sigma_-) \propto -|D_s||D_d|\sin(\delta_s - \delta_d)\sin 2\theta_e$. That is, the sign of the relative phase was obtained through the sine (rather than cosine) term and shown to be positive. The work is also interesting because it determines the sign of the relative phase using the same expression as above but observing the linear polarizations, parallel and perpendicular to the incident photon direction, of the fluorescent photons in coincidence with the photoelectrons detected at angle θ_e. It is noted also that the definition of left and right is made with the observer looking in the same direction as the incident beam is travelling i.e. opposite to that in general use in electron correlation studies [232].

These studies have a much broader attraction [185] because a broad range of spectroscopic features are enhanced by the 3p^5 4s^2 3d ^{1}P$_1$ resonance at 31.41 eV. There is coupling particularly between the 3p^5 4s^2 3d ^{1}P$_1$ and 3p^5 4s^4 P$_j$ nl states which enables autoionization into the Ca$^+$ ^{4}P continuum, and the 3d wavefunction for the 3p^5 4s^2 3d ^{1}P$_1$ level overlaps the 4d wavefunction of the Ca$^+$ ion. These aspects combine to increase the population of the nd levels of the Ca$^+$ ion. Alternatively these aspects are described as configuration mixing of the inner–shell excited levels. The study interprets the transition probabilities through these features when the Ca$^+$ ion is left in excited Rydberg levels converging to the ground level of the doubly charged ion. The angular distribution parameter β of electrons for a 5s Ca$^+$ level off–resonance indicates LS coupling does not apply; whereas for 3d and 4d Ca$^+$ levels on–resonance indicates p and f outgoing electrons have symmetry so that the spectator model (in which the spectator Auger electrons have p–wave symmetry) does not apply.

These studies are also noteworthy because they indicate the international scientific collaboration that has grown over the last decade. In this case there is the need arising from the high cost of obtaining photons from a synchrotron but the collaboration is of the nature that Professor Kleinpoppen will be long remembered for his striving to organize. The group comprised Ueda from Sendai (Japan), West and Ross from the UK, Beyer and Kleinpoppen from Stirling, Hamdy from Cairo, Kabachnik from Moscow.

5.2 ATOMIC OXYGEN [1S² 2S²2P⁴]

The other group studies on the Berlin synchrotron concern the photoionization of free oxygen atoms [178, 187] with measurements of the angular distribution of photoelectrons by ionizing the ground state $O(2p^4)$ to a final ionic $O^+(2p^3)$ $^4S_{3/2}$ level. The target atoms were J–polarized and the incident photons linearly polarized. The relative ease of using linear, rather than circularly, polarized photons from a synchrotron has given rise to measurements of angle–integrated magnetic linear dichroism (MLD) and linear magnetic dichroism in the photo electron angular distribution (LMDAD) [178, 184, 187]. The former varies the angle between E (the electric vector of the photon) and A (the alignment vector of the polarized target atom), while the latter varies A (parallel and aparallel) relative to the fixed E vector. The Berlin group [187], using MLD, studied polarized open–shell atomic oxygen with non–zero total angular momentum with the ejection of 2p electrons. The oxygen atoms were polarized by a hexapole magnet, which focused only the m_J positive values. The polarization of the beam was monitored by a Stern–Gerlach magnet to have m_J populations of 3, 7, 13, 24 and 53% for $m_J = -2, -1, 0, 1$ and 2 respectively. This performance is state–of–the–art. In contrast the photon beam was 99% linearly polarized.

This approach is an alternative to measurement of the spin of the photoelectrons to obtain information beyond the differential cross section and the asymmetry parameter. The reference frame for the collision was set by using a guiding magnetic field for the magnetic moment of the polarization of the atoms to be parallel (+) or aparallel (−) to the propagation vector of the polarized photons. The intensity I of photoelectrons, ejected at angle θ_e in the plane of the E vector of the photons, for aparallel and parallel atom polarization showed a *linear magnetic dichroism* for p-shell ionization in

$$O(1s^2 2s^2 2p^4)^3 P_2 + h\nu \rightarrow O^+(1s^2 2s^2 2p^3)^4 S_{3/2} + e^-(\varepsilon s, \varepsilon d) \quad (1.2)$$

whereas s–shell ionization to $O^+(1s^2 2s^2 2p^4)$ $^4P_2 + e^-$ showed no such dichroism, as expected since only relativistic effects cause dichroism for s electrons. Two interesting results were obtained [178, 187]. The data of figure 1.8 could be described only by the use of two angular distribution β parameters, in addition to the cross section, such that one β parameter was fitted to data from the sum of the parallel and antiparallel atom spins while the second was required for the difference of those spins. The dichroism data were interpreted using LS coupling and gave the ratio of the two radial components of the matrix elements for the 2p electron photoionized into the εs and εd continuum states as well as their phase difference. Over the photon energy range from 25 to 52 eV, the ratio of the dipole integrals was constant but the phase shifts change by about $30°$. The reason was attributed to the difference in Coulomb phase shifts for

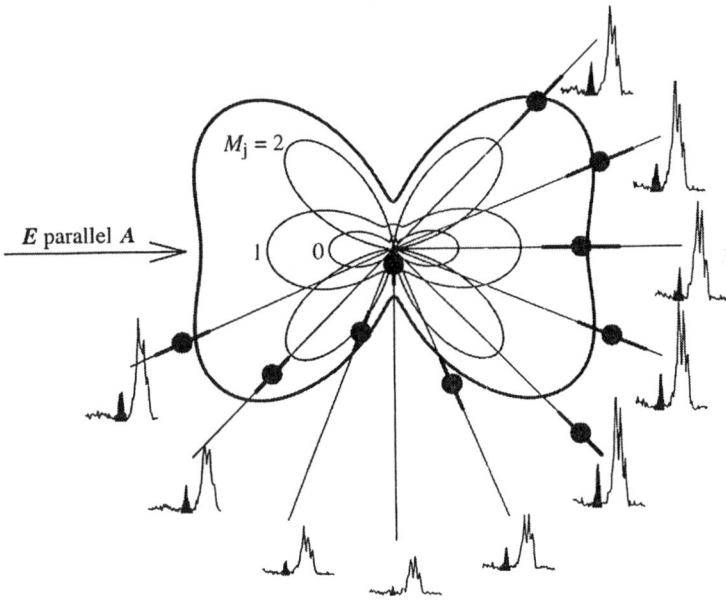

Figure 1.8 The measured angular distribution of photoelectrons from polarized oxygen atoms is shown by filled circles. The curve of best fit (thick line) is a superposition of two separate $(1 + \beta P_2 \cos \theta)$ distributions arising from the polarized oxygen atoms (filled spectra). The other (unfilled) parts of the spectra arise from vibrational multiplets of unpolarized oxygen molecules (after Ref. [178]).

the two angular momenta. This work showed that, within the assumption of LS coupling for the light oxygen atom, an essentially complete photoionization measurement was possible. The second interesting result [187] concerned the photoionization process

$$O(1s^2 2s^2 2p^4)^3 P_2 + h\nu \rightarrow O^+(1s^2 2s^2 2p^3)\,^4S_{3/2},\,^2P_{1/2,3/2},\,^2D_{3/2,5/2}, +e^- \tag{1.3}$$

which was studied using the angle–integrated magnetic linear dichroism approach by collecting the ions without separating the three states. As indicated in figure 1.9, the cross section and the asymmetry displayed Fano profiles for photoexcitation of the autoionizing 2s 2p^4(^{4}P$_e$) 3p (^{3}D^0, ^{3}S$_0$, ^{3}P$_0$) resonance near 25 eV. The signal depended on the angle between the photon and atom polarizations and showed very little asymmetry away from the resonances. But at resonance, LS coupling no longer applies because the different J components of the ^{2}P and ^{2}D lines show different behaviour. The separation of these contributions would be difficult because there is rapid change of the partial waves and branching ratios near the multiplets.

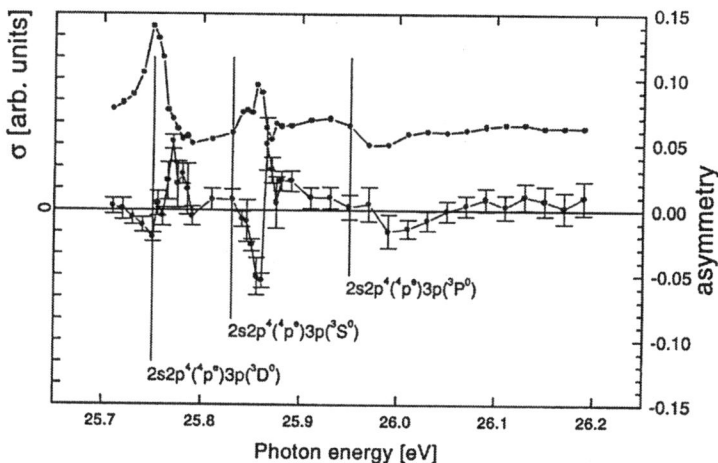

Figure 1.9 Resonance structure in the magnetic linear dichroism of free oxygen atoms. The upper line and data indicate the intensity and the lower line and data the asymmetry of the resonances (From Ref. [187]).

6. OUTER SHELL EXCITATION IN HEAVY PARTICLE COLLISIONS

In 1977 during a sabbatical at the University of Bielefeld Hans Kleinpoppen not only took part in ongoing experiments on atomic collisions processes but suggested and initiated several new ones. Unlike electronic collisions, slow atomic collisions proceed through well defined molecular states. Excitation and charge transfer processes are the result of crossings between the molecular states of the transiently formed quasi–molecule [233]. Naturally, the most fundamental atomic collision system H^+–H has attracted considerable interest. Of particular interest here is the excitation of the first excited H(2p) states, e.g.,

$$H^+ + H \to H^+ + H(2p) \tag{1.4}$$

that proceeds via a so–called rotational $2p\sigma$–$2p\pi$ rotational coupling at small internuclear distances (Fig. 1.10). This coupling, hence, leads to a preferred population of quasi–molecular p–states which asymptotically evolve into the magnetic H($2p_{\pm 1}$) sublevels. The integral alignment A_{20}, i.e., integrated over the scattering angle of the projectile, is defined as

$$A_{20} = \frac{\sigma_1 - \sigma_0}{\sigma} \tag{1.5}$$

where σ_m ($m = 0, \pm 1$) are the partial cross sections for excitation of the H($2p_m$) magnetic sublevels, and $\sigma = \sigma_0 + 2\sigma_1$. The measured A_{20} was found close to the expected value of +50% at low incident energies of 1–2 keV (Fig. 1.11), where

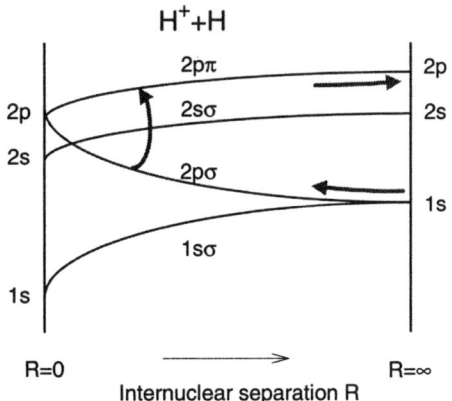

Figure 1.10 Correlation diagramme (schematic) for H^+–H versus internuclear spearation R. For clarity, the l–degeneracy has been removed (after Ref. [234]).

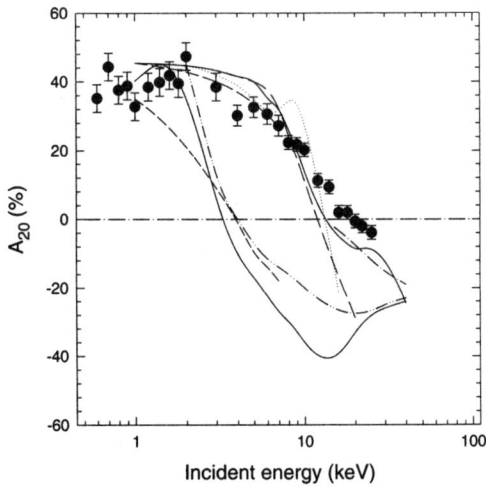

Figure 1.11 The integral alignment A_{20} for H^+–H collisions versus incident energy. Experimental points are compared with several theoretical predictions (after Ref. [140, 235]).

the quasi–molecular picture is expected to hold, followed by a pronounced decrease towards larger energies [140]. Fair agreement with state–of–the–art theoretical calculations was noted but some minor and thus far unresolved discrepancies remained [235]. The situation becomes more complicated for quasi–one–electron and two–electron systems like H–H and (H–He)$^+$, where other crossings, most notably the 2pσ–2sσ and similar radial couplings start playing a role [132, 141]. This is corroborated by angle–differential measurements of A_{20} employing the polarized photon–scattered projectile coincidence technique [66, 87, 117, 132]. These measurements were performed by employing an apparatus that made use of polarized photon–scattered projectile coincidence technique simultaneously measuring at four scattering angles θ_s and eight azimuthal angles ϕ of the scattered projectile. This allowed for a full polarization analysis, i.e., the simultaneous measurement of each of the three Stokes parameters P_1, P_2, and P_3 as a function of ϕ. The analysis followed

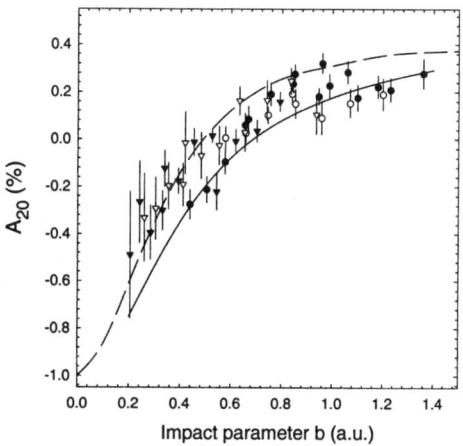

Figure 1.12 The differential alignment A_{20} for H$^+$–He collisions versus impact parameter b. The experimental data at incident energies of 1–4 keV are compared with different theoretical predictions (see text, after Ref. [132]).

along the lines of Blum and Kleinpoppen [62, 101] to obtain the orientation and alignment tensor components, $< T(1)_{11} >$ and $< T(1)_{2Q} > (Q = 0, 1, 2)$ of which the tensor A_{20} (which is equal to $< T(1)_{20} >$ within a normalization constant) displayed in Fig. 1.12 for H$^+$–He collisions is only one example. At distant collisions (i.e., large impact parameters respectively small scattering angles) A_{20} is positive indicating a preferred population of H(2p$_{\pm 1}$) substates [132]. At close collisions A_{20} becomes negative now indicating a preferred population of H(2p$_0$) substates (Fig. 1.12). The mechanisms responsible for this behaviour are the above mentioned 2pσ–2pπ rotational coupling that operates already for distant collision, while close collisions are influenced by the presence of a 2pσ–2pσ radial coupling. Due to this radial coupling the 2sσ and 2pσ are not pure states but contain admixtures of each other. This allows for a double 2pσ–2pπ–2sσ rotational coupling that operates at small internuclear separations [236]. The experimental results are in fair agreement with theoretical predictions [236, 237] and thus support this idea. Another observable is the orientation vector $< T(1)_{11} >$ that is related to the transferred angular momentum $< L_y >$ perpendicular to the scattering plane respectively to the relative phase χ between the excitation amplitudes f_0 and f_1. The phase χ was found to decrease with incident energy as is qualitatively expected if the time evolution of the adiabatic phase $(\hbar v_p)^{(} - 1) \int \Delta E(R) \, dR$ is taken into account. Here $\Delta E(R)$ is the energy of the molecular orbital along which the collision system proceeds, and v_p the collision velocity. For this collision system, a full description in terms of the four tensor components seemed unnecessary since full coherence, i.e., $P = (P_1^2 + P_2^2 + P_3^2)^{1/2} = 1$ can be assumed (although only to some extent established by the measurement). A two parameter description in terms of, e.g., λ and χ parameters hence should suffice. This is no longer true for more complicated atomic collision systems like, for example, H$^+$ + Ar → H(2p) + Ar$^+$. The final ionic state of the argon atom, Ar$^+(3p^{-1})$, by itself acts like a $3p_m$–state ($m = 0, \pm 1$). If the final argon state is not detected as it was the case in the experiment of Ref. [117], a full polarization analysis of the emitted photons is required to extract the maximum information. For this collision system we have a total of nine excitation amplitudes $f(m_1, m_2)$, where $m_1 = 0, \pm 1$ and $m_2 = 0, \pm 1$ refer to the magnetic substates of H(2p) and Ar$^+(3p^{-1})$, respectively. Reflection symmetry with respect to the scattering plane requires

$$f(m_1, m_2) = (-1)^{m_1+m_2} \times f(-m_1, -m_2) \quad (1.6)$$

which reduces the number of independent scattering amplitudes to five. This has two implications: (i) it is impossible to extract the full information from a standard polarized photon–scattered projectile coincidence experiment, and (ii) there will be a significant depolarization since the measured polarizations as well as the deduced tensor components are averages over the magnetic substates m_2 [117, 238]. For example, the alignment tensor $< T(1)_{22} >$ requiring a summation of m_2 is obtained as [117]

$$< T(1)_{22} > = - \mid f(1, 0) \mid^2 + 2 \Re \left[f(1, 1) f^*(1, -1) \right]. \quad (1.7)$$

As an interesting result, a positive $< T(1)_{22} >$ relates to a preferred population of $m_2 = \pm 1$–substates as was observed at some scattering angles [117].

Two–electron collision systems proceed through singlet and triplet spin channels. Although it is commonly assumed that spin interactions during ion–atom collisions are rather weak, the spin orbit coupling particularly in helium atoms leads to atomic states with energies that are significantly depending on the spin state. In He$^+$–H collisions, the reaction channel for direct excitation of H(2p)

$$He^+ + H \rightarrow He^+ + H(2p) \quad (1.8)$$

and for charge exchange excitation of He(2p)

$$He^+ + H \to He(2p) + H^+ \tag{1.9}$$

are in competition with each other [239]. The relative probability for these channels depends on the total energy of the final states that, for reaction (1.9) forming He(2^1P) or He(2^3P) states, differs by about 1 eV for the singlet and triplet channels. This yields to a different reaction path in the transiently formed quasi–molecule and to a different alignment for $H^+ + He \to H(2p) + He^+$ collisions, where only the singlet channel contributes, compared to $He^+ + H \to He^+ + H(2p)$ collisions, where owing to its larger statistical weight the triplet channel dominates [239].

Noticible spin effects have been also observed in other (quasi–) two–electron systems involving excitation of Na(3s) atoms to Na(3p) by Ne$^+$ [117] and He$^+$ [145] impact. Measurements using sodium atoms have the merit of a direct investigation of spin channels provided the sodium atom is spin polarised. These studies may be regarded as a straight–forward extension of electron impact excitation studies by electron impact [109]. The employed technique utilized collisions of unpolarised projectiles with spin–polarized Na(3s) atoms and detection of linearly and circularly polarised photons. It was shown [109] that the emitted circularly polarized light reflects the spin state of the excited Na(3p) atom after the collision. From the variation of the degree of circularly polarised light P_3 in Ne$^+$–Na(3s) collisions with collision energy a spin–dependent interaction was inferred. The observed spin depolarisation being largest for small collision velocities was explained by an interaction between the Na(3p) and Ne(3s) configurations. In the course of the collision, electron exchange between Na(3s) and Ne and back to Na(3p) may take place. During the collision the spin orientation of the active electron may be effected by the exchange energy, i.e., the energy difference E_{exc} between singlet and triplet levels, yielding to a spin polarisation that oscillates with a characteristic frequency of E_{exc}/h (h, Planck's constant). The experimental results were compared to a semi–quantitative model and fair agreement was obtained. Similarly, in He$^+$–Na(3s) collisions spin exchange is responsible for the observed spin depolarisation of the Na atom. In the absence of explicit spin–dependent forces, the processes that can be identified here are the direct and exchange excitation

$$Na(\uparrow) + He^+(\downarrow) \to \begin{cases} Na^*(\uparrow) + He^+(\downarrow) & \text{direct} \\ Na^*(\downarrow) + He^+(\uparrow) & \text{exchange} \end{cases} \tag{1.10}$$

and the interference process,

$$Na(\uparrow) + He^+(\uparrow) \to Na^*(\uparrow) + He^+(\uparrow) \quad \text{interference}. \tag{1.11}$$

During direct and the interference process the electron spin is conserved, while in exchange excitation the spin is reversed. A change of the excited Na*(3p) electron spin thus provides valuable information on electron exchange processes during atomic collisions. The experimental results for this collision system are

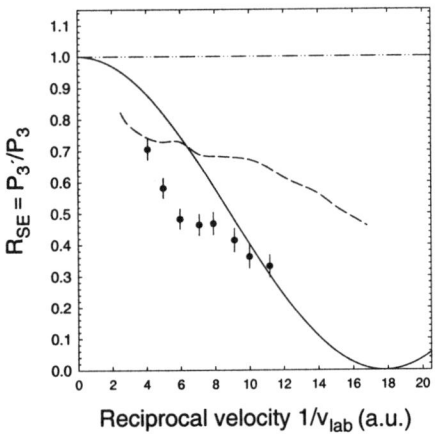

Figure 1.13 The spin exchange parameter $R_{SE} = P'_3/P_3$ versus reciprocal impact velcoity for He$^+$–Na(3s) → He$^+$–Na(3p) collisions (after Ref. [145]).

displayed in Fig. 1.13 versus the the reciprocal collision velocity. Fair agreement with a simple estimate taking the time evolution of the phase difference between singlet and triplet channels in the transiently formed quasi–molecule, as well as with more sophisticated calculations based on the quasi–molecular model of Mo and Riera [240] is noted. Heavy ion collisions involving hydrogen ions or atoms are well suited to investigate coherent excitation processes among states of different parity [238]. A particular interesting example is H(n=2) excitation in direct and charge exchange excitation processes, e.g,

$$H^+ + He \rightarrow H(n=2) + He^+.\tag{1.12}$$

The coherently excited H(n=2) atom formed during the interaction gives rise to pronounced forward–backward asymmetries with respect to the direction of the incident beam of photons emitted during the de–excitation process. Such a forward–backward asymmetry corresponds to a collision–induced dipole moment, i.e., an electron which is not centered at the origin of the atom. In the axially symmetric case it is related to the $\sigma_{sp0} = f_{2s0} f^*_{2p0}$ element of the

H($n = 2$) density matrix $\rho_{n=2}$,

$$\rho_{n=2} = \begin{pmatrix} \sigma_{2s_0} & 0 & \sigma_{sp_0} & 0 \\ 0 & \sigma_{2p_{-1}} & 0 & 0 \\ \sigma^*_{sp_0} & 0 & \sigma_{2p_0} & 0 \\ 0 & 0 & 0 & \sigma_{2p_{+1}} \end{pmatrix}.$$

In H$^+$ + He collision a pronounced dipole moment of the excited H(n=2)

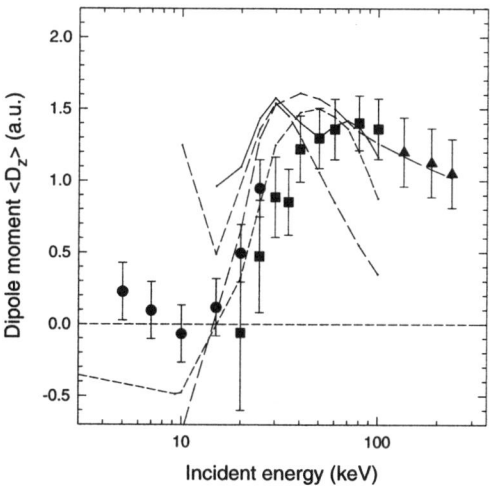

Figure 1.14 The collision–induced electric dipole moment of H(n=2) states for H$^+$–He collisions versus incident energy. The experimental results are compared with different theoretical predictions (after Ref. [151]).

atom was observed whose sign indicated that after the collision the electron was trailing behind the proton (Fig. 1.14). The measurements thus supported the simple picture of an electron trying to stay as long as possible between the two ion (H$^+$ and He$^+$) cores until it finally decides for either one. Large dipole moments were thus also found for other collision system involving H$^+$ ions, but not for systems employing neutral H projectiles [151, 165] [241].

7. INNER SHELL IONISATION

Vacancies in inner atomic shells produced by collisions with photons, electrons or ions may be aligned or oriented. In a typical collision experiment employing ground state atoms in a $^1S_0$ state, an electron removed from an inner atomic shell leaves the ion in a doublet state with a certain total angular momentum j. Provided we have $j > \frac{1}{2}$, the product ion may be aligned [231, 242].

This is, for example, the case for ionisation of 2p–electrons provided the electron hole is in the $2p_{3/2}$ state. First investigations of this so–called L3–shell ionisation for ionisation of argon by electron impact employed the Auger decay channel but found a rather weak alignment A_{20}

$$A_{20} = \frac{\sigma_{3/2} - \sigma_{1/2}}{\sigma_{3/2} + \sigma_{1/2}} \quad (1.13)$$

only [243]. A large alignment was, however, observed for electron impact ionisation of the xenon L3–shell [71, 74]. These measurements utilized the characteristic x–ray decay channel and were supporting a threshold law similar to that observed for outer shell excitation processes, e.g., Ref. [58]. These inner shell ionisation studies turned out to be rather interesting from another point of view. Over many years there has been a long and still continuing discussion on the threshold behaviour of electron impact ionisation cross sections. A Wannier–type threshold law [225, 226]

$$\sigma(E) \propto (E - I)^n \quad (1.14)$$

had been established for outer shell ionisation of, e.g., helium atoms [244]. In general this threshold law was observed over an energy range of, at most, few electron volts. In inner shell ionisation, however, such a threshold behaviour can persist over a much larger energy range [104, 245]. For Ar–K ionisation, the threshold law given by Eq. 1.14 was established over an energy range of almost 50 eV, while the exponent within error bars was in agreement with the original prediction of $n = 1.127$ [225, 226]. By contrast, for xenon L3–shell ionisation a linear threshold threshold ($n \approx 1$) was observed. The different behaviour of Ar–K and Xe–L3 shell ionisation could be explained by a model that takes screening effects into account [104, 246].

Characteristic x–ray production by electron impact is generally accompanied by emission of a continuous background that is caused by the acceleration (slowing–down) of electrons in the nuclear field of target atoms. The cross sections for the production of continuous x–ray (Bremsstrahlung) photons from free atoms display a pronounced Z^2 dependence on the atomic number Z (Figure 1.15). At low incident energies, marked deviations from this predictions were observed, however, and attributed to two–photon bremsstrahlung processes not accounted for in standard theories [78]. These energetic photons arise from the slowing–down of electrons in the nuclear field of target atoms. The angular distribution $I(\theta)$ of bremsstrahlung photons owing to its dipole character should, hence, follow a dipole–type pattern. Taking retardation effects due to the finite speed c of light into account, $I(\theta)$ may be expressed as [248, 249]

$$I(\theta)/I(90^o) = \frac{(1 - \beta \cos \theta)^2 - P_1(\cos \theta - \beta)^2}{(1 - P_1 \beta^2)(1 - \beta \cos \theta)^4}. \quad (1.15)$$

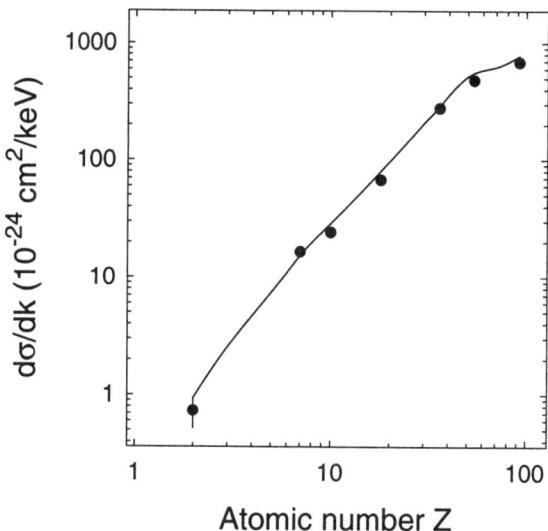

Figure 1.15 The differential cross section $d\sigma/dk$ for bremsstrahlung production versus atomic number Z. The experimental data, taken at an incident electron energy $T = 2.5$ keV and a photon energy $k = 2$ keV are compared with a theoretical prediction of Kissel et al. [247] (after Ref. [78]).

where $\beta = v/c$, v is the electron velocity and c the speed of light. Retardation leads to an angular distribution of the emitted photons that is no longer symmetric around $\theta = 90°$ but shifted towards forward angles. If the incident electron decelerates along a straight line, i.e., without deflection, as is the case in the so–called high–energy limit where all the incident electron energy is radiated, the polarization P_1 should amount to $P_1 = 100$ %. A large, albeit non–complete degree of linear polarization that presumably is caused by the deflection of the projectile electron and, hence, a deviation from a straight line trajectory was indeed observed. The experimental results were in fair agreement with more recent theoretical predictions and the pronounced shift of the maximum angle of intensity due to retardation effects confirmed [71, 74].

A large inner shell alignment was also observed for heavy particle impact. Measurements of the gold L3–shell ionisation by proton impact established a significant alignment being negative at large incident velocities $v_p \gg v_e$ (v_p and v_e are projectile and bound electron orbital velocity, respectively) where the interaction is dominantly in longitudinal direction. At medium velocities

$v_p \approx v_e$, where the interaction is strongest and preferentially in transverse directions, the alignment becomes positive. At low energies, the projectile carries little momentum into the collision making it difficult to transfer sufficient momentum and energy to bound electrons. Here the interaction is dominated by close collisions acting preferentially in longitudinal directions. A large negative alignment is, hence, here observed. At even lower velocities, the deflection of the projectile in the nuclear field of the target atom becomes noticable. For the large scattering angles encountered in such very slow collisions, the quantization axis is no longer defined by the direction of the incoming projectile but rotates in course of the collision. Still, the alignment is preserved but only with respect to the direction of the outgoing projectile. In non–coincidence experiments where the direction of the outgoing projectile is not specified an average over all possible trajectories and, in particular over azimuthal angle occurs. This leads to a significant decrease of the observed angle–integrated alignment, as was observed by experiment [64, 85].

The impact of a charged particles on neutral atoms may lead not only to singly but also to multiply charged ions. During electron impact singly charged ions are produced with reasonable cross sections while direct multiple ionization processes are relatively weak. During inner shell ionization, multiple ionization of atoms may result from the decay of the inner shell vacancy by, for example, Auger or Coster–Kronig transitions. The combination of these processes in, for example, L–shell ionization of argon atoms leads to about 70–80 % doubly, 20–30 % triply, and 3–5 % quadruply charged argon ions [114, 250]. Performing experiments by which one of the ejected electrons is detected in coincidence with charged–state analysed product ions, multiply–charged ions resulting from multiple outer shell ionisation and from inner shell ionization were distinguished. For atoms as heavy as xenon, multiple ionization processes may become pronounced and ionic charge states as large as Xe^{8+} were found for 6 keV electron impact ionization of xenon [127, 143].

8. TWO–PHOTON STUDIES AND THE EINSTEIN–PODOLSKY–ROSEN PARADOX

Much of Hans Kleinpoppen's scientific work is related to another fundamental process: the photon decay of metastable hydrogen atoms [38, 118, 120, 122, 125, 129, 135, 138, 144, 152, 158, 166, 181–183, 186]. First investigations were aimed at a direct detection of the two–photon decay mechanism [38] that had been predicted by Maria Göppert–Mayer as early as 1931 [251]. This early experiment, although of limited accuracy, already showed a clear deviation of the coincident two–photon angular correlation function $I(\theta)$ from a symmetric angular distribution while it was in accordance with the quantum mechanical prediction $I(\theta) \propto 1 + \cos 2\theta$. Here θ is the angle between the two–photon detectors. This first experiment was already at variance with a simple local realistic theory that, in essence, was a follow up of a discussion based on the Einstein–Podolsky–Rosen debate about the completeness of quantum mechanics (e.g., Refs. [135,252,253]). In a *local* and *realistic* theory the world is made of objects with physical properties that exist independently of any observations made on them nor on other objects. In quantum mechanics as a *non–local* and *non–realistic* theory, however, a measurement of a certain physical quantity taken at a certain point A may be influenced by a measurement taken at another object at another point B. The angular distribution of the two decay measurement established a departure from simple local realistic theories but could not rule out more sophisticated versions of it. Rather it was argued that these as well as other experimental observations may point to an incompleteness of quantum mechanics and to the necessity of introducing so called *hidden variables* ultimately leading to Bell's inequalities [254]. There have been var-

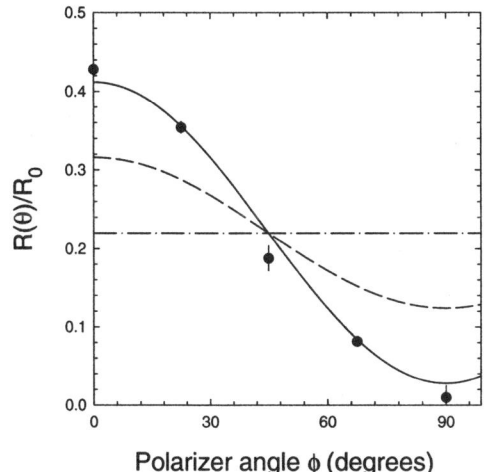

Figure 1.16 Linear polarization correlation function $R(\theta)/R_0$ versus relative polarizer angle ϕ for the two–photon decay of metastable hydrogen. The experimental results (•) are compared with non–local quantum mechanical predictions (solid line) as well as with predictions based on two different local realistic theories (dashed and dash–dotted lines, see text) (after Ref. [186]).

ious tests of hidden variables theories. Fig. 1.16 displays first measurements in which the linear polarization correlation function of both photons during the two–photon decay of metastable hydrogen was measured. Excellent agreement with the predictions according to quantum mechanics have been obtained, if the finite acceptance angle and imperfect polarizers was taken into account. By contrast, poor agreement with local realistic theories clearly outside the experimental error bars was noted [118,186]. In following years further experimental improvements allowing for even more precise and more sophisticated test of quantum mechanics ruling out predictions of local realistic theories in various forms. Among these are circular polarization experiments, polarization measurements employing three polarizers [138], tests of the Breit–Teller hypothesis of spin indepencence of the two–photon emission of metastable hydrogen atoms [118], and Fourier analysis of the two–photon radiation of metastable hydrogen atoms [182]. From the latter investigation a coherence length of 350 nm corresponding to a coherence time $t_{coh} = 1.2 \times 10^{-15}$ sec of the two–photon pair was deduced.

We mention in passing that the one–photon decay of metastable hydrogen taking place in an external electric field results in the emission of completely linearly polarized light. The polarization vector of the emitted Lyman–α radiation is perpendicular to the electric field [144], just the opposite of what is expected on the grounds of standard text book theory of quantum mechanics. This reversal of polarization is due to the Lamb–splitting and causes the $H(2s_{1/2})$ wave function to behave like $H(2p_{\pm 1})$ rather than like a linear combination of $H(2s)$ and $H(2p_0)$. Text book physics is recovered if the electric field is switched on fast enough to overcome the Lamb–splitting [255].

9. CONCLUSION

Professor Kleinpoppen has exerted a wide influence on many aspects of physics. An indication is given by the many successful conferences he has organized, by his writing of chapters in various books, the editing of conference proceedings, the co–editing a number of books, and by a book that has been devoted to him by his colleagues and friends on the occasion of his 60th birthday [256]. The contributions of Professor Kleinpoppen as a facilitator in such development are legendary and may be discerned from the bibliography. These activities are indicated in references [16, 43, 65, 67, 72, 77, 82, 96, 139, 154, 169–173, 179]. The active and positive role that Professor Kleinpoppen has maintained for many years in communicating advances in the field, both for his own studies and for the field as a whole, are indicated in a short note to Nature [131] in 1987.

However the best recognition of his contributions is seen in the development of 'the complete scattering experiment' in atomic and molecular scattering physics and in the photoionization processes of atoms. This volume of conference proceedings provides excellent examples of the growth and significance of those contributions.

References

[1] W. Hanle, H. Kleinpoppen, and A. Scharmann, *Dichroismus and Luminesenzpolarization an Verstreckten Polyvinylalkohol-Folien (Dichroism and Luminescence Polarization of Stretched Polyvinylalcohol-films)*, Z. Naturforschung **13a**, 64 (1958).

[2] H. Kleinpoppen, *Vermessung der Lamb-Verschiebung des $3^2S_{1/2}$-Zustandes gegenuber dem $3^2P_{1/2}$-Zustand am Wasserstoffatom, (Measurement of the $3^2S_{1/2}$-$3^2P_{1/2}$ Lamb Shift of Atomic Hydrogen)*, Z. Physik **191**, 164 (1961).

[3] H. Kleinpoppen, H. Krüger, and R. Ulmer, *Excitation and Polarization of Balmer-alpha-Radiation in Electron-Hydrogen Atom Collisions*, Phys. Letts. **2**, 78 (1962).

[4] D. Haidt, H. Kleinpoppen, and H. Krüger, *Polarization of the HeII 4686 Line in Electron-Helium Atom Collisions*, Phys. Letts. **16**, 122 (1965).

[5] H. Kleinpoppen and V. Raible, *On the Narrow Resonance in the Scattering of Electrons by Atomic Hydrogen*, Phys. Letts. **18**, 24 (1965).

[6] H. Hafner, H. Kleinpoppen, and H. Krüger, *Polarization of Alkali Line Radiation Excited by Electron Impact*, Phys. Letts. **18**, 270 (1965).

[7] D. Haidt and H. Kleinpoppen, *Elektronens to β anregung und Polarization der HeII-Fowler-alpha-Line. (Electron Impact Excitation and Polarization of the HeII-Fowler-alpha-line)*, Z. f. Physik **196**, 72 (1966).

[8] H. Kleinpoppen and H. Krüger, in *Conference Proceedings on Excitation Electronique d'une Vapeur Atomique, Application a la Spectroscopie (Grenoble) A 41* (1966).

[9] D. Hils, H. Kleinpoppen, and H. Koschmieder, *Re-measurement of the Total Cross section for Excitation of the Hydrogen $2^2S_{1/2}$ State by Electron Impact*, Proc. Phys. Soc. **89**, 35 (1966).

[10] H. Kleinpoppen and J. D. Schumann, *Ein Statisches Vierpolfeld zur dreidimensionalen Stabilisierung von Elektronen, (A Static Quadrupole Field for Three-Dimensional Stabilizing of Electrons)*, Z. Angewandte Physk **22**, 152 (1967).

[11] H. Hafner and H. Kleinpoppen, *Polarization of Atomic Line Radiation. I. Excitation by Electron Impact*, Z. Physik **198**, 315 (1967).

[12] H. Kleinpoppen and R. Neugart, *Polarization of Atomic Line Radiation. II. Optical Excitation and its Relationship to Electron Impact Excitation*, Z. Physik **198**, 321 (1967).

[13] H. Beyer and H. Kleinpoppen, *Measurement of the $4^2S_{1/2}$-$4^2P_{1/2}$ Lamb Shift of the He^+ Ion*, Z. f. Physik **201**, 177 (1967).

[14] H. Kleinpoppen and E. Kraiss, *Cross Section and Polarization of the Balmer-alpha-Line by Electron Impact Excitation*, Phys. Rev. Lett. **20**, 361 (1968).

[15] H. Kleinpoppen, *Atomic Collisions in the Growth Points in Physics*, Physics Bulletin **20**, 213 (1969).

[16] in *Physics of the One-and Two-Electron Atoms, Conference Proceeding of the International Symposium in Honour of A. Sommerfeld*, edited by F. Bopp and H. Kleinpoppen (North Holland Publishing Company, 1969).

[17] H. Kleinpoppen, *Polarization of Atomic Line Radiation, in Physics of the One- and Two-Electron Atoms* (North Holland Publishing Company, 1969), p. 612.

[18] H. Kleinpoppen, *Analysis of Scattering and Excitation Amplitudes in Polarized Electron-Atom Collisions. I. Elastic Scattering on One-Electron Atoms and Excitation Processes $^2S_{1/2} \sim ^2P_{1/2,3/2}$*, Phys. Rev. A **3**, 2015 (1971).

[19] H. J. Beyer. and H. Kleinpoppen, *Fine Structure Measurement in He^+, n = 4L. Intervals S-F*, J. Phys. B **4**, L129 (1971).

[20] H. J. Beyer, H. Kleinpoppen, and J. M. Woolsey, *Cascading Anticrossings from n = 5, He^+*, Phys. Rev. Lett. **28**, 263 (1972).

[21] H. Beyer and H. Kleinpoppen, *Fine Structure Measurement in He^+, n = 4 II. Intervals S-D and Lamb Shift $P_{3/2}$-$D_{3/2}$*, J. Phys. B **5**, L12 (1972).

[22] D. Hils, H. Kleinpoppen, M. V. McCusker, and S. J. Smith, *Total Differential and Differential Direct Cross Sections for Low-Energy Electron Potassium Scattering*, Phys. Rev. Lett. **29**, 398 (1972).

[23] T. Ottley, D. Denne, and H. Kleinpoppen, *Near-Threshold Polarization of the 6^1S_0–6^3P_1 Line of Mercury by Electron Impact*, Phys. Rev. Lett. **29**, 1646 (1972).

[24] H. Koschmieder, V. Raible, and H. Kleinpoppen, *Resonance Structure in the Excitation Cross Section of the Atomic Hydrogen 2S State by Electron Impact*, Phys. Rev. A **8**, (1973).

[25] M. Eminvan, K. MacAdam, J. Slevin, and H. Kleinpoppen, *Measurements of Complex Excitation Amplitudes in Electron-Helium Collisions by Angular Correlations Using a Coincidence Method*, Phys. Rev. Lett. **31**, 576 (1973).

[26] H. J. Beyer, H. Kleinpoppen, and J. M. Woolsey, *Measurements of Non-Linear Stark Effect in Hydrogen-Like He^+, n = 4*, J. Phys. B **6**, 1949 (1973).

[27] H. Kleinpoppen, in *Tenth Coral Gables Conference on Fundamental Interactions* (Plenum Publishing Company, New York, 1973), p. 229.

[28] K. Blum and H. Kleinpoppen, *Analysis of Scattering and Excitation Amplitudes in Polarized Electron-Atom Collisions II. Scattering on Two-Electron Atoms*, Phys. Rev. A **9**, 1902 (1974).

[29] V. Raible, H. Koschmieder, and H. Kleinpoppen, *Structure in Near Threshold Excitation of Argon and Neon*, J. Phys. B **7**, L14 (1974).

[30] T. W. Ottley, D. R. Denne, and H. Kleinpoppen, *Polarization of the 6^1S_0-6^1P_1, 1850 Å Line of Mercury Excited by Electron Impact*, J. Phys. B **7**, L179 (1974).

[31] M. Eminyan, K. MacAdam, J. Slevin, and H. Kleinpoppen, *Electron-Photon Angular Correlations in Electron-Helium Collisions: Measurements of Complex Excitation Amplitudes*, J. Phys. B **7**, 1519 (1974).

[32] K. J. Kollath and H. Kleinpoppen, *Measurement of the Non-Linear Stark Effect of the $2^2P_{1/2}$ State of Hydrogen*, Phys. Rev. A **10**, 1519 (1974).

[33] H. Kleinpoppen, *Spin and Coherence Effects in Electron-Atom Collision Processes*, Int. Journal of Quantum Chemistry Symposium **8**, 59 (1974).

[34] H. Kleinpoppen, in *Electron-Photon Coincidences and Polarization of Impact Radiation, Fourth International Conference on Atomic Physics (Heidelberg 1974)*, edited by G. zu Putlitz, E. W. Weber, and U. A. Winnacker (Plenum Press, New York, 1975), p. 449.

[35] K. Blum and H. Kleinpoppen, *On the Theory of Electron-Photon Coincidences from Electron-Atom Excitation Processes*, J. Phys. B **8**, 922 (1975).

[36] T. W. Ottley and H. Kleinpoppen, *Resonances Near Threshold in the Electron Impact Excitation and Polarization of the 6^1S_0–6^3P_1 Line of Mercury*, J. Phys. B **8**, 621 (1975).

[37] M. Standage, M. Eminyan, H. Kleinpoppen, K. MacAdam, and J. Slevin, *Electron- Photon Angular-Correlations in Electron Helium Collisions for 3'P Excitations*, J. Phys. B **8**, 2058 (1975).

[38] D. O'Connell, K. J. Kollath, A. J. Duncan, and H. Kleinpoppen, *The Two-Photon Decay of Metastable Atomic Hydrogen*, J. Phys. B **8**, L214 (1975).

[39] K. Blum and H. Kleinpoppen, *Spin Analysis of Electron Atom Collisions, Proceedings of Thomas Symposium*, Int. Journal Quantum Chem. **9**, 415 (1975).

[40] H. J. Beyer and H. Kleinpoppen, *Fine Structure Measurements of the n=5 He States*, J. Phys. B **8**, 2449 (1975).

[41] K. Blum and H. Kleinpoppen, *Some Remarks on the Theory of Coincidence Experiments*, J. Phys. B **8**, 922 (1975).

[42] M. Eminyan, H. Kleinpoppen, J. Slevin, and M. Standage, in *Proceedings of Stirling Symposium 1974, and Festschrift for Ugo Fano* (Plenum Press, New York, 1976), p. 455.

[43] in *Proceedings of the Stirling Symposium 1974, and Festschrift for Ugo Fano*, edited by H. Kleinpoppen and M. R. C. McDowell (Plenum Press, New York, 1976).

[44] M. Standage and H. Kleinpoppen, *Photon Vector Polarization and Coherence Parameters in an Electron Photon Coincidence Experiment on Helium*, Phys. Rev. Lett. **36**, 577 (1976).

[45] H. Kleinpoppen, K. Blum, and M. Standage, in *International Conference of Electronic and Atomic Collisions*, edited by J. S. Risley and R. Gaballe (Washington Press, Washington, 1976), p. 641.

[46] K. Blum and H. Kleinpoppen, in *Electron and Photon Interactions with Atoms, Proceedings of the Stirling Symposium 1974, and Festschrift for Ugo Fano* (Plenum Press, New York, 1976), p. 501.

[47] H. Kleinpoppen, *Direct Measurement of Quantum Mechanical Phases of Atomic Excitation Amplitudes*, Comments Atom. Molec. Phys. **1**, 35 (1976).

[48] H. Kleinpoppen and I. McGregor, in *Physics of Ionized Gases* (SPIG 1976, Stefan Institute, 1976), p. 71.

[49] K. Blum and H. Kleinpoppen, *Coherent and Incoherent Excitation of Atoms by Electron Impact*, Int. Jour. Quant. Chem. Symp. **10**, 331 (1976).

[50] H. Kleinpoppen, *Analysis of Electron-Atom Collision Processes*, Advances in Quantum Chemistry **10**, 77 (1977).

[51] K. Blum and H. Kleinpoppen, *Even-odd Parity Coherence Effects in Electron Hydrogen Scattering*, J. Phys. B **10**, 3283 (1977).

[52] H. J. Beyer and H. Kleinpoppen, *Fine Structure and Stark Shifts from Anti-crossing Studies*, Int. Jour. Quantum Chemistry **11**, 271 (1977).

[53] H. J. Beyer and H. Kleinpoppen, *Measurement of Intervals between $^{1,3}D$ and High L States of He for n=7 to 10 by Electric Field Induced Anti-crossings*, J. Phys. B **11**, 979 (1978).

[54] A. A. Zaidi, I. McGregor, and H. Kleinpoppen, *Effect of the Polarization of the 6^3P_1 to 6^1S_0 line of Mercury Excited by Electron Impact*, J. Phys. B **11**, L283 (1978).

[55] D. Hils and H. Kleinpoppen, *The Ionization of Polarized Potassium Atoms by Polarized Electrons*, J. Phys. B **11**, L283 (1978).

[56] S. H. Al-Shama and H. Kleinpoppen, *Near Threshold Polarization and Excitation Function of Xe Atoms*, J. Phys. B **11**, L367 (1978).

[57] W. Hanle and H. Kleinpoppen, in *Progress in Atomic Spectroscopy*, edited by W. Hanle and H. Kleinpoppen (Plenum Press, New York, 1978, 1979), Vol. A and B.

[58] H. Kleinpoppen and A. Scharmann, in *Progress in Atomic Spectroscopy, Part A*, edited by W. Hanle and H. Kleinpoppen (Plenum Press, New York, 1978), p. 329.

[59] H. J. Beyer and H. Kleinpoppen, in *Progress in Atomic Spectroscopy, Part A*, edited by W. Hanle and H. Kleinpoppen (Plenum Press, New York, 1978), p. 607.

[60] K. Blum, E. E. Fitchard, and H. Kleinpoppen, *Coherent Excitation of the n=3 levels of Atomic Hydrogen in a Field Free Region*, Z. f. Physik A **287**, 137 (1978).

[61] H. J. Beyer and H. Kleinpoppen, in *Proc. VIth International Conference on Atomic Physics, Riga* (Plenum Press, New York, 1978), p. 435.

[62] K. Blum and H. Kleinpoppen, *Electron-Photon Angular Correlations in Atomic Physics*, Physics Report **52**, 205 (1979).

[63] H. Kleinpoppen, in *Coherence and Correlation in Atomic Collisions*, edited by H. Kleinpoppen and J. F. Williams (Plenum Press, New York, 1979), p. 15.

[64] W. Jitschin, H. Kleinpoppen, R. Hippler, and H. O. Lutz, *L-Shell Alignment of Heavy Atoms Induced Proton Impact Ionization*, J. Phys. B **12**, 4077 (1979).

[65] H. Kleinpoppen, in *Invited Papers of XIth ICPEAC Conference, Kyoto* (1979), p. 559.

[66] R. Hippler, H. Kleinpoppen, and H. O. Lutz, in *Invited Papers of XIth ICPEAC Conference, Kyoto* (1979), p. 611.

[67] *Coherence and Correlation in Atomic Collisions, Plenum Monograph series of Physics of Atoms and Molecules*, edited by H. Kleinpoppen and J. F. Williams (Plenum Press, New York, 1980).

[68] H. Kleinpoppen, in *Coherence and Correlation in Atomic Collisions*, edited by H. Kleinpoppen and J. F. Williams (Plenum Press, New York, 1980), p. 697.

[69] I. McGregor, A. Chutjian, and H. Kleinpoppen, in *Some New Aspects of Electron-Photon Coincidence Measurements, in Coherence and Correlation in Atomic Collisions*, edited by H. Kleinpoppen and J. F. Williams (Plenum Press, New York, 1980), p. 109.

[70] D. Hils, K. Rubin, and H. Kleinpoppen, in *Coherence and Correlation in Atomic Collisions*, edited by H. Kleinpoppen and J. F. Williams (Plenum Press, New York, 1980), p. 689.

[71] M. Aydinol, I. McGregor, R. Hippler, and H. Kleinpoppen, in *Coherence and Correlation in Atomic Collisions*, edited by H. Kleinpoppen and J. F. Williams (Plenum Press, New York, 1980), p. 205.

[72] H. Kleinpoppen, *Summary of the Symposium on Amplitudes and State Parameters in Atomic Collisions XIth ICPEAC*, Comments Atomic Molecular Collisions **9**, 201 (1980).

[73] A. Zaidi, I. McGregor, and H. Kleinpoppen, *A Coincident Photon Coherence Analysis of the Mercury 6^1S_0-6^3P_1-6^1P_0 Excitation De-excitation Process in Terms of Spin Orbit Effects*, Phys. Rev. Lett. **45**, 1168 (1980).

[74] M. Aydinol, I. McGregor, R. Hippler, and H. Kleinpoppen, *Angular Distribution of X-Radiation Following Electron Bombardment of Free Atoms*, J. Phys. B **13**, 989 (1980).

[75] D. Hils, W. Jitschin, and H. Kleinpoppen, *Highly Polarized Sodium Atomic Beam*, Applied Phys. **25**, 39 (1981).

[76] R. Hippler, I. McGregor, M. Aydinol, and H. Kleinpoppen, *Ionization of Xenon L Subshells by Low-Energy Electron Impact*, Phys. Rev. A **23**, 1730 (1981).

[77] H. Kleinpoppen, *Elementarprozesse in der Elektronenstreuung an Atomen. I. Vom Franck-Hertz-Experiment zu Elektron-Photon-Winkelkorrelationen (Fundamental Processes in Electron Scattering on Atoms, Part 1, From the Franck-Hertz-Experiment to Electron-Photon Angular Correlations)*, Physikalische Blätter **37**, 131 (1981).

[78] R. Hippler, I. McGregor, K. Saeed, and H. Kleinpoppen, *Z Dependence of Bremsstrahlung Radiation from Free Atoms*, Phys. Rev. Lett. **46**, 1622 (1981).

[79] R. Hippler, I. McGregor, and H. Kleinpoppen, *Comment on L-Subshell Ionization Cross Section of Xenon by Electron Impact Near Threshold Region*, J. Phys. B **14**, L537 (1981).

[80] *Innershell and X-Ray Physics of Atoms and Solids*, edited by D. Fabian, H. Kleinpoppen, and L. Watson (Plenum Press, New York, 1981).

[81] A. A. Zaidi, S. M. Khalid, I. McGregor, and H. Kleinpoppen, *Determination of a Complete Set of Parameters Incorporating Spin-Orbit Effects from an Electron-Photon Coincidence Experiment on the Excitation of the 6^3P_1 State of Mercury by Electron Impact*, J. Phys. B **14**, L503 (1981).

[82] H. Kleinpoppen, *Elementarprozesse in der Elektronenstreuung an Atomen. II. Strenung Polarisierter Electronen and Polarisierten Atomen (Fundamental Processes in Electron-Atom Scattering, Part II, Scattering of Polarized Electrons on Polarized Atoms)*, Physikalische Blätter **37**, 289 (1981).

[83] M. Aydinol, R. Rippler, I. McGregor, and H. Kleinpoppen, in *Inner-Shell and X-Ray Physics of Atoms and Solids* (Plenum Press, New York, 1981), p. 323.

[84] R. Hippler, I. McGregor, K. Saeed, and H. Kleinpoppen, in *Inner-Shell and X-Ray Physics of Atoms and Solids* (Plenum Press, New York, 1981), p. 379.

[85] W. Jitschin, H. O. Lutz, and H. Kleinpoppen, in *Inner-Shell and X-Ray Physics of Atoms and Solids* (Plenum Press, New York, 1981), p. 89.

[86] W. Jitschin, A. Kaschuba, H. Kleinpoppen, and H. O. Lutz, *Proton Induced Alignment of the L3-Subshell in Heavy Atoms*, Z. Physik A **304**, 69 (1982).

[87] R. Hippler, G. Malunat, M. Faust, H. Kleinpoppen, and H. O. Lutz, *Determination of Alignment Tensor Components for the Charge Exchange Excitation of H(2P) in Proton-Argon Collisions*, Z. f. Physik A **304**, 63 (1982).

[88] H. Klar and H. Kleinpoppen, *Angular Distribution of Photoelectrons from Polarized Atoms Exposed to Polarized Radiation*, J. Phys. B **15**, 933 (1982).

[89] G. Tepehan, H. J. Beyer, and H. Kleinpoppen, *Anticrossing Measurements of Fine Structure Intervals at n=4 of He*, J. Phys. B **15**, 3141 (1982).

[90] G. Tepehan, H. J. Beyer, and H. Kleinpoppen, *Low Field Stark Shifts of n=4 States in Helium*, J. Phys. B **15**, 3159 (1982).

[91] D. Hils, W. Jitschin, and H. Kleinpoppen, *The Spin Dependent Ionization Asymmetry in Electron Sodium Collisions*, J. Phys. B **15**, 3347 (1982).

[92] R. Hippler, K. Saeed, I. McGregor, and H. Kleinpoppen, *Energy Dependence of Characteristic and Bremsstrahlung Cross Sections of Argon Induced by Electron Bombardment*, Z. Physik A **307**, 83 (1982).

[93] H. J. Beyer, H. Kleinpoppen, I. McGregor, and L. C. McIntyre, *Some Consequences of Atomic Orientation in Electron-Atom Scattering by Consideration of Attractive and Repulsive Forces*, J. Phys. B **15**, L545 (1982).

[94] I. McGregor, D. Hils, R. Hippler, N. Malik, J. F. Williams, A. Zaidi, and H. Kleinpoppen, *Electron-Photon Angular Correlations from Electron Impact Excitation of Heavy-Rare Gas Atoms*, J. Phys. B **15**, L411 (1982).

[95] W. Jitschin, A. Kaschuba, H. Kleinpoppen, and H. O. Lutz, *L3-Subshell Alignment of Heavy Atoms by Light Ion Impact*, Z. f. Physik A **304**, 69 (1982).

[96] H. Kleinpoppen, *Spin-Orbit Couplings and Quantum Beats in Electron-Photon Angular Correlations from Impact Excitation of Atoms*, Comments Atomic and Molecular Physics **12**, 111 (1982).

[97] K. Blum and H. Kleinpoppen, in *Proceedings of International Workshop on Electron-Atom and Electron-Molecule Collisions*, edited by J. Hinze (Plenum Press, New York, 1983).

[98] S. M. Khalid and H. Kleinpoppen, *Elastic Scattering of Electrons from Spin-1/2 Targets Including Spin-Orbit Interactions*, Phys. Rev. A **27**, 236 (1983).

[99] S. M. Khalid and H. Kleinpoppen, *Alignment and Orientation of Charge Exchange Excitation with Polarized Electrons*, Z. Physik A **311**, 57 (1983).

[100] W. Jitschin, R. Hippler, R. Shanker, H. Kleinpoppen, H. O. Lutz, and R. Schuch, *L X-Ray Anisotrophy and L3-Subshell Alignment of Heavy Atoms Induced by Ion Impact*, J. Phys. B **16**, 1417 (1983).

[101] K. Blum and H. Kleinpoppen, *Angular Correlation Studies of Heavy-Particle Impact Excitation of Atoms*, Phys. Reports **96**, 251 (1983).

[102] H. J. Beyer and H. Kleinpoppen, in *Fundamental Processes in Energetic Atomic Collisions*, edited by H. O. Lutz, J. S. Briggs, and H. Kleinpoppen (Plenum Press, New York, 1983), p. 531.

[103] K. Blum and H. Kleinpoppen, *Spin-Dependent Phenomena in Inelastic Electron-Atom Collisions*, Adv. Atom. and Molec. Phys. **19**, 187 (1983).

[104] R. Hippler, H. Klar, K. Saeed, I. McGregor, A. J. Duncan, and H. Kleinpoppen, *Threshold Behaviour of Ar-K and Xe L3 Ionization by Electron Impact*, J. Phys. B **20**, L617 (1983).

[105] H. O. Lutz, J. S. Briggs, and H. Kleinpoppen, in *Fundamental Processes in Energetic Atomic Collisions* (Plenum Press, New York, 1983).

[106] H. Kleinpoppen, in *Fundamental Processes in Energetic Atomic Collisions*, edited by H. O. Lutz, J. S. Briggs, and H. Kleinpoppen (Plenum Press, New York, 1983), p. 611.

[107] H. Kleinpoppen, in *Fundamental Processes in Energetic Atomic Collisions*, edited by H. O. Lutz, J. S. Briggs, and H. Kleinpoppen (Plenum Press, New York, 1983), p. 489.

[108] H. J. Beyer and H. Kleinpoppen, in *Progress in Atomic Spectroscopy, Part C* (Plenum Press, New York, 1983).

[109] W. Jitschin, S. Osimitsch, H. Reihl, H. Kleinpoppen, and H. O. Lutz, *Electron Exchange in the Na 3p Electron Impact Excitation*, J. Phys. B **17**, 1899 (1984).

[110] S. M. Khalid and H. Kleinpoppen, *Theory of Electron-Photon Coincidence using Spin-Polarized Electrons and Spin-Polarized Atoms*, J. Phys. B **17**, 243 (1984).

[111] H. Kleinpoppen, *Synopsis of International Symposium on Polarization and Correlation in Electron-Atom Collisions, Angular Correlations as a Major Topic*, Comments Atomic and Molecular Physics **14**, 321 (1984).

[112] H. Kleinpoppen and I. McGregor, in *Atomic Physics 8* (Plenum Press, New York, 1984), p. 431.

[113] N. Anderson, I. V. Hertel, and H. Kleinpoppen, *Shape and Dynamics of States Excited in Electron-Atom Collisions: A Comment on Orientation and Alignment of Parameters by Consideration of Attractive and Repulsive Forces*, J. Phys. B **17**, L901 (1984).

[114] R. Hippler, K. Saeed, A. J. Duncan, and H. Kleinpoppen, *Electron Spectroscopy of Multiple Ionization of Argon by Electron Impact*, Phys. Rev. A **30**, 3329 (1984).

[115] R. Hippler, H. Klar, K. Saeed, I. McGregor, A. J. Duncan, and H. Kleinpoppen, in *Electronic and Atomic Collisions* (Elsevier Science, 1984), p. 789.

[116] G. G. Tepehan, H. J. Beyer, and H. Kleinpoppen, *Measurement of the 5^1D-5^3D Interval in Helium*, J. Phys. B. **18**, 1125 (1985).

[117] R. Hippler, M. Faust, R. Wolf, H. Kleinpoppen, and H. O. Lutz, *Polarization Studies of H(2p) Charge-Exchange Excitation: H^+-Ar Collisions*, Phys. Rev. A **31**, 1399 (1985).

[118] W. Perrie, A. J. Duncan, H. J. Beyer, and H. Kleinpoppen, *Polarization Correlation of the Two Photons Emitted by Metastable Atomic Deuterium: A Test of Bell's Inequality*, Phys. Rev. Lett. **54**, 1790 (1985).

[119] H. Kleinpoppen, J. S. Briggs, and H. O. Lutz, in *Fundamental Processes in Atomic Collison Physics* (Plenum Press, New York, 1985).

[120] A. J. Duncan, W. Perrie, H. J. Beyer, and H. Kleinpoppen, in *Fundamental Processes in Atomic Collision Physics* (Plenum Press, New York, 1985), p. 555.

[121] W. Jitschin, S. Osimitsch, H. Reihl, D. Mueller, H. Kleinpoppen, and H. O. Lutz, *Spin Exchange in the Excitation of Spin-Polarized Na Atoms by Ne-Ion Impact*, Phys. Rev. A **34**, 3684 (1986).

[122] H. Kleinpoppen, in *Foundations of Quantum Mechanics, Proceedings 2nd Int. Symposium, Tokyo* (1986), p. 59.

[123] H. A. Silim, A. H. El-Farrash, and H. Kleinpoppen, *Lifetime Measurement of the 3^3P State of Helium*, Z. Phys. D **5**, 61 (1987).

[124] H. J. Beyer and H. Kleinpoppen, *Progress in Atomic Spectroscopy, Part D* (Plenum Press, New York, 1987).

[125] H. Kleinpoppen, *On the State Vector Model of the Two-Photon Emission of Metastable Atomic Hydrogen*, Atom. Phys. **10**, 141 (1987).

[126] H. A. Silim, H. J. Beyer, A. El-Sheikh, and H. Kleinpoppen, *Measurement of the Spin-Exchange Amplitudes in the Electron Impact Excitation of the 3^3P State of Helium*, Phys. Rev. A **35**, 4455 (1987).

[127] N. A. Chaudhry, N. A. Duncan, R. Hippler, and H. Kleinpoppen, *Coincidence Spectroscopy of Highly Charged Xenon Ions by Electron Impact*, Phys. Rev. Lett. **59**, 2036 (1987).

[128] H. Kleinpoppen, S. Trajmar, J. C. Nickel, and G. Csanak, *Synopsis of International Symposium on Correlation and Polarization in Electron-Atom Collisions*, Comments Atom. Mol. Phys. **20**, 277 (1987).

[129] T. Haji-Hassan, A. J. Duncan, W. Perrie, H. J. Beyer, and H. Kleinpoppen, *Experimental Investigation of the Possibility of Enhanced Photon Detection in the EPR Type Experiments*, Phys. Lett. **123A**, 110 (1987).

[130] R. Hippler, W. Harbich, H. Madeheim, H. Kleinpoppen, and H. O. Lutz, *Cross Sections for Charge-Exchange Excitation to H(2p) in Photon-Rare-Gas Collisions (1 - 25 keV)*, Phys. Rev. A **35**, 3139 (1987).

[131] H. Kleinpoppen, *Electron-Scattering Experiments*, Nature **330**, 20 (1987).

[132] R. Hippler, M. Faust, R. H. Wolf, H. Kleinpoppen, and H. O. Lutz, *Polarization Studies of H(2p) Charge Exchange Excitation: H^+–He Collisions*, Phys. Rev. A **36**, 4644 (1987).

[133] *Fundamental Processes of Atomic Dynamics*, edited by J. S. Briggs, H. Kleinpoppen, and H. O. Lutz (Plenum Press, New York, 1988).

[134] H. Kleinpoppen, in *Fundamental Processes in Atomic Dynamics* (Plenum Press, New York, 1988), p. 393.

[135] A. J. Duncan and H. Kleinpoppen, in *Quantum Mechanics Versus Local Realism: The Einstein, Podolsky and Rosen Paradox (ed F. Selleri)* (Plenum Press, New York, 1988), p. 175.

[136] H. Kleinpoppen, in *Laser Spectroscopy*, edited by J. Heldt and R. Lawruszczuk (World Scientific, Singapore, 1988), p. 286.

[137] H. J. Beyer, K. Blum, H. A. Silim, M. C. Standage, and H. Kleinpoppen, *Polarization Correlation Measurements on the 3^1P State of Helium for Positive and Negative Scattering Angles*, J. Phys. B **21**, 2953 (1988).

[138] T. Haji-Hassan, A. J. Duncan, W. Perrie, and H. Kleinpoppen, *Polarization Correlation Analysis of the Radiation from a Two-Photon Deuterium*

Source Using Three Polarizers: A Test of Quantum Mechanics Versus Local Realism, Phys. Rev. Lett. **62**, 237 (1988).

[139] H. Kleinpoppen, in *Introduction, International Symposium on Correlation and Polarization in Electronic and Atomic Collisions*, edited by A. Crowe and M. R. H. Rudge (World Scientific, Singapore, 1988), p. 1.

[140] R. Hippler, H. Madeheim, W. Harbich, H. Kleinpoppen, and H. O. Lutz, *Alignment of H(2p) in H^+ -H, H_2 Collisions*, Phys. Rev. A **38**, 1662 (1988).

[141] R. Hippler, H. Madeheim, H. Kleinpoppen, and H. O. Lutz, *Alignment of H(2p) in $H + H \to H + H(2p)$ Collisions*, J. Phys. B **10**, L257 (1989).

[142] H. Kleinpoppen, in *Synchroton Radiation Sources and their Application*, edited by G. N. Greaves and I. H. Munro (Edinburgh University Press, Edinburgh, 1989), p. 256.

[143] M. A. Chaudhry, A. J. Duncan, R. Hippler, and H. Kleinpoppen, *Partial Doubly Differential Cross Sections for Multiple Ionization of Argon, Krypton and Xenon Atoms by Electron Impact*, Phys. Rev. A **36**, 530 (1989).

[144] W. Harbich, R. Hippler, H. Kleinpoppen, and H. O. Lutz, *Complete Linear Polarization of Lyman-alpha Radiation from Metastable Hydrogen Atoms in External Electric Fields*, Phys. Rev. A **39**, 3388 (1989).

[145] S. Osimitsch, W. Jitschin, H. Reihl, H. Kleinpoppen, H. O. Lutz, O. Mo, and A. Riera, *Alignment and Spin Exchange in the Na 3p Excitation by He -Ion Impact*, Phys. Rev. A **40**, 2958 (1989).

[146] H. A. Silim, H. J. Beyer, , and H. Kleinpoppen, in *Bulletin of the Faculty of Science* (Mansoura University, 1989), Vol. 16, p. 1.

[147] H. Kleinpoppen, H. J. Beyer, and M. A. Chaudhry, in *Atomic Physics 11* (World Scientific, Singapore, 1989).

[148] H. Handy, H. J. Beyer, and H. Kleinpoppen, *Excitation Cross Section and Polarization of the Ca ll $\lambda = 393.3$ nm Line by Electron Impact of Ca atoms*, J. Phys. B **23**, 1671 (1990).

[149] M. A. Chaudhry, A. J. Duncan, R. Hippler, and H. Kleinpoppen, *Angular Variation of the Partial Doubly Differential Cross Sections for Multiple Ionization of Argon Atoms by Electron Impact*, Phys. Rev. A **41**, 4056 (1990).

[150] H. Kleinpoppen, H. J. Beyer, H. Hamdy, J. B. West, and E. I. Zohny, in *Proceedings of the UK USSR Seminar, Leningrad*, edited by M. Y. Amusia and J. B. West (1990), p. 161.

[151] R. Hippler, O. Plotzke, W. Harbich, H. Madeheim, H. Kleinpoppen, and H. O. Lutz, *Coherent Excitation of H(n=2) in H^+, H - He Collisions*, Z. Physik D **18**, 61 (1991).

[152] T. Haji-Hassan, A. J. Duncan, W. Perrie, H. Kleinpoppen, and E. Merzbacher, *Circular Polarization Correlation of the Two Photon Emitted in Decay Metastable Atomic Deuterium*, J. Phys. B **24**, 5053 (1991).

[153] H. J. Beyer, H. Hamdy, J. B. West, and H. Kleinpoppen, *Fluorescence from Excited States of Ca^+ and Sr^+ Induced by Synchrotron Radiation*, J. Phys. B **24**, 4957 (1991).

[154] H. Kleinpoppen, *Lehrbuch der Experimental Physik, Vol. IV* (Walter de Gruyter Company, Berlin, New York, 1992), p. 1.

[155] K. Ueda, J. B. West, K. J. Ross, H. Hamdy, H. J. Beyer, and H. Kleinpoppen, *Angle-resolved Photo-electron Spectroscopy of Ca in the 3p-3d Giant Resonance*, Phys. Rev. A **48**, R 863 (1993).

[156] A. Zaidi and H. Kleinpoppen, *A Novel Technique for Elimination of the Hyperfine Effect on the Polarization of Atomic Line Radiation*, J. Phys. B **26**, 1669 (1993).

[157] K. Ueda, J. B. West, K. Ross, H. Beyer, H. Hamdy, and H. Kleinpoppen, *Measurement of the Partial Cross Section and Asymmetry Parameters in the 4p-4d Giant Resonance Region of Sr*, J. Phys. B **26**, L347 (1993).

[158] A. J. Duncan, H. Kleinpoppen, and Z. A. Sheikh, in *Proc Int. Conf on Bell's Theorem and the Foundations of Modern Physics, Cesena, Italy 1991* (World Scientific, Singapore, 1993), p. 161.

[159] H. Kleinpoppen and H. Hamdy, in *Foundations of Quantum Mechanics in the Light of New Technology on Quantum Control and Measurement*, edited by H. Ezawa and Y. Mureyama (Elsevier Science, 1992), p. 51.

[160] A. H. Al-Nasir, M. A. Chaudhry, A. J. Duncan, R. Hippler, and H. Kleinpoppen, *Doubly Differential Cross Sections for the Ionization of the Hydrogen Molecule by the Impact of 100 eV Electrons*, Phys. Rev. A **47**, 2922 (1993).

[161] H. Hamdy, H. J. Beyer, and H. Kleinpoppen, *Electron Impact Coherence and Polarization Parameters for the Excitation of the 5^1P_1 in Strontium*, J. Phys. B **26**, 4237 (1993).

[162] H. Kleinpoppen, in *Foundation of Physics, Cologne 1993*, edited by P. Busch, P. Lahti, and P. Mittelstaedt (World Scientific, Singapore, 1993), p. 215.

[163] H. Kleinpoppen and H. Hamdy, *How Perfect are Complete Atomic Collision Experiments?*, Adv. At. Mol. Opt. Phys. **32**, 223 (1994).

[164] H. J. Beyer, H. Hamdy, E. I. M. Zohny, K. R. Mahmoud, M. A. K. El-Fayoumi, H. Kleinpoppen, J. Abdallah, R. E. H. Clark, and G. Csanak, *Electron Impact Coherence Parameters for the Excitation of the 5^1P State of Sr*, Z. Physik D **30**, 91 (1994).

[165] B. Siegmann, G. G. Tepehan, R. Hippler, H. Madeheim, H. Kleinpoppen, and H. O. Lutz, *The Collisions-Induced Electric Dipole Moment of H(n=2) in H-He, Ne and Ar Collisions*, Z. Physik D **30**, 223 (1994).

[166] R. Sheikh, A. J. Duncan, H. J. Beyer, and H. Kleinpoppen, *Depolarization of Atomic Two-Photon Radiation*, Z. Physik D **30**, 132 (1994).

[167] H. J. Beyer, J. B. West, K. Ross, K. Ueda, N. M. Kabachnik, H. Hamdy, and H. Kleinpoppen, *A New Approach to the Complete Photoionization Experiment by Means of a Coincidence Measurement between Autoionized Electrons and Polarized Fluorescent Photons in the Region of the 3p-3d Resonance in Calcium*, J. Phys. B **28**, L47 (1995).

[168] M. A. H. Bukhari, H. J. Beyer, M. A. Chaudhry, D. M. Campbell, A. J. Duncan, and H. Kleinpoppen, *Excitation of Spin-Polarized Sodium and Potassium Atoms by Electron Impact*, J. Phys. B **28**, 1889 (1995).

[169] *Polarized Electron/Polarized Photon Physics*, edited by H. Kleinpoppen and W. R. Newell (Plenum Press, New York, 1995).

[170] H. Kleinpoppen, in *Polarized Electron/Polarized Photon Physics*, edited by H. Kleinpoppen and W. R. Newell (Plenum Press, New York, 1995).

[171] M. A. Chaudhry and H. Kleinpoppen, in *Polarized Electron/Polarized Photon Physics*, edited by H. Kleinpoppen and W. R. Newell (Plenum Press, New York, 1995), p. 159.

[172] D. M. Campbell and H. Kleinpoppen, in *Proceedings of the Peter Farago Symposium on Selected Topics on Electron Physics* (Plenum Press, New York, 1996).

[173] D. M. Campbell and H. Kleinpoppen, in *Introduction to the Proceedings of the Peter Farago Symposium on Selected Topics on Electron Physics*, edited by M. Campbell and H. Kleinpoppen (Plenum Press, New York, 1996).

[174] J. B. West, K. Ueda, N. M. Kabachnik, K. J. Ross, H. J. Beyer, and H. Kleinpoppen, *Circular Dichroism in the Polarization of the Fluorescence Resulting from the Decay of Photoionized Ca Atoms*, Phys. Rev. A **53**, R 9 (1996).

[175] A. H. Alnasir, M. A. Chaudhry, A. J. Duncan, R. Hippler, D. M. Campbell, and H. Kleinpoppen, *Doubly Differential Cross Sections for the Ionization of the SF_6 Molecule by Electron Impact*, J. Phys. B **29**, 1849 (1996).

[176] K. Ueda, J. B. West, N. M. Kabachnik, Y. Sato, K. J. Ross, H. J. Beyer, H. Hamdy, and H. Kleinpoppen, *Evolution from Spectator to Normal Auger Lines Through the Thresholds Observed in the Ca 3P Excitation Region*, Phys. Rev. A **54**, 490 (1996).

[177] H. J. Beyer, J. B. West, K. J. Ross, K. Ueda, N. M. Kabachnik, H. Hamdy, and H. Kleinpoppen, *A New Approach to the Complete Experiment of Atomic Photoionization*, J. Electron Spectros. & Related Phenom. **79**, 342 (1996).

[178] O. Plotzke, G. Prümper, B. Zimmermann, U. Becker, and H. Kleinpoppen, *Magnetic Dichroism in the Angular Distribution of Atomic Oxygen 2P Photoelectrons*, Phys Rev. Lett. **77**, 2642 (1996).

[179] D. M. Campbell and H. Kleinpoppen, in *Preface to Selected Topics on Electron Physics*, edited by D. M. Campbell and H. Kleinpoppen (Plenum Press, New York, 1996), p. xiii.

[180] H. Kleinpoppen, in *Selected Topics on Electron Physics*, edited by D. M. Campbell and H. Kleinpoppen (Plenum Press, New York, 1996), p. 95.

[181] B. Siegmann, R. Hippler, H. Kleinpoppen, and H. O. Lutz, in *Selected Topics on Electron Physics*, edited by D. M. Campbell and H. Kleinpoppen (Plenum Press, New York, 1996), p. 295.

[182] A. J. Duncan, Z. A. Sheikh, and H. Kleinpoppen, in *Selected Topics on Electron Physics*, edited by D. M. Campbell and H. Kleinpoppen (Plenum Press, New York, 1996), p. 457.

[183] A. J. Duncan, Z. A. Sheikh, H. J. Beyer, and H. Kleinpoppen, *Two-Photon Polarization Fourier Spectroscopy of Metastable Atomic Hydrogen*, J. Phys. B **30**, 1347 (1997).

[184] G. Prümper, B. Zimmermann, O. Plotzke, U. Becker, and H. Kleinpoppen, *Multiplet- Dependent Magnetic Dichroism in the Atomic Oxygen 2P Spectrum Source*, Europhysics Letts. **38**, 19 (1997).

[185] K. Ueda, J. B. West, K. J. Ross, N. M. Kabachnik, H. J. Beyer, H. Hamdy, and H. Kleinpoppen, *Electronic Decay of the Photo-excited Ca 3P-3D Resonance*, J. Phys. B **30**, 2093 (1997).

[186] H. Kleinpoppen, A. J. Duncan, H. J. Beyer, and Z. A. Sheikh, *Coherence and Polarization Analysis of the Two-Photon Radiation of Metastable Atomic Hydrogen*, Physica Scripta T **72**, 7 (1997).

[187] G. Prümper, B. Zimmermann, B. Langer, O. Plotzke, M. Martins, K. Wielizcek, A. Hempelmann, M. Wiedenhoft, U. Becker, and H. Kleinpoppen, *Angle-Integrated Magnetic Linear Dichroism in Valence Photoionization of Free Oxygen Atoms*, J. Phys. B **30**, L 683 (1997).

[188] H. Kleinpoppen, *Atoms, in Textbook on Experimental Physics* (Walter de Gruyter Company, Berlin, New York, 1998), p. 1.

[189] S. Datz, G. W. F. Drake, T. F. Gallagher, H. Kleinpoppen, and G. zu Putlitz, *Atomic Physics*, Rev. Mod. Phys. **71**, S 223 (1999), centenary Issue 1999.

[190] H. Kleinpoppen, *Lehrbuch der Experimental Physik (in German), Vol. IV*, 2nd ed. (Walter de Gruyter Company, Berlin, New York, 2000), in print.

[191] G. Prümper, B. Zimmermann, B. Langer, J. Viefhaus, R. Hentges, N. A. Cherepkov, B. Schmidtke, M. Drescher, U. Heinzmann, U. Becker, and H. Kleinpoppen, *Sudden Interchannel Coupling in the Tl 6p Ionization Above the 5d Threshold*, 2000, submitted.

[192] A. J. Duncan, H. Kleinpoppen, and M. A. Scully, *Polarization and Correlation Effects in the Two-Photon-Radiation of Metastable Atomic Hydrogen*, 2000, submitted.

[193] M. A. Chaudhry, A. Al-Nasir, A. J. Duncan, R. Hippler, and H. Kleinpoppen, *Multiple Ion and Ejected Electron Coincidences of Direct and Dissociative Ionization of Sulphur Dioxide by Electron Impact*, to be published.

[194] H. Hamdy, H. J. Beyer, R. Clarke, D. Cartright, G. Csanak, and H. Kleinpoppen, *Theoretical and Experimental Investigations on Coherence Parameters of Ca and Sr Atoms by Electron Impact*, to be published.

[195] A. H. Nasir, M. A. Chaudhry, R. Hippler, A. J. Duncan, and H. Kleinpoppen, *Doubly Differential Cross section for the Ionization of SO_2 Molecule by Electron Impact*, to be published.

[196] I. C. Percival and M. Seaton, Proc. Camb. Phil. Soc. **53**, 654 (1957).

[197] I. C. Percival and M. Seaton, Phil. Trans. Roy. Soc. (London) A **251**, 113 (1958).

[198] P. G. Burke, S. Ormonde, and W. Whitaker, Proc. Phys. Soc. (London) **92**, 319 (1967).

[199] W. E. Lamb and R. C. Retherford, Phys. Rev. **72**, 241 (1947).

[200] W. E. Lamb and R. C. Retherford, Phys. Rev. **79**, 549 (1950).

[201] W. E. Lamb and R. C. Retherford, Phys. Rev. **81**, 222 (1951).

[202] I. Rabi, S. Millman, P. Kusch, and J. Zacharias, Phys. Rev. **55**, 526 (1939).

[203] J. Brossel and F. Bitter, Phys. Rev. **86**, 308 (1952).

[204] W. Hanle, Z. Physik **30**, 93 (1924).

[205] H.-J. Beyer, in *Progress in Atomic Spectroscopy, Part A*, edited by W. Hanle. and H. Kleinpoppen (Plenum Press, New York, 1978), p. 529.

[206] A. M. Ermolaev, in *Progress in Atomic Spectroscopy, Part A*, edited by W. Hanle. and H. Kleinpoppen (Plenum Press, New York, 1978), p. 149.

[207] M. J. Seaton, in *Atomic and Molecular Processes*, edited by D. R. Bates (Academic Press, New York, 1966).

[208] P. G. Burke and H. M. Schey, Phys. Rev. **126**, 163 (1962).

[209] P. G. Burke and J. F. B. Mitchell, J. Phys. B **7**, 214 (1974).
[210] B. Bederson, Comments At. Mol. Opt. Phys. **1**, 41 (1969).
[211] B. Bederson, Comments At. Mol. Opt. Phys. **1**, 65 (1969).
[212] B. Bederson, Comments At. Mol. Opt. Phys. **2**, 160 (1970).
[213] K. Rubin, B. Bedersen, M. Goldstein, and R. E. Collins, Phys. Rev. **182**, 201 (1969).
[214] I. V. Hertel and W. Stoll, J. Phys. B **7**, 570 (1974).
[215] I. V. Hertel and W. Stoll, J. Phys. B **7**, 583 (1974).
[216] I. Hertel and W. Stoll, Adv. At. Mol. Phys. **13**, 113 (1977).
[217] J. Kessler, *Polarized Electrons* (Springer Verlag, Heidelberg, 1976).
[218] U. Fano, Rev. Mod. Phys. **29**, 74 (1957).
[219] J. Macek and D. H. Jaecks, Phys. Rev. A **4**, 2288 (1971).
[220] U. Fano and J. H. Macek, Rev. Mod. Phys. **45**, 553 (1973).
[221] R. H. McKnight and D. H. Jaecks, Phys. Rev. A **4**, 2281 (1971).
[222] H. G. Berry, Rep. Prog. Phys. **40**, 155 (1977).
[223] N. Andersen, J. W. Gallagher, and I. V. Hertel, Phys. Rep. **165**, 1 (1988).
[224] I. Bray and A. T. Stelbovics, Adv. At. Mol. Opt. Phys. **35**, 209 (1995).
[225] G. H. Wannier, Phys. Rev. **90**, 817 (1953).
[226] G. H. Wannier, Phys. Rev. **100**, 1180 (1955).
[227] G. F. Hanne, Phys. Rep. **95**, 95 (1983).
[228] F. J. D. Paixao, N. T. Padial, G. Csanak, and K. Blum, Phys. Rev. Lett. **45**, 1164 (1980).
[229] U. Heinzmann, J. Phys. B **13**, 4367 (1980).
[230] C. Heckenkamp, F. Schäfers, G. Schönhense, and U. Heinzmann, Phys. Rev. Lett. **52**, 421 (1984).
[231] W. Mehlhorn, Phys. Letters **26 A**, 166 (1968).
[232] K. Bartschat, J. Phys. B **32**, 355 (1999).
[233] U. Fano and W. Lichten, 1967.
[234] R. Hippler, in *Fundamental Processes in Atomic Collision Physics*, edited by H. Kleinpoppen, J. S. Briggs, and H. O. Lutz (Plenum Press, New York, 1985), p. 181.
[235] R. Hippler, C. D. Lin, B. Siegmann, N. Toshima, and J. B. Wang, Can. J. Phys. **74**, 959 (1996).
[236] J. Macek and C. Wang, Phys. Rev. A **34**, 176 (1986).
[237] W. Fritsch, 1985, unpublished.
[238] R. Hippler, J. Phys. B **26**, 1 (1993).

[239] R. Hippler, H. Madeheim, H. Lutz, M. Kimura, and N. Lane, Phys. Rev. A **40**, 3446 (1989).

[240] O. Mo and A. Riera, J. Phys. B **21**, 119 (1988).

[241] G. G. Tepehan, B. Siegmann, H. Madeheim, R. Hippler, and M. Kimura, J. Phys. B **27**, 5527 (1994).

[242] B. Cleff and W. Mehlhorn, J. Phys. B **9**, 593 (1974).

[243] W. Sandner and W. Schmitt, J. Phys. B **11**, 1833 (1978), e.g.

[244] F. H. Read, in *Atomic Physics 7*, edited by D. Kleppner and F. M. Pipkin (Plenum, New York, 1981), p. 429, e.g.

[245] W. Hink, L. Kees, H.-P. Schmitt, and A. Wolf, in *Inner Shell and X-ray Physics of Atoms and Solids*, edited by D. J. Fabian, H. Kleinpoppen, and L. M. Watson (Plenum, New York, 1981), p. 327.

[246] R. Hippler, in *Coherence in Atomic Collision Physics*, edited by H. J. Beyer, K. Blum, and R. Hippler (Plenum, New York, 1988), p. 137.

[247] L. Kissel, C. MacCallum, and R. H. Pratt, *Albuquerque: Sandia National Laboratories, Report SAND81-1337*, 1981.

[248] A. Sommerfeld, Ann. Physik. (Leipzig) **11**, 257 (1931).

[249] H. Kulenkampff, Z. Physik **157**, 282 (1959).

[250] R. Hippler, J. Bossler, and H. O. Lutz, J. Phys. B **17**, 2453 (1984).

[251] M. Göppert-Mayer, Ann. Phys. **9**, 173 (1931).

[252] *Quantum Mechanics versus Local Realism – The Einstein–Podolski–Rosen Paradox*, edited by F. Selleri (Plenum, New York, 1988).

[253] A. J. Duncan, in *Coherence in Atomic Collision Physics*, edited by H. J. Beyer, K. Blum, and R. Hippler (Plenum, New York, 1988), p. 321.

[254] J. F. Bell, Physics **1**, 195 (1964).

[255] O. Plotzke, U. Wille, R. Hippler, and H. O. Lutz, Phys. Rev. Lett. **65**, 2982 (1990).

[256] *Coherence in Atomic Collision Physics*, edited by H. J. Beyer, K. Blum, and R. Hippler (Plenum, New York, 1988).

I
ELECTRON SCATTERING

Chapter 2

COMPLETE EXPERIMENTS IN ELECTRON–ATOM COLLISIONS – BENCHMARKS FOR ATOMIC COLLISION THEORY

Klaus Bartschat
Department of Physics and Astronomy, Drake University, Des Moines, IA 50311, U.S.A.

1. INTRODUCTION

Electron collisions with atoms and ions, involving elastic scattering, excitation, and ionization, are of tremendous importance not only as a fundamental branch of atomic physics, but also because of the urgent practical need for accurate data in many applications. These data serve as input for modeling processes in air pollution research, astronomy, electrical discharges, laser developments and applications, magnetically confined thermonuclear fusion devices, planetary atmospheres, and surface science, to name just a few.

In light of the importance of these processes, the need for benchmark comparisons between experimental data and theoretical predictions is obvious. Such comparisons can be made at various levels of detail, ranging from rather global observables such as rate coefficients, i.e., total, angle-integrated cross sections that are further integrated over a range of collision energies, to the very detailed parameters measured in "complete experiments" [1,2] which resolve the scattering angle of the projectile, its spin, and even the polarization of the light emitted from excited targets in possibly spin-resolved electron–photon coincidence experiments.

In this article, we present some key examples to illustrate the role that complete, or almost complete, experiments have played in pushing the theoretical models used to describe electron–atom collisions beyond their original limits. As a result, some numerical approaches have become so sophisticated that elastic scattering and excitation of relatively simply quasi-one-electron and quasi-two-electron targets, such as Na and He, can now be described with very high accuracy. In some cases, the reliability of theoretical predictions has even been assessed higher than the corresponding experimental data [3], particularly for

62   COMPLETE SCATTERING EXPERIMENTS

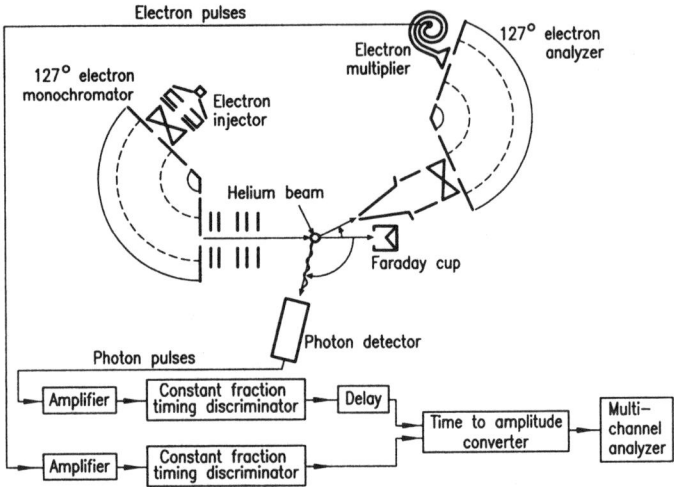

Figure 2.1   Angular correlation setup of the Stirling group

absolute cross sections whose measurements often suffer from normalization problems.

The outline of this contribution is as follows. We concentrate on electron-impact excitation processes and begin with the simplest example of P-state excitation in helium from the ground state $(1s^2)^1$S. The level of complexity is then increased for the case of excited D-states or S $\rightarrow$ P transitions in targets with unsaturated spin, with the specific example of sodium. The situation becomes even more complex for targets such as the heavy noble gases or mercury, where explicitly spin-dependent forces such as the spin-orbit interaction have to be taken into account both in the target description and the projectile–target interaction. Finally, a brief look at simultaneous electron-impact ionization–excitation shows a possibility for future benchmark experiments that will further challenge and stimulate theoretical developments.

2.   S $\rightarrow$ P EXCITATION IN HELIUM

This system was studied in a pioneering experiment by Hans Kleinpoppen's group in Stirling [4]. They began by measuring the angular correlation of the light emitted in the reaction plane in coincidence with the scattered electron. The setup of the experiment is shown in Fig. 2.1. Figure 2.2 shows the results for electron impact excitation of the He 2^1P state at an incident electron energy of 60 eV. Note how the intensity pattern observed in the scattering plane is reproduced quite well in the first Born approximation (FBA) at the small scattering angle of 16°, while a dramatic discrepancy between experiment and

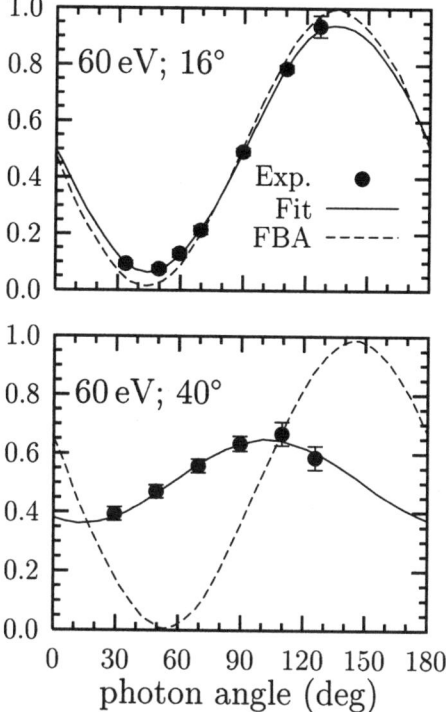

Figure 2.2 Electron–photon coincidence rates after impact excitation of He (2^1P) at 60 eV incident energy and scattering angles of 16° and 40° [4].

the FBA prediction is observed at 40°. This gives a first impression about how such experiments can serve as a detailed test of theoretical predictions.

The apparatus was later modified by Standage and Kleinpoppen [5] to allow for polarization measurements of the emitted light. They observed the radiation emitted perpendicular to the scattering plane, again in coincidence with the scattered projectile, but they also measured the three relative Stokes parameters

$$P_1 \equiv \frac{I(0°) - I(90°)}{I(0°) + I(90°)}, \tag{2.1}$$

$$P_2 \equiv \frac{I(45°) - I(135°)}{I(45°) + I(135°)}, \tag{2.2}$$

$$P_3 \equiv \frac{I(RHC) - I(LHC)}{I(RHC) + I(LHC)}. \tag{2.3}$$

Here $I(\beta)$ is the light intensity transmitted by a linear polarization analyzer oriented at an angle β with respect to the incident beam direction while $I(RHC)$ and $I(LHC)$ are the intensities transmitted by filters for right-circularly and left-circularly polarized light, respectively. Their schematic setup and results

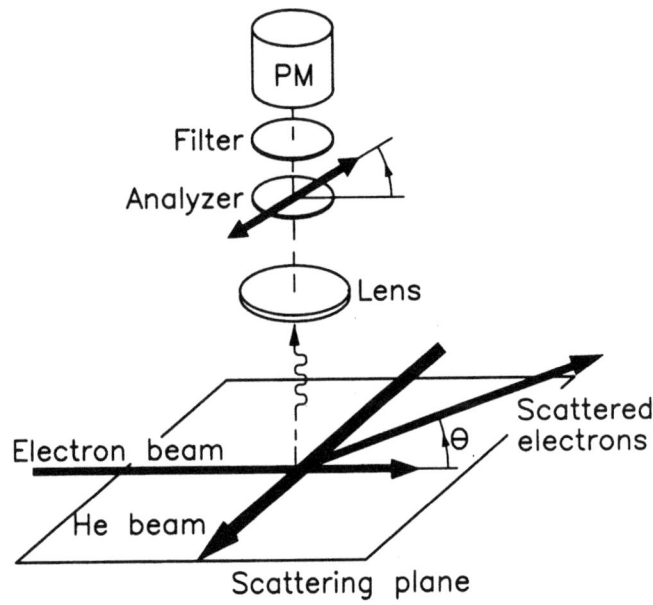

Figure 2.3 Electron–photon coincidence setup

for the circular polarization P_3 in the $3^1P \to 2^1S$ transition of helium are shown in Fig. 2.4.

It turns out that the above measurements, together with the *absolute* differential cross section σ indeed correspond to a "complete experiment", since they allow for the determination of the two independent scattering amplitudes that describe this process for each combination of scattering angle and collision energy. In fact, the excited P-state for this case can be written as a fully coherent linear combination

$$|\Psi(L)\rangle = \sum_M f_M(\theta)|L, M\rangle, \qquad (2.4)$$

where $f_M(\theta)$ is the scattering amplitude for excitation of a magnetic sublevel with quantum number M.

A standard parameterization used for many years was the set of complex amplitudes $(f_0^c, f_1^c = -f_{-1}^c)$, where the superscript "c" refers to the "collision system" with the quantization (z) axis chosen along the incident beam direction. Although this system is convenient for numerical calculations, Andersen and collaborators [7] demonstrated in a very convincing way that the general analysis becomes much simpler in the so-called "natural frame" with the quantization axis chosen perpendicular to the reaction plane. Most importantly, the

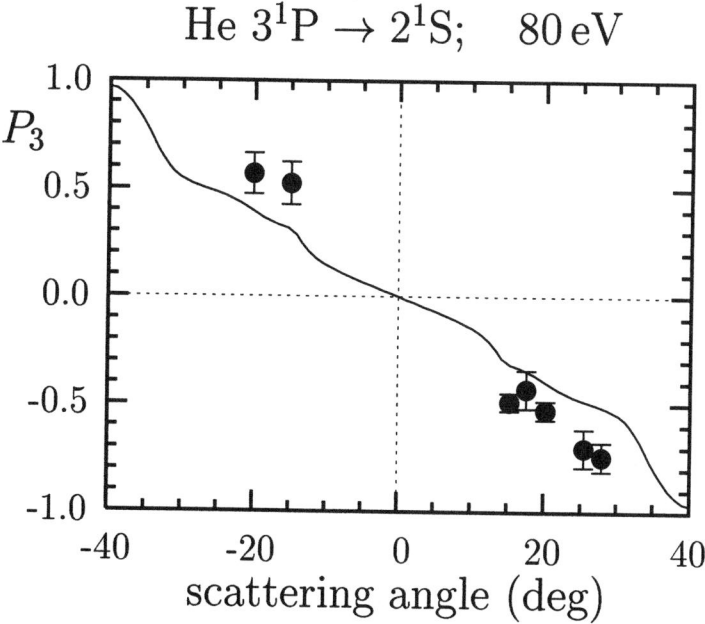

Figure 2.4 Circular light polarization P_3 observed in the He $3^1P \to 2^1S$ transition after impact excitation by 80 eV incident electrons [5]. The theoretical curve represents a multi-channel eikonal approximation calculation [6].

underlying physics can be clearly explained in this frame as well, and hence we will use it for most of the formal analysis presented below. In the non-relativistic limit, the independent amplitudes in the "natural frame" are given by $f^n_{\pm 1} = \mp \frac{1}{\sqrt{2}} f^c_0 - i f^c_1$ and $f^n_0 \equiv 0$.

Figure 2.5 shows the relationship between the Stokes (P_1, P_2, P_3, P_4) parameters that describe the polarization of the emitted light and the charge cloud characteristics for an excited P state. The linear polarization P_4, corresponding to P_1 measured with a photon detector in the scattering plane perpendicular to the incident beam direction, provides additional information if the charge cloud has a finite height h. For $S \to P$ excitation, this is the case if spin-flips perpendicular to the scattering are possible and hence the $M = 0$ sublevel (in the natural frame) can be excited. Alternatively, $P_4 \neq 0$ is possible if states with higher angular momentum are excited, as will be shown for the case of D-states below.

As seen from Fig. 2.5, the (relative) length, width, and height, as well as the alignment angle γ are obtained from various linear polarization measurements, while the circular polarization P_3 is related to the angular momentum transfer $L_\perp$. For the particular case of $S \to P$ excitation without spin-dependent

$$P_1 = \frac{I(0°) - I(90°)}{I(0°) + I(90°)}; \quad P_2 = \frac{I(45°) - I(135°)}{I(45°) + I(135°)}; \quad P_3 = \frac{I(\sigma^-) - I(\sigma^+)}{I(\sigma^-) + I(\sigma^+)}$$

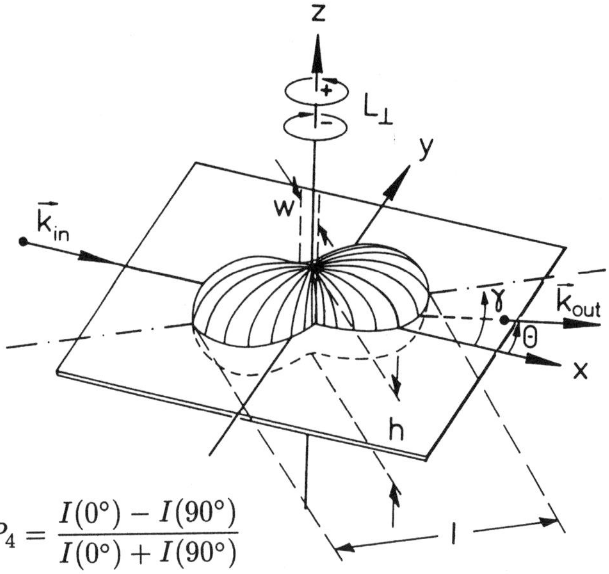

$$P_4 = \frac{I(0°) - I(90°)}{I(0°) + I(90°)}$$

$$L_\perp = L_\perp^+(1-h) = -P_3(1-h) = \eta_2(1-h)$$
$$\gamma = \tfrac{1}{2}\arctan(P_2/P_1) = \tfrac{1}{2}\arctan(\eta_1/\eta_3)$$
$$P_\ell = \sqrt{P_1^2 + P_2^2} = \sqrt{\eta_3^2 + \eta_1^2} = l - w$$
$$h = \frac{(1+P_1)(1-P_4)}{4 - (1-P_1)(1-P_4)}$$

Figure 2.5 Physical interpretation of Stokes parameter measurements for excitation of atomic P states. The linear polarizations (P_1, P_2, P_4) are given in terms of intensities $I(\beta)$ defined as the intensity transmitted through a linear polarizer whose transmission direction makes an angle β with the incident beam direction. Furthermore, the circular polarization P_3 is defined via the intensities of right-hand and left-hand circular-polarized light with negative (σ^-) and positive (σ^+) helicities, respectively. The set $(\eta_3, \eta_1, -\eta_2)$ [9] is a frequently used alternative notation to (P_1, P_2, P_3).

effects, the relationship to the scattering amplitudes is given by

$$\sigma = |f_{+1}^n|^2 + |f_{-1}^n|^2, \tag{2.5}$$
$$L_\perp = \left[|f_{+1}^n|^2 + |f_{-1}^n|^2\right]/\sigma, \tag{2.6}$$
$$2\gamma = \pi - \delta = \pi - \arg(f_{+1}^n f_{-1}^{n\,*}). \tag{2.7}$$

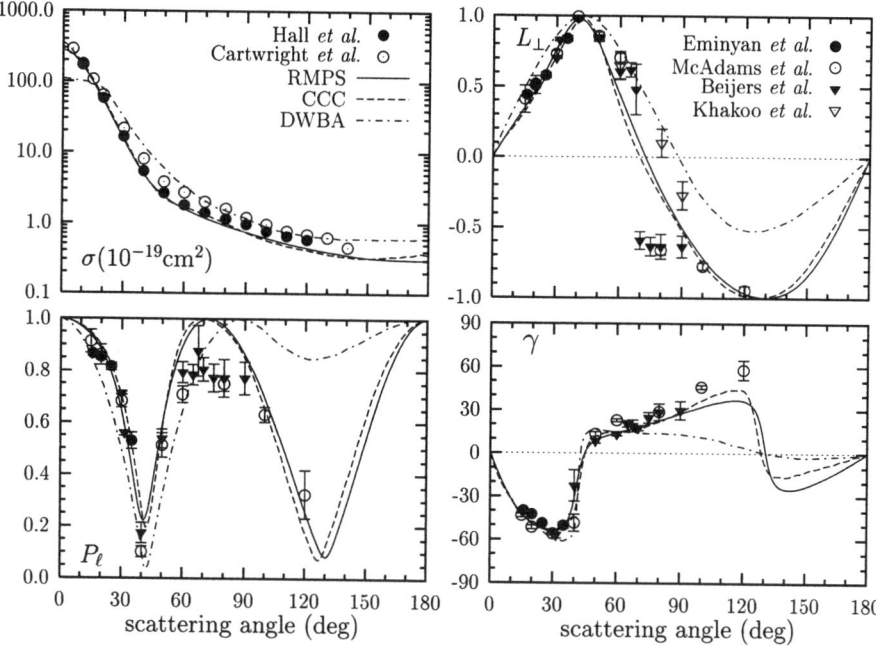

Figure 2.6 Differential cross section σ and electron impact coherence parameters $L_\perp$, γ, and P_ℓ for electron impact excitation of the 2^1P^o state in helium from the ground state 1^1S at an incident electron energy of 50 eV. The theoretical curves correspond to: ———, RMPS [10]; – – –, CCC [11]; – · – ·, DWBA [12]. The sources for the experimental data are: cross sections [13,14]; EICPs [4, 15–17]

We also note that an angular correlation measurement represents an alternative method to determine the alignment angle γ from the minimum of the distribution and the degree of linear polarization, $P_\ell = \sqrt{P_1^2 + P_2^2}$, from the amplitude of the correlation signal.

Figures 2.6 and 2.7 demonstrate how well the S → P excitation process is now understood, both experimentally and theoretically, and how the detailed electron-impact coherence parameters (EICP) can serve as benchmark tests of theoretical models. There is very good agreement between various experimental datasets and theoretical predictions from the "R-matrix with pseudostates" (RMPS) [10] and the "convergent-close-coupling" (CCC) [11] models for excitation of the 2^1P in helium at an incident electron energy of 50 eV. A much simpler first-order distorted-wave approach, formulated many years ago by Madison and Shelton [22], does very well in reproducing the differential cross section (DCS) and also the EICPs for small scattering angles. Not surprisingly, however, it fails for the large scattering angles when the very detailed sublevel and phase resolved information is compared. Similar problems, at compara-

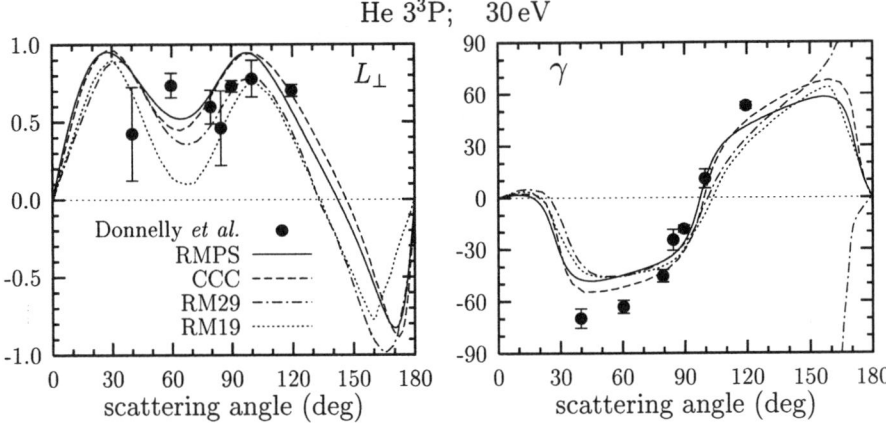

Figure 2.7 Electron impact coherence parameters $L_\perp$ and γ for electron impact excitation of the 3^3P^o state in helium from the ground state 1^1S at an incident electron energy of 30 eV. The experimental data of Donnelly *et al.* [18] are compared with results from various theoretical models: ———, RMPS; – – –, CCC [19]; – · – ·, RM29 [20]; · · · · · ·, RM19 [21].

ble impact energies, are found with predictions from the "first-order may-body theory" (FOMBT) [23].

At lower collision energies, such as 30 eV, simpler close-coupling-type methods including only discrete target states can also be expected to predict the outcome of the collision process with reasonable accuracy. This is shown in Fig. 2.7 where experimental EICP results [18] for excitation of the 3^3P^o state in helium are compared with results from RMPS, CCC [19], 29-state [20], and 19-state [21] R-matrix calculations. The overall agreement between experiment and theory is satisfactory, slightly favoring (as expected) the CCC and RMPS models. Note that the observed radiation ($3^3P^o \rightarrow 2^3S^e$) is depolarized in this case, but the EICPs directly after excitation can be recovered [7].

The charge clouds representing the $(1s2p)^1P^o$ state of helium after impact excitation by electrons at an incident energy of 50 eV are shown for selected scattering angles in Fig. 2.8, respectively. The scattering amplitudes were obtained from an RMPS calculation [10]. For small scattering angles the charge cloud is indeed well represented by an $M = 0$ state aligned along the momentum transfer direction, as predicted by the FBA, whereas the situation becomes much more complex for larger scattering angles. Using software packages such as *Mathematica*, it is now possible to produce even movies (angle-to-time conversion) to visualize the charge cloud in a very effective way [24]. For examples, see http://bartschat.drake.edu/dloveall.

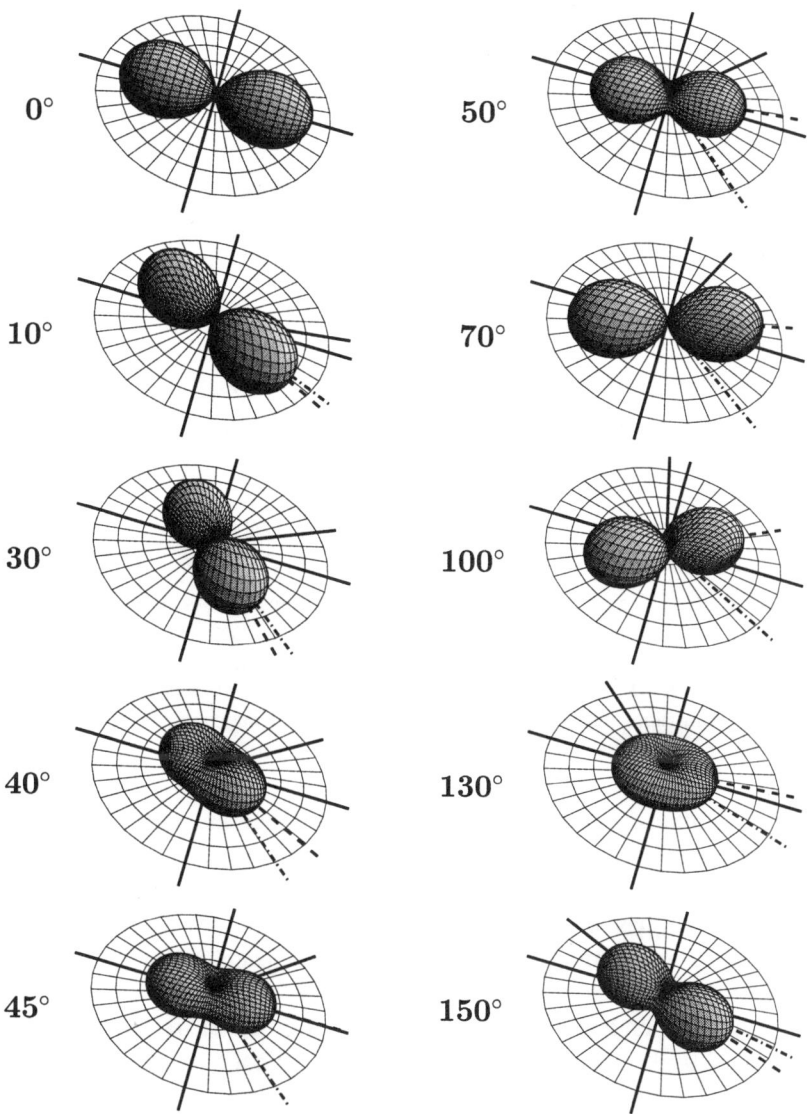

Figure 2.8 Charge cloud for electron impact excitation of the $(1s2p)^1P^o$ state of helium at an incident electron energy of 50 eV at various scattering angles between 0° and 150°. The scattering amplitudes were obtained from an RMPS calculation [10]. For the black-and-white printing, the thick lines were chosen as follows: solid, axes in the plane and scattering angle; dashed, alignment angle; dot-dashed, momentum transfer direction. In the color pictures created by the actual application [24], the solid thick lines represent the axes in the plane (red), the scattering angle (blue), the alignment angle (orange), and the momentum transfer direction (green).

3. S → D EXCITATION IN HELIUM

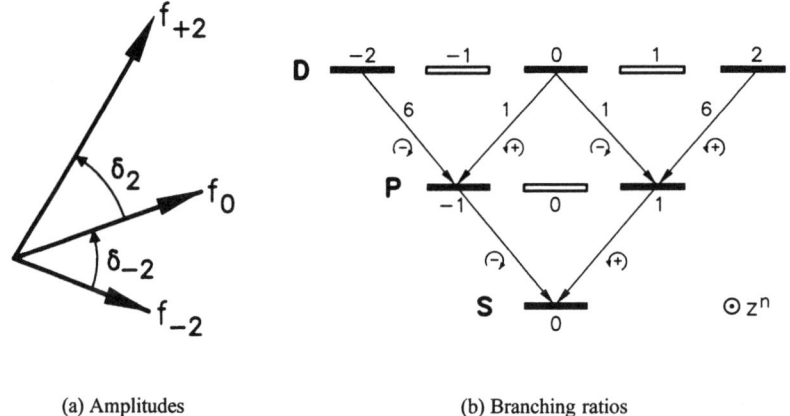

(a) Amplitudes (b) Branching ratios

Figure 2.9 Left: The amplitudes f_{+2}, f_0 and f_{-2} for impact S → D excitation without accounting for electron spin.
Right: Branching ratios for decay of an atomic D state through photon emission along the quantization axis. S → D impact excitation will only populate the magnetic sublevels with $M = 0, \pm 2$. Transitions through the $M = 0$ sublevel of the P state are not visible for photon observation along the z direction.

Having summarized the good experimental and theoretical understanding of S → P excitation processes in helium, we now move on to the significantly more complicated case of electron induced S → D transitions.

This problem has been discussed extensively in the literature over the years [25–27]. The summary here is based upon a recent review [28] to which we refer for further details.

The three independent amplitudes (in the natural frame) are shown in the left part of Fig. 2.9(a). A complete analysis of the optical decay pattern requires a triple-coincidence experiment involving the scattered particle and the two photons from the subsequent D → P and P → S transitions. The decay pattern for observation of both photons along the positive z direction is detailed in the right part of Fig. 2.9(b), where the helicities of the individual photons and the branching ratios for the sublevel transitions are indicated as well. A characteristic distinction of this observation direction is the fact that photons from transitions through the $M = 0$ sublevel of the P state, $|2\,0\rangle \to |1\,0\rangle$ and $|1\,0\rangle \to |0\,0\rangle$, are not emitted and detected in the z direction, due to the properties of $\Delta M = 0$ electric dipole radiation.

Because of the difficulty of triple–coincidence experiments, double–coincidences between the scattered particle and one of the emitted photons [12,

26, 29–31], or the two photons without observation of the scattered particle [32,33] have been observed in the majority of experimental studies until today. However, a triple-coincidence experiment was recently performed [34], thereby demonstrating the possibility (and experimental difficulty) of such studies.

In Fig. 2.9(b), two decay paths can be distinguished by the sublevel ($M_L = +1$ or $M_L = -1$) of the intermediate P level. (Recall that transitions to or from the state with $M_L = 0$ are not observable along the $\pm z$ directions.)

These paths will be labeled by the helicity ($+$ or $-$) of the subsequent P $\to$ S cascade photon. The two channels may be studied individually if observed in coincidence with the corresponding cascade photon, detected along the positive z direction through a circular polarization analyzer. If only the D $\to$ P photons are observed without regard of the cascade, the total signal is the weighted, incoherent sum of the two channels. In detail,

$$(1 - h)\mathbf{P} = w^+ (1 - h^+) \mathbf{P}^+ + w^- (1 - h^-) \mathbf{P}^-, \qquad (2.8)$$

or, splitting into linear and circular polarizations,

$$(1 - h) P_\ell e^{2i\gamma} = w^+ (1 - h^+) P_\ell^+ e^{2i\gamma^+} + w^- (1 - h^-) P_\ell^- e^{2i\gamma^-}; \qquad (2.9)$$
$$(1 - h) P_3 = w^+ (1 - h^+) P_3^+ + w^- (1 - h^-) P_3^-.$$

Since $\mathbf{P}^+$ and $\mathbf{P}^-$ in general are not parallel, $\mathbf{P}$ is not a unit vector. The degree of polarization $P = |\mathbf{P}|$ is thus not unity, although the excitation process is fully coherent.

Equation 2.9 corresponds to a triangle in the complex plane shown in Fig. 2.10. For this triangle, the lengths of all sides can be evaluated from scattered-particle–one-photon coincidence data, and the same holds for the direction 2γ of the sum vector. This information, however, is not sufficient to uniquely determine the directions of the "+" and "−" components, since, as the figure suggests, the mirror triangle has the same properties.

However, additional information about the excited D state, namely the height parameter h, can still be obtained using the scattered-particle–one-photon coincidence technique. This is achieved by photon observation in a second direction, traditionally chosen along the y axis, i.e., in the collision plane perpendicular to the incident beam direction. The Stokes vector in this direction is $(P_4, 0, 0)$. Here, light emitted with linear polarization perpendicular to the scattering plane arises from $\Delta M = 0$ optical decays of the $|2\,0\rangle$ state through the $|1\,0\rangle$ state which has negative reflection symmetry with respect to the scattering plane.

As mentioned above, the triangle in Fig. 2.9(a) cannot be defined uniquely in terms of the four Stokes parameters (P_1, P_2, P_3, P_4). By transforming the problem into the "atomic frame", obtained through rotation of the natural frame by the alignment angle γ around the quantization axis, and defining (up to

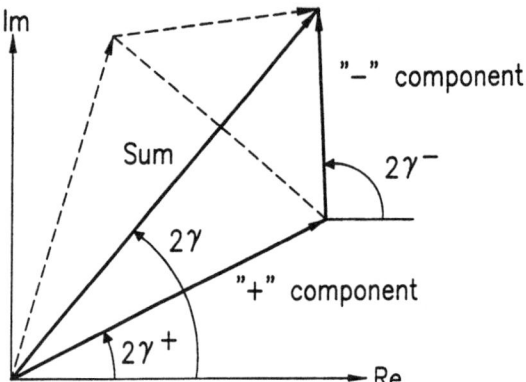

Figure 2.10 Vector diagram corresponding to equation 2.9. Instead of the values indicated for $2\gamma^{\pm}$, the ones obtained using the long dashed lines in the kite would also fulfill the equation.

mod $\pi/4$)

$$\hat{\eta} = -\frac{1}{4}(\hat{\delta}_2 + \hat{\delta}_{-2}) = \frac{1}{2}(\hat{\gamma}^+ + \hat{\gamma}^-) = \frac{1}{2}(\gamma^+ + \gamma^-) - \gamma = \eta - \gamma, \quad (2.10)$$

it can be shown that two D states which only differ by the sign of $\hat{\eta}$ will yield *an identical set of Stokes parameters*. The two solutions merge to one if, and only if, all three amplitudes are real in the atomic frame [28]. In general, however, the proper sign of $\hat{\eta}$ is revealed by a coincidence analysis with the cascade photons.

Before we show an example for such a truly complete experiment, let us consider again the benchmark character of scattered-electron–polarized-photon coincidence studies of this problem. For electron-impact excitation of the $(1s^2)^1S \rightarrow (1s3d)^1D$ transition in helium, Fig. 2.11 shows the state of the art in 1994, a mere five years ago: whereas there is very good agreement between experiments from three different groups for the circular polarization P_3 of the light emitted perpendicular to the reaction plane in the optical transition $(1s3d)^1D \rightarrow (1s2p)^1P$ transition after impact excitation by 40 eV incident electrons, theoretical predictions from various first-order distorted-wave-type approximations are dramatically different from each other and show little similarity to the experimental data.

Fig. 2.12 shows the progress that has been made by accounting for the coupling to both higher discrete and continuum states of the target. This coupling effect is known to be particularly important for optically forbidden transitions in neutral targets. Its proper representation by including a large number of square-integrable pseudo-states in the close-coupling + correlation expansion is the principal advantage of both the CCC and RMPS methods over standard discrete-state-only treatments. Clearly, the agreement between experimental

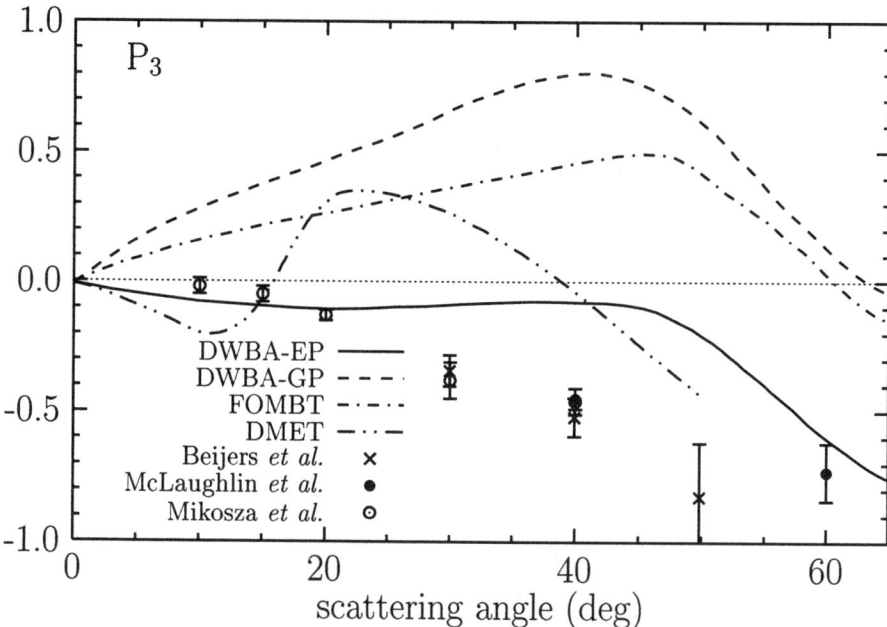

Figure 2.11 Circular light polarization P_3 for electron-impact excitation of the He(1s3d)^{1}D state at an incident electron energy of 40 eV. Experimental data of Beijers *et al.* [12] (×), McLaughlin *et al.* [28] (•), and Mikosza *et al.* [31] (○) are compared with various predictions from first-order models (DWBA [35], FOMBT [36], DMET [37]).

data for all four Stokes parameters $(P_1 - P_4)$ and the CCC/RMPS predictions is very satisfactory.

The results for the Stokes parameters (P_1, P_2, P_3, P_4) can then be transformed into various sets of parameters describing the charge cloud of the excited state. Since all the details can be found in the review by Andersen and Bartschat [28], we concentrate in Fig. 2.13 on the form of the charge cloud predicted by the CCC calculation for scattering angles of 40°, 60°, 80°, 100°, and 120°. The full picture is shown on the left, whereas the form derived from a one-photon angular distribution is shown on the right. Note the almost pure (four-leaf) cloverleaf shape around 100°, emitting nearly unpolarized light perpendicular to the scattering plane. This example shows very clearly how, in this case, the light polarization measurement yields much more information than an angular correlation analysis alone — in contrast to the case of S → P excitation discussed above, where the sign of P_3 is the only quantity that is not accessible in a correlation measurement.

As pointed out before, the Stokes parameters (P_1, P_2, P_3, P_4) obtained from a scattered-electron–polarized-photon double-coincidence experiment do not

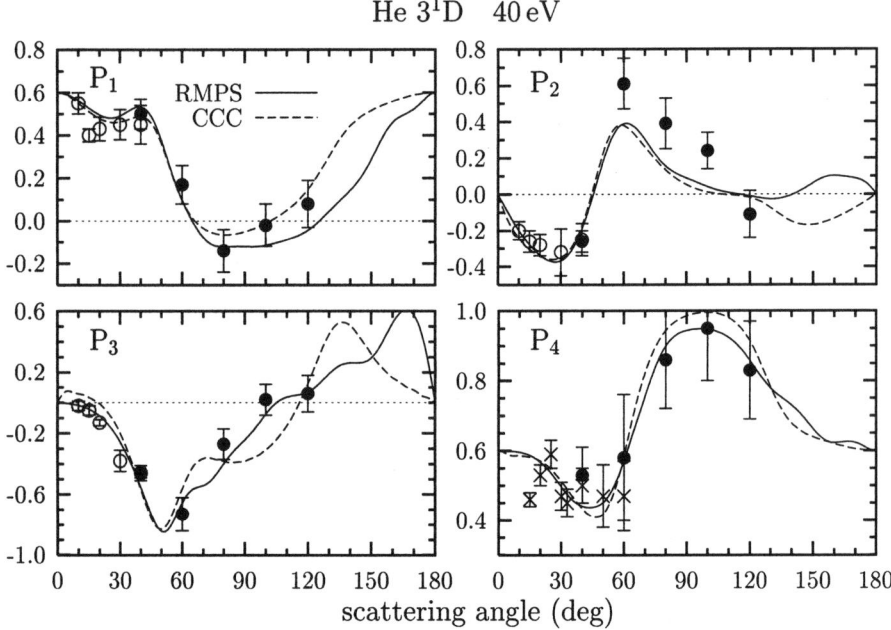

Figure 2.12 Stokes parameters (P_1, P_2, P_3, P_4) for electron impact excitation of the He $1^1S \to 3^1D$ transition at an incident electron energy of 40 eV. Experimental data of Beijers *et al.* [12] (×), McLaughlin *et al.* [29] (•), and Mikosza *et al.* [31] (○) are compared with predictions from CCC [19] and RMPS [38] calculations.

provide the full information about the excited state. Instead, some information about the cascade photon is needed through a triple-coincidence experiment, but the required information is limited to removing the ambiguity in the sign of $\hat{\eta}$, i.e., distinguishing between the "true" and the "ghost" solutions for the relative phases $\delta_{\pm 2}$ introduced in Fig. 2.9(a). Experimental results [34] for electron impact excitation of the He 3^1D state for an incident energy of 60 eV are displayed in Fig. 2.14. At a scattering angle of 40°, a triple-coincidence measurement was able to *experimentally* distinguish between the true and the "ghost" solution, both of which are consistent with the data obtained from a standard scattered-particle–one-photon double-coincidence setup. Interestingly, the good agreement between the corresponding CCC predictions and the two sets of data might suggest a simpler, though less satisfactory from a fundamental point of view, method of removing the ambiguity: using a reliable theory to distinguish between the two possibilities.

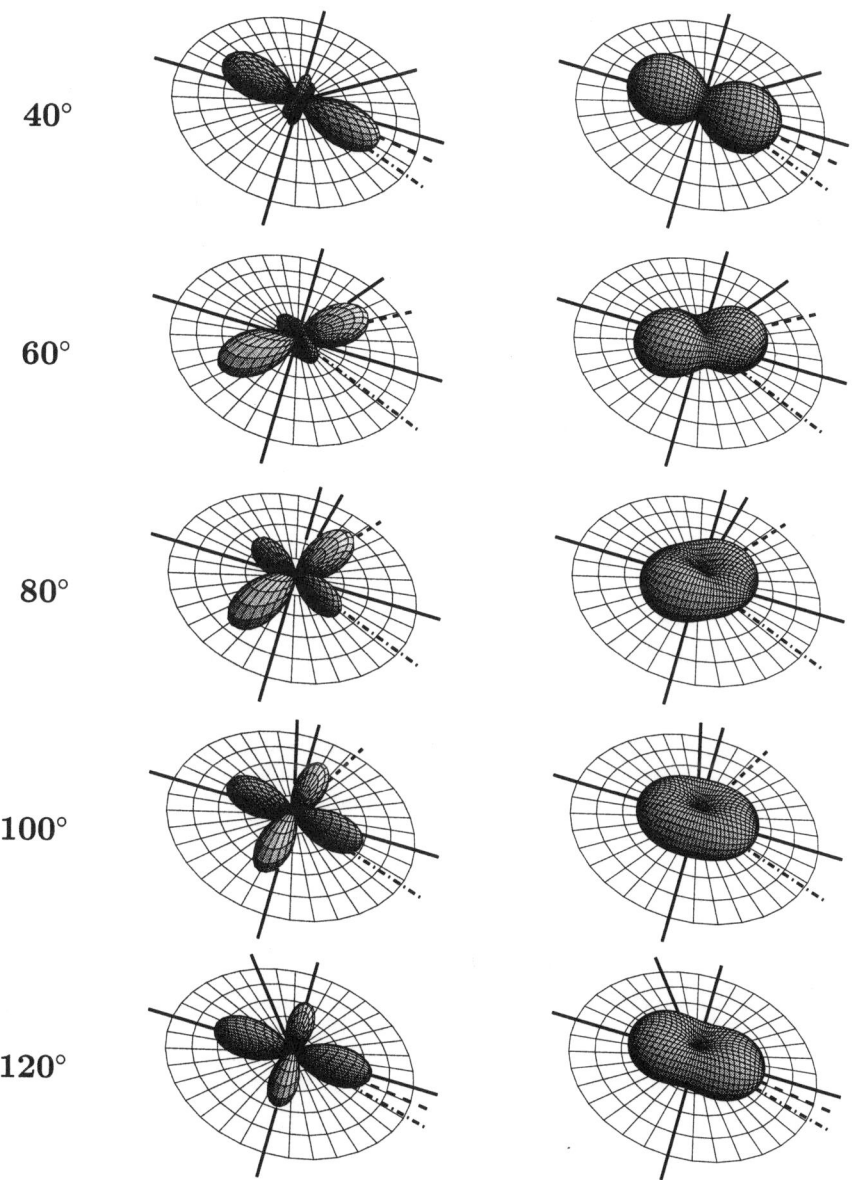

Figure 2.13 The form of the charge cloud predicted by the CCC calculation for electron impact excitation of the He $1^1S \rightarrow 3^1D$ transition at an incident electron energy of 40 eV. The full picture for selected scattering angles is shown on the left, whereas the result derived from a one-photon angular distribution is shown on the right [39]. The dashed and dot-dashed lines represent the alignment angle and the momentum transfer direction, respectively.

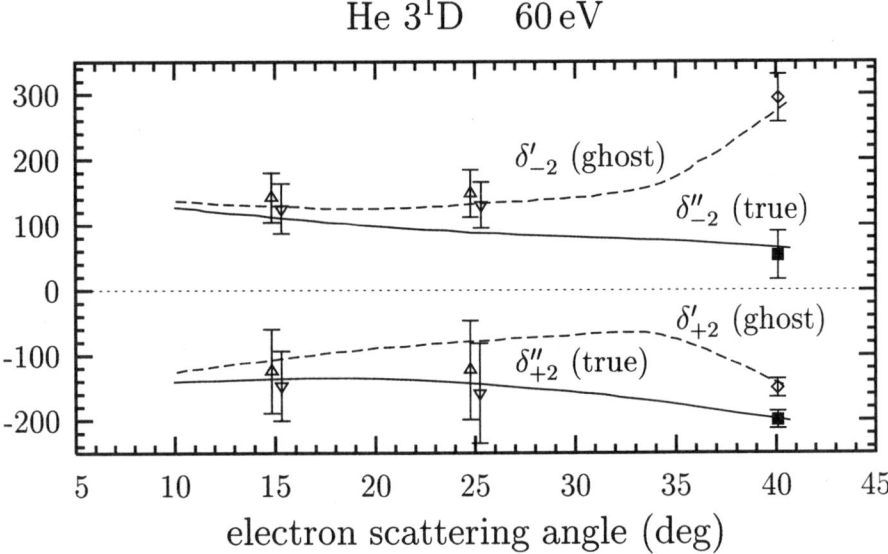

Figure 2.14 The true and ghost solutions for the relative phase $\delta_{\pm 2}$ for electron impact excitation of the He $1^1S \to 3^1D$ transition at an incident electron energy of 60 eV [34], as obtained in scattered-electron–single-photon (double) coincidence measurements. By using a scattered-electron–two-photon (triple) coincidence setup, it was possible to distinguish the two solutions experimentally for a scattering angle of 40°. Also shown are the corresponding predictions obtained from the CCC calculation of Fursa and Bray [19].

4.   S → P EXCITATION IN SODIUM

We now move on to the case of light (quasi-)one-electron targets. In contrast to excitation processes involving at least one singlet state (such as the $(1s^2)^1S$ ground state of helium), the projectile and target spins of $s = \frac{1}{2}$ double the number of independent scattering amplitudes, because we now have the possibility of triplet (t) and singlet (s) scattering, i.e., two independent channels for the combined spin S of the projectile + target system. [1] Fig. 2.15(a) illustrates the situation for the most important case, $(ns)^2S \to (n'p)^2P^o$, where the four independent amplitudes in the natural frame are denoted as $f_{\pm 1}^{t,s}$.

Neglecting an overall phase, one thus needs to determine seven independent parameters for a complete experiment. In addition to the differential cross section σ_u, six dimensionless parameters must be defined, three to characterize the relative lengths of the four vectors, and three to define their relative phase

---

[1] An early analysis of this problem was given by Hans Kleinpoppen (Phys. Rev. A **3** (1971) 2015) in terms of direct and exchange amplitudes in the collision frame.

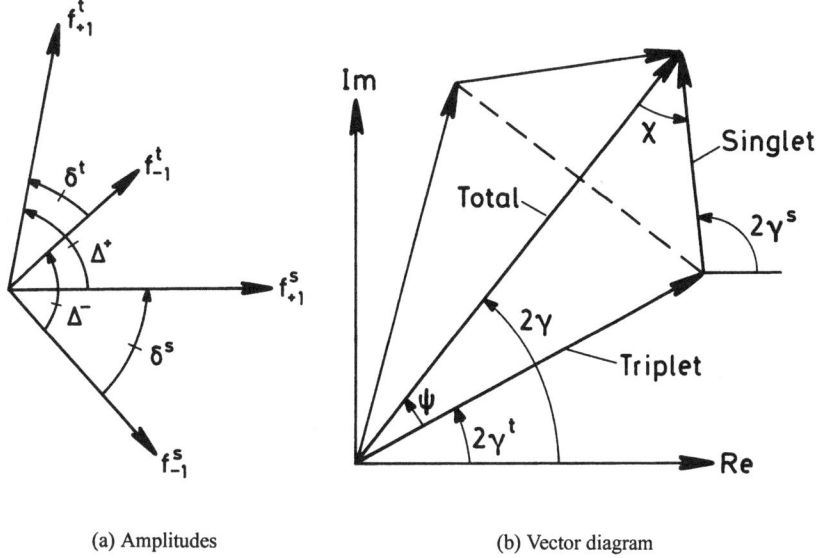

(a) Amplitudes (b) Vector diagram

Figure 2.15 Left: Schematic diagram of triplet (t) and singlet (s) scattering amplitudes in the natural frame for $^2S \to {}^2P$ transitions by electron impact. Note that $\Delta^+ + \delta^s = \Delta^- + \delta^t$. Right: Vector diagram corresponding to equation 2.12. Note the analogy to Fig. 2.10.

angles. In the spirit of the parameters introduced for S → P excitation without the necessity to separate the total spin channels of the system, Andersen and Bartschat [40] suggested the following complete set:

$$(\sigma_u; w^t, L_\perp^t, L_\perp^s; \gamma^t, \gamma^s, \Delta^+). \tag{2.11}$$

In addition to the differential cross section σ_u for scattering of unpolarized beams, the set contains angular momentum transfer and alignment parameters for the two spin channels, a weight parameter (w^t) to account for the relative strength of singlet and triplet scattering, and a phase parameter (Δ^+) to relate the relative phases between the singlet and the triplet amplitudes.

Experimentally, the singlet and triplet spin channels can be separated by using spin-polarized projectile and target beams. Extensive studies were performed at NIST [41–43], but again a few ambiguities remained in the determination of the relative phase between the complex amplitudes. Similar to the case of D-state excitation discussed above, the (reduced, i.e., fine-structure corrected) Stokes vector $\vec{P}$ of an unpolarized beam experiment is given by the weighted sum of the singlet and triplet (unit) Stokes vectors $\vec{P}^{s,t}$ as

$$\vec{P} = 3w^t \vec{P}^t + w^s \vec{P}^s, \tag{2.12}$$

78 COMPLETE SCATTERING EXPERIMENTS

from which the set of parameters $(L_\perp, \gamma, P_\ell)$ for an unpolarized beam experiment may be evaluated.

As illustrated in Fig. 2.15(b), the linear polarization part of 2.12 can be cast into a complex form that corresponds to addition of the two vectors $\vec{P}_\ell^s$ and $\vec{P}_\ell^t$, multiplied by weighting factors $3w^t$ and w^s, respectively, to form the resulting vector $\vec{P}_\ell$. Hence, elementary geometry can be applied to obtain two pairs of solutions, $(\gamma^t, \gamma^s)_{\text{true}}$ (the true solution) and $(\gamma^t, \gamma^s)_{\text{ghost}}$ (the other possibility) as

$$\begin{aligned} \gamma^t &= \gamma \mp \chi/2, \\ \gamma^s &= \gamma \pm \psi/2, \end{aligned} \quad (2.13)$$

where the angles χ and ψ are defined in the figure. Provided experimental data are available for the parameter set $(P_\ell, \gamma, w^t, L_\perp^t, L_\perp^s)$ at a given collision energy and scattering angle, two sets of possible angles (γ^t, γ^s) can be determined. As pointed out by Hertel et al. [44], this ambiguity is mathematically identical to the one for S $\rightarrow$ D excitation processes discussed above. The physical origin is the same in the two cases, namely the incoherent addition in the experiment of two channels that are, in principle, distinguishable.

Andersen and Bartschat [40] offered suggestions about how the ambiguities could, in principle, be removed. Due to the lack of experimental data, however, they designed inversion procedures based on the available data and theoretical predictions from a "close-coupling plus optical potential" (CCO) calculation [45] to identify the true solutions. The results are shown in Figs. 2.16 and 2.17

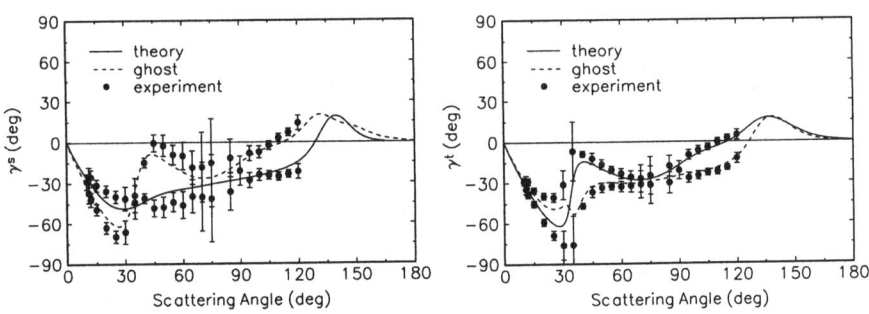

Figure 2.16 Alignment angles γ^t and γ^s calculated from the NIST data [42] for $(w^t, L_\perp^t, L_\perp^s)$ and theoretical results for (P_ℓ, γ) from scattering amplitudes of Bray [45] for electron impact excitation of the $3\,^2$P state of sodium at an incident electron energy of 4.1 eV; •, two sets of inverted experimental data as well as true (———) and ghost (– – –) theoretical solutions [40].

for electron impact excitation of the $3\,^2$P state of sodium at an incident electron energy of 4.1 eV. Since the true and the ghost solutions evaluated from theory are

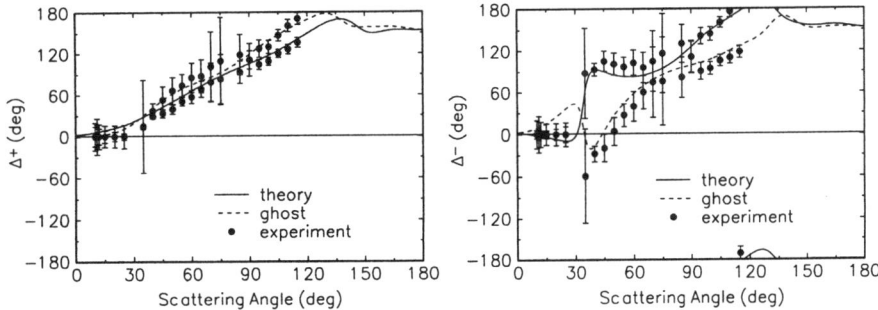

Figure 2.17 Singlet-triplet phase angles Δ^+ and Δ^- calculated from data for $(w^t, L^t_\perp, L^s_\perp)$ [42], the corresponding alignment angles (γ^t, γ^s) presented in Fig. 2.16, and theoretical T parameter results [46] for electron impact excitation of the $3\,^2\text{P}$ state of sodium at an incident electron energy of 4.1 eV; •, two sets of inverted experimental data as well as the true (———) and one ghost (– – –) solution [40].

in very good agreement with the experimental values, the "true" experimental data were selected as those that follow the true theoretical solution.

As an example for the importance of such detailed benchmark measurements, followed by a thorough analysis of the data, Fig. 2.18 shows the full set of results obtained at a collision energy of 10 eV [47]. Looking at the parameter $L^s_\perp$ reveals that only the most sophisticated CCC theory [48] can reproduce the experimental results for this observable; in contrast, the parameter $L^t_\perp$ and all the parameters for unpolarized beams (left column) are much less sensitive to the quality of the theoretical model.

5. S → P EXCITATION IN MERCURY

Electron-impact excitation of the $6^1\text{S}_0 \to 6^3\text{P}^o_1$ transition in mercury is certainly one of the show cases for (almost) complete experiments. This process is described by six independent scattering amplitudes, thereby requiring the determination of one *absolute* differential cross section, five *relative* magnitudes, and five *relative* phases. The large number of independent parameters reflects the additional degrees of freedom that the problem presents, due to the need for including explicitly spin-dependent effects such as the spin-orbit interaction in both the target description and the projectile–target interaction. Fig. 2.19 shows the notation of the six amplitudes, as introduced by Andersen and Bartschat [51]; the subscript indicates the magnetic sublevel while the arrow indicates the initial spin projection relative to the quantization axis.

As one might expect, the detailed analysis becomes very complicated but, compared to an early attempt in the collision frame [52], is relatively straightforward. A complete experiment is, in principle, possible by combining mea-

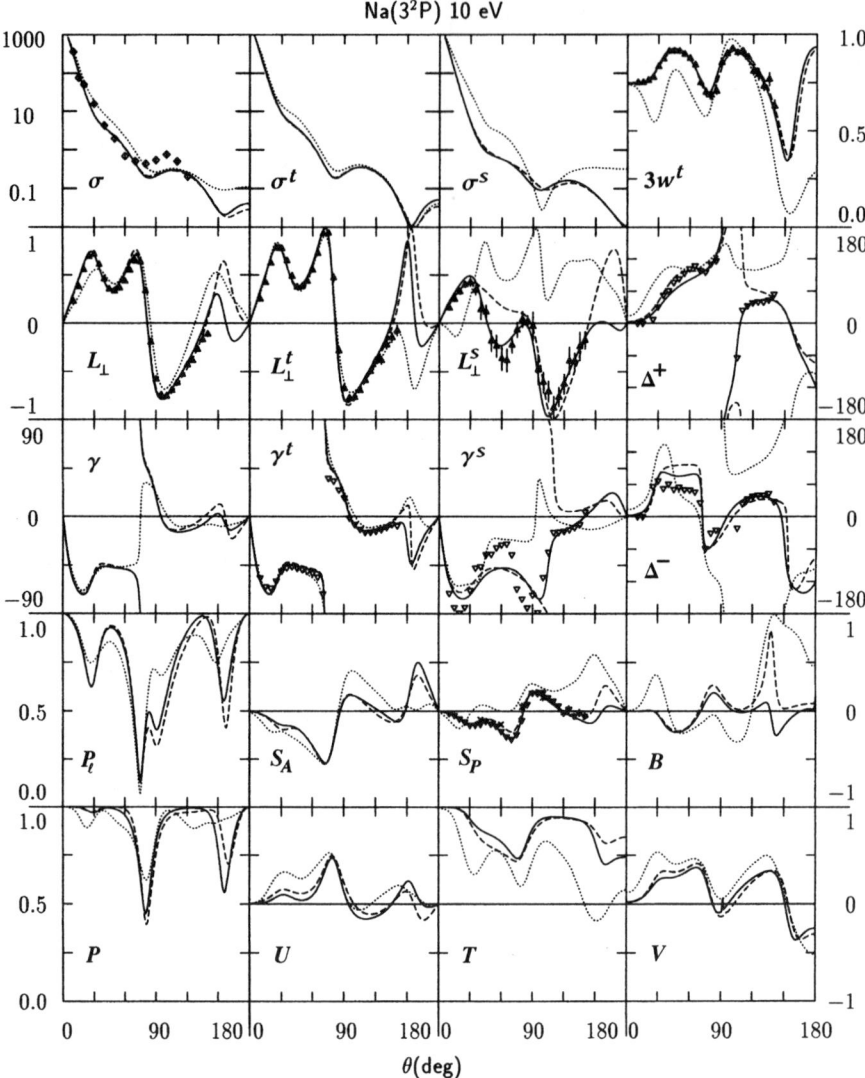

Figure 2.18 Survey of alignment and orientation parameters for excitation of the Na $3\,^2$P state by spin-polarized electrons at an incident electron energy of 10 eV [47]. The differential cross sections are given in units of a_0^2/sr. The experimental data of the NIST and Münster groups have been transformed to the parameter set (11); they are compared with CCC (——— [48]), CCO (— — — [45]), and DWBA2 (······ [49]) results. The differential cross sections are from Srivastava and Vuskovic [50].

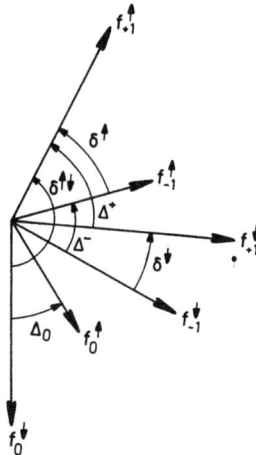

Figure 2.19 Schematic diagram of scattering amplitudes in the natural frame for $J = 0 \to J = 1$ transitions by electron impact [51]. Note that $\Delta^+ + \delta^\downarrow = \Delta^- + \delta^\uparrow$.

surements of the "generalized *STU* parameters" [53], that describe the change of an initial electron spin polarization through the collision process, and the "generalized Stokes parameters" [51], that represent scattered-electron–polarized-photon coincidence rates depending on the spin polarization of the incident electron beam.

Interesting features can be seen in Fig. 2.20 after transformation of the generalized Stokes parameter results into the spin-dependent coherence parameters represented by angular momentum transfer, alignment angle, and height parameters for different spin orientations, as well as parameters that connect the amplitude triplets for the two initial spin projections. The differential cross sections for the two spin directions are fairly similar at all angles. The coherence parameters, however, are vastly different for the two incident spin projections. Note, for example, that $L_\perp^{+\uparrow} > 0$ while $L_\perp^{+\downarrow} < 0$ at small scattering angles. While the positive value of $L_\perp^{+\uparrow}$ is in agreement with well-established propensity rules [54], the negative value of $L_\perp^{+\downarrow}$ might seem surprising. Both $L_\perp^{+\uparrow}$ and $L_\perp^{+\downarrow}$ are non-zero for forward scattering, and $L_\perp^{+\uparrow}(0°) = -L_\perp^{+\downarrow}(0°)$ by symmetry requirements. The alignment angles $\gamma^\uparrow$ and $\gamma^\downarrow$ show no similarities, with the directions of the major axes often being perpendicular to each other ($\gamma_\perp^{+\uparrow}(0°) = -\gamma_\perp^{+\downarrow}(0°) \neq 0$). There is a large difference between the height parameters $h^\uparrow$ and $h^\downarrow$ (which measure the relative importance of spin-flips) in this angular range, with $h^\downarrow$ assuming a maximum value of 75% near a scattering of 40°. This means that spin-flips are very likely for incoming spin down electrons, but those spin down electrons, whose spin is not flipped, tend to transfer a *negative* angular momentum to the atom.

82   COMPLETE SCATTERING EXPERIMENTS

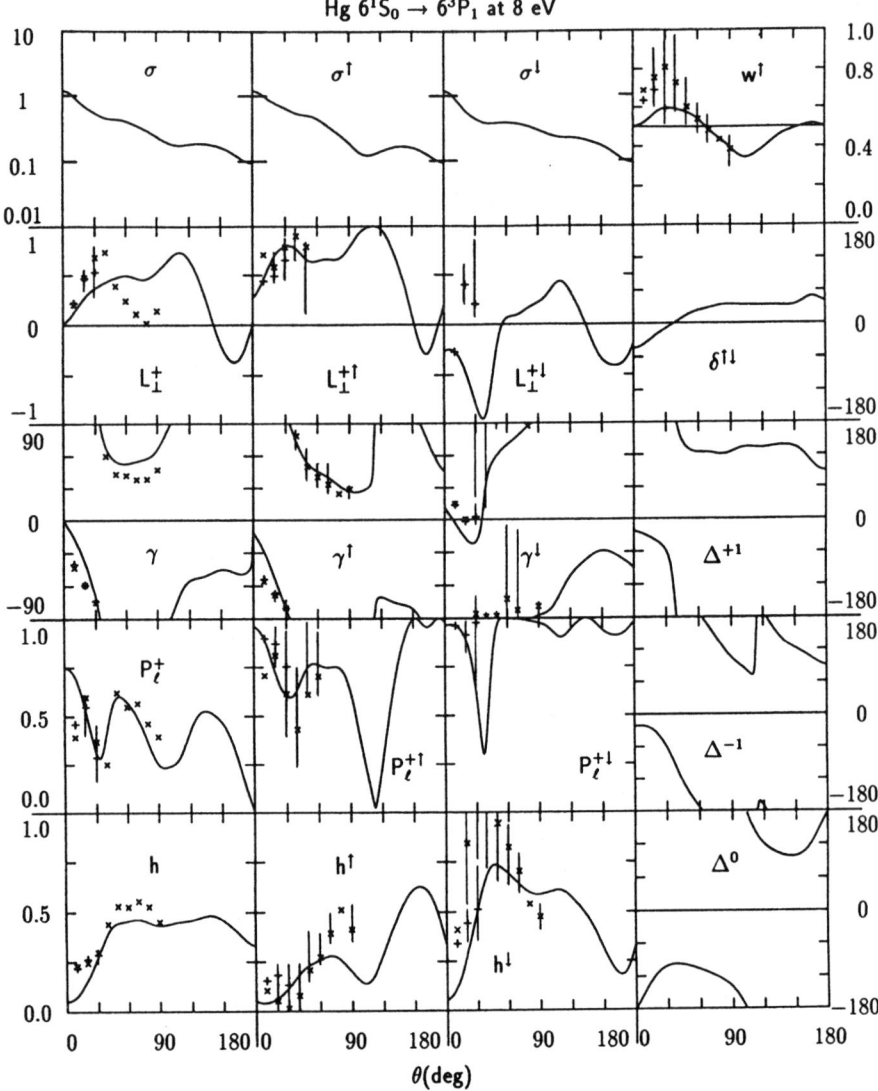

Figure 2.20 Spin-dependent coherence parameters for electron impact excitation of Hg(6s6p)$^3P_1$ at an incident electron energy of 8 eV [54]. ×, prediction based on data of Goeke *et al.* [55] and Sohn and Hanne [56]; +, prediction based on unpublished data of Sohn and Hanne. The experimental data are compared with the results from a five-state Breit-Pauli R-matrix calculation [57].

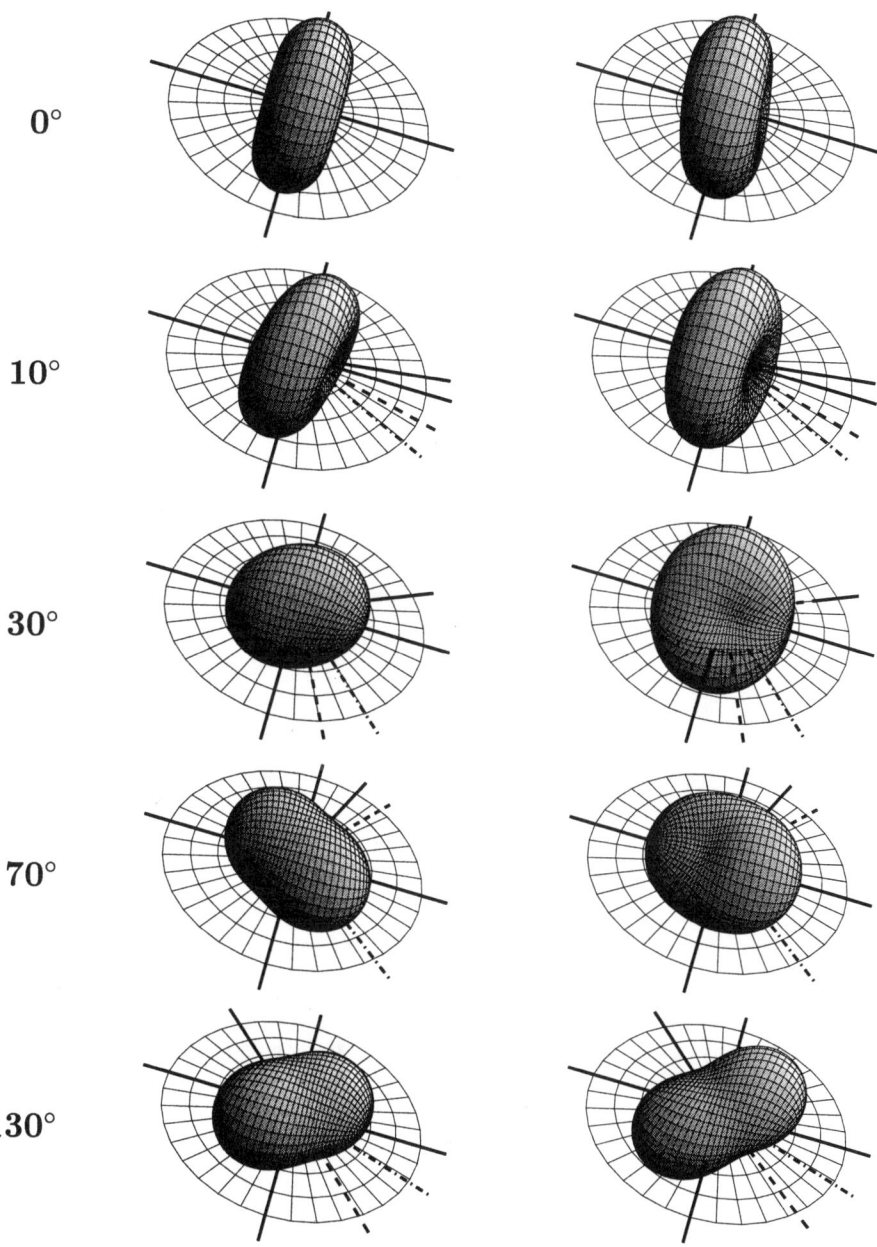

Figure 2.21 Charge cloud m representing the excited 6p electron after impact excitation of Hg(6s6p)^{3}P$_1$ at an incident electron energy of 8 eV [39]. The left column is for unpolarized incident electrons, the right column for an in-plane transversal spin polarization $P_y = 1$. The predictions are based upon a five-state Breit-Pauli R-matrix calculation [57]. The dashed and dot-dashed lines represent the alignment angle and the momentum transfer direction, respectively.

Fig. 2.21 shows the charge clouds representing the excited 6p electron. The graphs on the left correspond to unpolarized incident electrons, whereas those on the right correspond to an in-plane transversal spin polarization $P_y = 1$. Note how the charge cloud tilts and twists out of the reaction plane in the latter case, since the (axial) polarization vector breaks the planar symmetry [58].

6. S → P SIMULTANEOUS IONIZATION–EXCITATION IN HELIUM

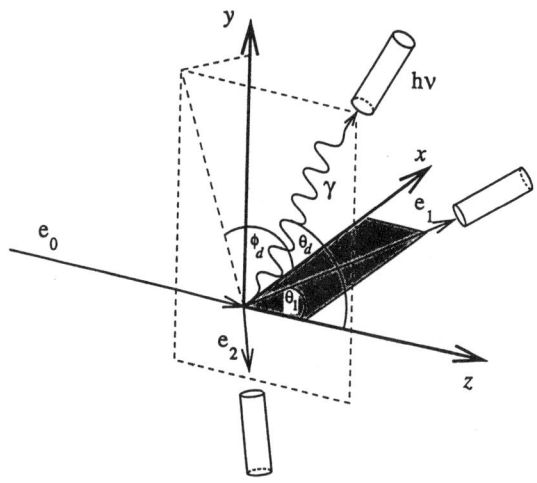

Figure 2.22 Scheme of an (e, 2eγ) process for ionization–excitation [59].

Electron-impact ionization of atoms has been of great interest for many years. In most cases, experimental efforts concentrated on the measurement of ionization probabilities, with the most detailed information being obtained from triple-differential cross section (TDCS) studies, for which the kinematics of the incident, scattered, and ejected electrons are fully defined. In the case of ionization–excitation, however, this information is generally not sufficient. Consider, for example, the situation shown in Fig. 2.22: A possibly spin-polarized electron beam is used to ionize a possibly spin-polarized target. In addition, the residual ion may end up in an excited state which subsequently can decay by photon emission.

Even if the projectile and target spins are essentially irrelevant (as in ionization of helium atoms in their ground state), the full information about an ionization–excitation process, e.g.,

$$e(\mathbf{k}_i) + He(1s^2)^1S \rightarrow He^+(2p)^2P + e_e(\mathbf{k}_e) + e_s(\mathbf{k}_s)$$
$$\downarrow$$
$$He^+(1s)^2S + h\nu \, (30.4 \text{ nm}), \quad (2.14)$$

can only be obtained in a triple-coincidence experiment between the two outgoing electrons and the photon which, in turn, also needs to be analyzed with respect to its polarization. In equation 2.14, $\mathbf{k}_i$ is the momentum of the incident electron while $\mathbf{k}_s$ and $\mathbf{k}_e$ denote the momenta of the scattered and ejected electrons, respectively. Although the two outgoing electrons are, in principle, indistinguishable, the faster (slower) one of these is usually referred to as the scattered (ejected) electron.

Although triple-coincidence setups between one electron and two photons [34] or all three outgoing electrons in double-ionization studies [60] have been developed, the above two-electron–one-photon coincidence experiment has not been performed to date. Instead, the investigations have been simplified

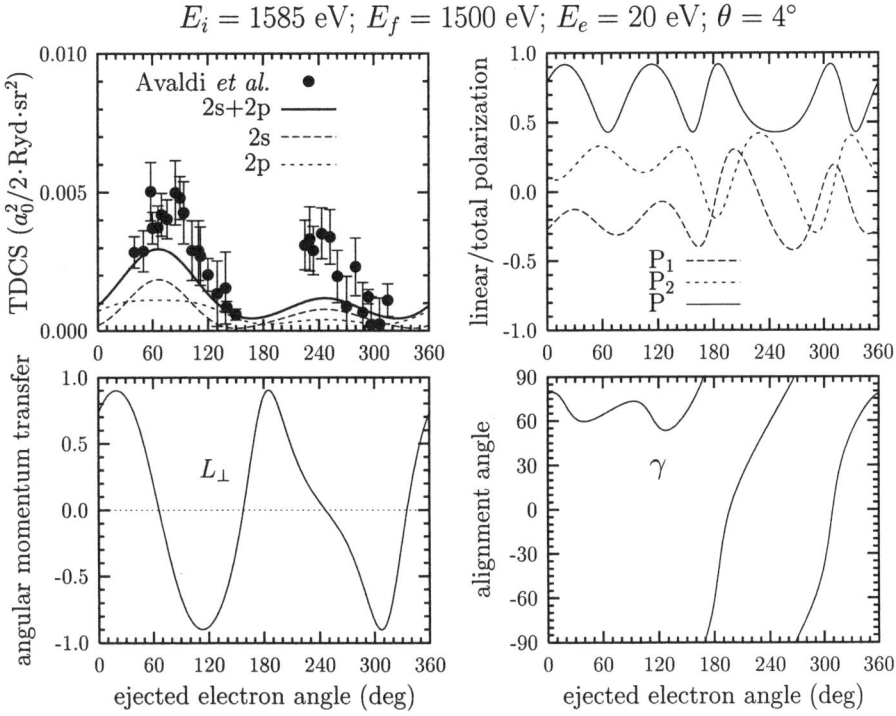

Figure 2.23 TDCS, light polarizations, angular momentum transfer, and alignment angle for ionization–excitation of $He(1s^2) \to He^+(2p)$ [70]. A fast incident electron of energy $1,585$ eV is observed at a scattering angle of $4°$ and an energy of $1,500$ eV, corresponding to an energy of 20 eV for the slow electron. The experimental data for the TDCS ($n = 2$) are taken from Avaldi et al. [69].

by observing only the photon to measure angle-integrated optical ionization–excitation cross sections [61–63] and alignment parameters [64], only one electron in double-differential cross section (DDCS) studies [65], one photon

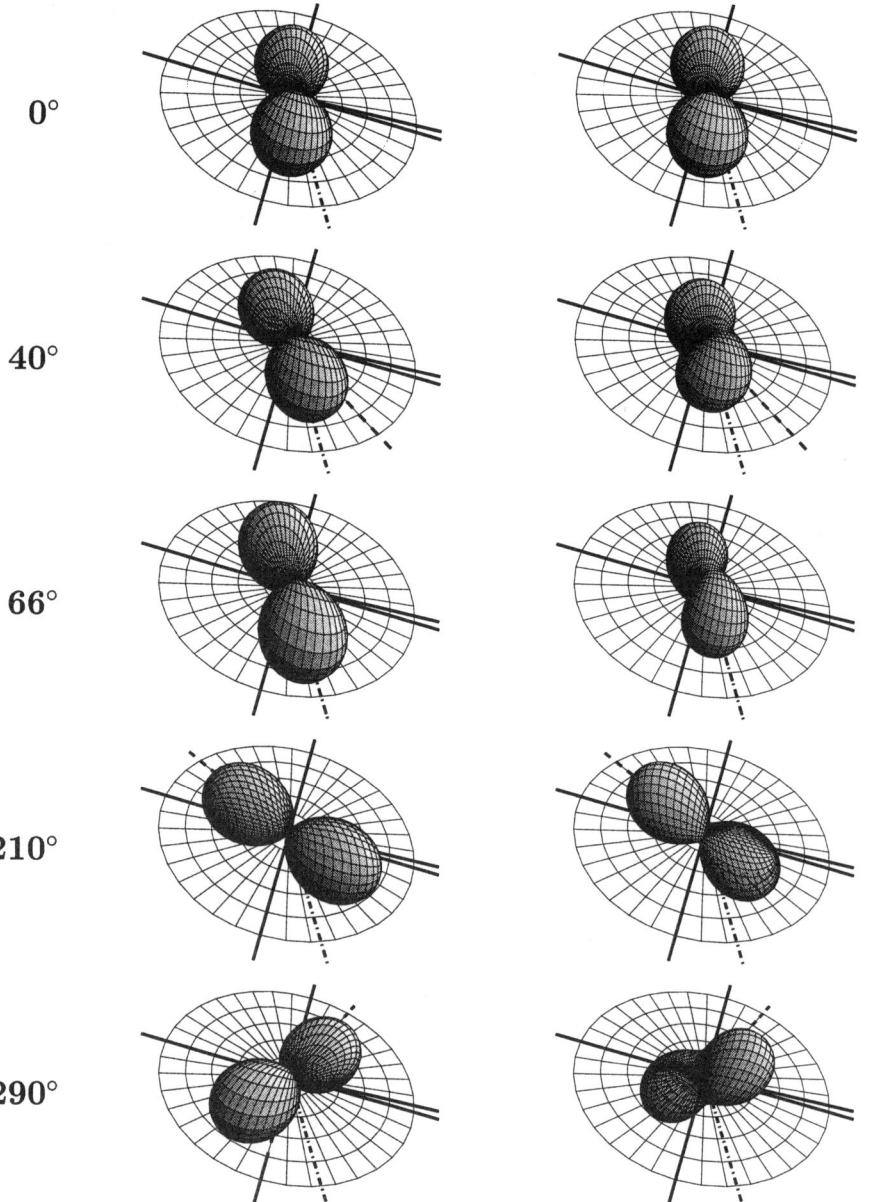

Figure 2.24 Charge cloud describing the He$^+$(2p)^{2}P state for the kinematical situation shown in Fig. 2.23 for various angles of the ejected electron, if this electron is observed in the reaction plane (left) or at 45° out of the plane [39]. The dashed and dot-dashed lines represent the ejected electron angle and the momentum transfer direction, respectively.

and one electron in coincidence [66, 67], or both electrons in coincidence for the TDCS [68, 69].

While all these experiments complete each other, a more stringent test of ionization theories could be performed by testing the predictions shown in Fig. 2.23. It shows predictions for the parameters that could be observed in a "complete" (e,2eγ) experiment for ionization–excitation of He($1s^2$) to the ionic state He$^+$($2p$)^{2}P in a coplanar geometry for the two outgoing electrons, as calculated in a "PWBA-RMPS" model, corresponding to a plane-wave description of the fast electron combined with an RMPS-type treatment of the initial bound state and the interaction between the residual ion and the ejected electron [70, 71]. Note that the predicted degree of light polarization, $P = \sqrt{P_1^2 + P_2^2 + P_3^2}$, is less than unity, due to fine-structure depolarization effects in the linear polarizations P_1 and P_2. Furthermore, if the electron is ejected parallel or anti-parallel to the direction of the momentum transfer (66° or 246° in the case shown), the symmetry properties of the PWBA force the angular momentum transfer $L_\perp (= -P_3)$ to vanish and the charge cloud to be aligned along this direction.

On the other hand, the symmetry may be broken further in non-planar observation geometries, thereby resulting in twisted and tilted charge clouds similar to those discussed above for electron-impact excitation of mercury with (in-plane) spin-polarized electrons. An example for such a twisted charge cloud is shown in the right part of Fig. 2.24, where the ejected electron is observed at an angle of 45° above the plane defined by the incident and the scattered electron directions.

7. SUMMARY

In this article, we have shown how complete, or nearly complete, experiments in electron–atom collisions can provide benchmarks for testing numerical methods to simulate these processes. In addition, the detailed interpretation of the results generally leads to a much deeper understanding of the collision dynamics than the restricted information available from experiments in which the dependence on the preparation of the initial state is averaged over and several possible outcomes are summed incoherently. Starting from the simple case of S → P transitions in the light helium target, the number of independent amplitudes grows significantly if higher orbital angular momenta of the target states are involved or spin exchange and explicitly relativistic effects such as the spin-orbit interaction have to be accounted for. Nevertheless, a systematic parameterization of such experiments is now available and should lead to further progress, particularly in the description of highly correlated processes such as simultaneous ionization–excitation where three (or more) electrons undergo significant changes in their quantum states as a result of the collision.

Acknowledgments

I would like to express my sincere thanks to Professor Hans Kleinpoppen for his pioneering contributions to the field of complete experiments and, even more importantly, for his long-term interest in and support of my research. Much of the work presented here was performed in collaboration with Nils Andersen over the past decade, and it would have been impossible without the contributions of numerous other colleagues from Adelaide, Belfast, Bielefeld, Boulder, Brisbane, Gaithersburg, Lincoln, Münster, Newcastle, Perth, Rolla, Toronto, Windsor, and Drake University. Finally, financial support from the United States National Science Foundation under grant PHY-9605124 is gratefully acknowledged.

References

[1] B. Bederson, Comments At. Mol. Phys. **1**, 41 (1969).

[2] B. Bederson, Comments At. Mol. Phys. **2**, 65 (1969).

[3] F. J. de Heer, International Nuclear Data Committee INDC(NDS) 385 .

[4] M. Eminyan *et al.*, J. Phys. B: At. Mol. Opt. Phys. **8**, 2058 (1975).

[5] M. C. Standage and H. Kleinpoppen, Phys. Rev. Lett. **36**, 577 (1976).

[6] M. R. Flannery and K. J. McCann, J. Phys. B: At. Mol. Opt. Phys. **5**, 1716 (1975).

[7] N. Andersen, I. V. Hertel, and H. Kleinpoppen, J. Phys. B: At. Mol. Opt. Phys. **17**, L901 (1984).

[8] N. Andersen, J. W. Gallagher, and I. V. Hertel, Phys. Rep. **165**, 1 (1988).

[9] K. Blum, *Density Matrix Theory and Applications*, 2 ed. (Plenum Press, New York and London, 1996).

[10] K. Bartschat *et al.*, J. Phys. B: At. Mol. Opt. Phys. **29**, 2875 (1996).

[11] D. V. Fursa and I. Bray, Phys. Rev. A **52**, 1279 (1995).

[12] J. P. Beijers, D. H. Madison, J. van Eck, and H. D. M. Heideman, J. Phys. B: At. Mol. Opt. Phys. **20**, 167 (1987).

[13] R. I. Hall *et al.*, J. Physique **34**, 827 (1973).

[14] D. C. Cartwright, G. Csanak, S. Trajmar, and D. F. Register, Phys. Rev. A **45**, 1602 (1992).

[15] R. R. McAdams, M. T. Hollywood, A. Crowe, and J. F. Williams, J. Phys. B: At. Mol. Opt. Phys. **13**, 3691 (1980).

[16] M. Eminyan, K. B. MacAdam, J. Slevin, and H. Kleinpoppen, J. Phys. B: At. Mol. Opt. Phys. **7**, 1519 (1974).

[17] M. A. Khakoo, K. Becker, J. L. Forand, and J. W. McConkey, J. Phys. B: At. Mol. Opt. Phys. **19**, L209 (1986).

[18] B. P., Donnelly, P. A. Neill, and A. Crowe, J. Phys. B: At. Mol. Opt. Phys. **21**, L321 (1988).

[19] D. V. Fursa and I. Bray, J. Phys. B: At. Mol. Opt. Phys. **30**, 757 (1997).

[20] W. C. Fon, K. P. Lim, K. A. Berrington, and T. G. Lee, J. Phys. B: At. Mol. Opt. Phys. **28**, 1569 (1995).

[21] W. C. Fon, K. A. Berrington, and A. E. Kingston, J. Phys. B: At. Mol. Opt. Phys. **24**, 2161 (1991).

[22] D. H. Madison and W. N. Shelton, Phys. Rev. A **7**, 499 (1973).

[23] D. C. Cartwright and G. Csanak, Phys. Rev. A **38**, 2740 (1988).

[24] D. Loveall, M. B. Hamley, B. J. Miller, and K. Bartschat, Comp. Phys. Commun. **124**, 90 (2000), see also http://bartschat.drake.edu/dloveall.

[25] G. Nienhuis, in *Coherence and Correlation in Atomic Collisions*, edited by H. Kleinpoppen and J. F. Williams (Plenum Press, New York, 1980).

[26] N. Andersen *et al.*, J. Phys. B: At. Mol. Opt. Phys. **16**, 817 (1983).

[27] N. Andersen and K. Bartschat, Adv. At. Mol. Opt. Phys. **36**, 1 (1996).

[28] N. Andersen and K. Bartschat, J. Phys. B: At. Mol. Opt. Phys. **30**, 5071 (1997).

[29] D. T. McLaughlin, B. P. Donnelly, and A. Crowe, Z. Phys. D **29**, 259 (1994).

[30] D. T. McLaughlin, B. P. Donnelly, and A. Crowe, Phys. Rev. A **49**, 2545 (1994).

[31] A. G. Mikosza, R. Hippler, J. B. Wang, and J. F. Williams, Z. Phys. D **30**, 129 (1994).

[32] A. G. Mikosza, R. H. J. B. Wang, and J. F. Williams, Phys. Rev. Lett. **71**, 235 (1993).

[33] A. G. Mikosza, R. Hippler, J. B. Wang, and J. F. Williams, Phys. Rev. A **53**, 3287 (1996).

[34] A. G. Mikosza, J. B. Wang, and J. F. Williams, Phys. Rev. Lett. **79**, 3375 (1997).

[35] K. Bartschat and D. H. Madison, J. Phys. B: At. Mol. Opt. Phys. **21**, 168 (1988).

[36] D. C. Cartwright and G. Csanak, J. Phys. B: At. Mol. Opt. Phys. **21**, 1717 (1988).

[37] E. J. Mansky and M. R. Flannery, J. Phys. B: At. Mol. Opt. Phys. **24**, L551 (1991).

[38] K. Bartschat, J. Phys. B: At. Mol. Opt. Phys. **32**, L355 (1999).

[39] N. Andersen and K. Bartschat, *Polarization, Alignment, and Orientation in Atomic Collisions* (Springer, New York, 2000), in press.

[40] N. Andersen and K. Bartschat, Comm. At. Mol. Phys. **29**, 157 (1993).

[41] J. J. McClelland, M. H. Kelley, and R. J. Celotta, Phys. Rev. Lett. **55**, 688 (1985).

[42] J. J. McClelland, M. H. Kelley, and R. J. Celotta, Phys. Rev. A **40**, 2321 (1989).

[43] R. E. Scholten et al., J. Phys. B: At. Mol. Opt. Phys. **24**, L653 (1991).

[44] I. V. Hertel, M. H. Kelley, and J. J. McClelland, Z. Phys. D **6**, 163 (1987).

[45] I. Bray, Phys. Rev. Lett. **69**, 1908 (1992).

[46] T. Hegemann, M. Oberste-Vorth, R. Vogts, and G. F. Hanne, Phys. Rev. Lett. **66**, 2968 (1991).

[47] N. Andersen and K. Bartschat, Phys. Rep. **279**, 251 (1997).

[48] I. Bray, Phys. Rev. A **49**, 1066 (1994).

[49] D. H. Madison, K. Bartschat, and R. P. McEachran, J. Phys. B: At. Mol. Opt. Phys. **25**, 5199 (1992).

[50] S. K. Srivastava and L. Vuskovic, J. Phys. B: At. Mol. Opt. Phys. **13**, 2633 (1980).

[51] N. Andersen and K. Bartschat, J. Phys. B: At. Mol. Opt. Phys. **27**, 3189 (1994), corrigendum: *J. Phys. B***29** (1996) 1149.

[52] K. Bartschat, K. Blum, G. F. Hanne, and J. Kessler, J. Phys. B: At. Mol. Opt. Phys. **14**, 3761 (1981).

[53] K. Bartschat, Phys. Rep. **180**, 1 (1989).

[54] N. Andersen et al., Phys. Rev. Lett. **76**, 208 (1996).

[55] J. Goeke, G. F. Hanne, and J. Kessler, J. Phys. B: At. Mol. Opt. Phys. **22**, 1075 (1989).

[56] M. Sohn and G. F. Hanne, J. Phys. B: At. Mol. Opt. Phys. **25**, 4627 (1992).

[57] N. S. Scott, P. G. Burke, and K. Bartschat, J. Phys. B: At. Mol. Opt. Phys. **16**, L361 (1983).

[58] A. Raeker, K. Blum, and K. Bartschat, J. Phys. B: At. Mol. Opt. Phys. **26**, 1491 (1993).

[59] R. Schwienhorst, A. Raeker, K. Bartschat, and K. Blum, J. Phys. B: At. Mol. Opt. Phys. **29**, 2305 (1996).

[60] I. Taouil, A. Lahmam-Bennani, A. Duguet, and L. Avaldi, Phys. Rev. **81**, 4600 (1998).

[61] E. W. O. Bloemen et al., J. Phys. B: At. Mol. Opt. Phys. **14**, 717 (1981).

[62] J. J. Forand, K. Becker, and J. W. McConkey, J. Phys. B: At. Mol. Opt. Phys. **18**, 1409 (1985).

[63] H. Merabet *et al.*, Phys. Rev. A **60**, 1187 (1999).

[64] A. Götz, W. Mehlhorn, A. Raeker, and K. Bartschat, J. Phys. B: At. Mol. Opt. Phys. **29**, 4699 (1996).

[65] R. Müller-Fiedler, K. Jung, and H. Ehrhardt, J. Phys. B: At. Mol. Opt. Phys. **19**, 1211 (1986).

[66] P. A. Hayes and J. F. Williams, Phys. Rev. Lett. **77**, 3098 (1996).

[67] M. Dogan, A. Crowe, K. Bartschat, and P. J. Marchalant, J. Phys. B: At. Mol. Opt. Phys. **31**, 1611 (1998).

[68] C. Dupré *et al.*, J. Phys. B: At. Mol. Opt. Phys. **25**, 259 (1992).

[69] L. Avaldi *et al.*, Phys. Rev. A **31**, 2981 (1998).

[70] K. Bartschat, J. Phys. IV France **9**, 6 (1999).

[71] A. S. Kheifets, I. Bray, and K. Bartschat, J. Phys. B: At. Mol. Opt. Phys. **32**, L433 (1999).

Chapter 3

DIFFERENTIAL CROSS SECTION AND SPIN ASYMMETRIES FOR COLLISIONS BETWEEN ELECTRONS AND ORIENTED CHIRAL MOLECULES

A. Busalla
Institut für Theoretische Physik I, Universität Münster, Wilhelm–Klemm–Str. 9, D–48149 Münster, Germany

M. Musigmann
Institut für Theoretische Physik I, Universität Münster, Wilhelm–Klemm–Str. 9, D–48149 Münster, Germany

K. Blum
Institut für Theoretische Physik I, Universität Münster, Wilhelm–Klemm–Str. 9, D–48149 Münster, Germany

D. G. Thompson
Department of Applied Mathematics and Theoretical Physics, Queen's University of Belfast, Belfast BT7 1NN, UK

1. INTRODUCTION

In recent years collisions between electrons and chiral molecules have been investigated by several theoretical and experimental groups. Nearly exclusively, all these studies concentrated on scattering processes between spin–polarised electrons and optically active molecules. For example, longitudinally polarised electrons behave as *screws* (with its screw sense defined by the relative orientation of spin and momentum vector) and the interaction between these *handed* electrons and chiral molecules has been investigated. In particular, first experi-

94   COMPLETE SCATTERING EXPERIMENTS

mental results for randomly oriented molecules have been obtained by J. Kessler and his group [1, 2]. Numerical calculations have been performed for unoriented and oriented molecules (I. M. Smith, D. G. Thompson, and K. Blum [3] and refs. therein). For a review we refer to K. Blum and D. G. Thompson [4].

In the present article we will study collisions between *unpolarised* projectiles and chiral molecules. We will show that the differential cross sections for the molecule (M) and its optical isomer (M') differ *if the molecules are oriented*.

This effect is due to the different positions of some of the atoms within M and M' (to be defined below). We will first derive general equations for elastic and inelastic collisions between electrons or atoms and oriented space–fixed chiral molecules which might be used as a basis for further research. In particular, a symmetry relation will be derived which relates the differential cross section for M and M' to each other. The general theory is then applied to elastic electron collisions with oriented H_2S_2– and $CHBrClF$–molecules, and numerical results are given in order to get some estimates for the magnitudes of the effects.

In section 3 we will study spin asymmetries. We will first present a general analysis, and disentangle the various effects, and then present numerical results for H_2S_2 and $CHBrClF$.

2.   DISCUSSION OF CROSS SECTION

2.1   GENERAL THEORY

We define a laboratory–fixed system XYZ in such a way that the Z–axis is parallel to the incoming beam axis, and the XZ–plane is the scattering plane (collision system). In addition, we define a molecule–fixed system denoted by xyz. We consider a collision with a space–fixed chiral molecule M whose spatial orientation relative to the collision system is specified by the three Euler angles $\alpha\beta\gamma$. Here, β is the angle between z and Z, α is the azimuth angle of the molecular z–axis with respect to the collision system, and the third Euler angle γ is the angle between the molecular x–axis and the zZ–plane. If the x–axis lies within the zZ–plane then γ is zero.

The differential cross section for an elastic collision between a projectile and the molecule M will be denoted by $\sigma(M, \theta, \alpha\beta\gamma)$ where θ is the scattering angle. It is always possible to expand σ in terms of the complete set of rotation matrix elements $D_{qQ}^{(K)}(\alpha\beta\gamma)$ [5]

$$\sigma(M,\theta,\alpha\beta\gamma) = \sum_{KqQ} \frac{2K+1}{8\pi^2} I_{qQ}^{(K)}(M,\theta)\, D_{qQ}^{(K)*}(\alpha\beta\gamma) \qquad (3.1)$$

where the star denotes the complex conjugate element. Throughout this article we will use the conventions of R. N. Zare [6] for the rotation matrices

$$D_{qQ}^{(K)}(\alpha\beta\gamma) = e^{-i\alpha q} d_{qQ}^{(K)}(\beta) e^{-i\gamma Q}. \tag{3.2}$$

The expansion coefficients $I_{qQ}^{(K)}(M,\theta)$ (steric factors) are defined by the expression

$$I_{qQ}^{(K)}(M,\theta) = \int d\Omega\, D_{qQ}^{(K)}(\alpha\beta\gamma)\, \sigma(M,\theta,\alpha\beta\gamma) \tag{3.3}$$

where $d\Omega = d\alpha\, d\beta\, \sin\beta\, d\gamma$ denotes the element of the solid angle. The steric factors are independent of the molecular orientatione and contain the information on the collision dynamics whereas the collision geometry is expressed in eq. (3.1) by the rotation matrices. From the symmetry properties of the rotation matrices it follows for eq. (3.3)

$$I_{qQ}^{(K)}(M,\theta) = (-1)^{Q-q} I_{-q-Q}^{(K)*}(M,\theta). \tag{3.4}$$

The steric factors have turned out to be a convenient tool in molecular collisions and in particular in the field of *stereodynamics* (see for example H. J. Loesch [7], A. Busalla and K. Blum [8]. For a recent review we refer to A. Busalla and K. Blum [9]). In particular, it has been shown that the same set of steric factors characterise also collisions with rotating, state–selected molecules (A. Busalla and K. Blum [9]).

The steric factors are real for molecules possessing a symmetry plane (see below) but complex in general. Dividing the steric factors into real (Re) and imaginary (Im) parts and using eq. (3.2) we write eq. (3.1) in a form which contains explicitly only real parameters:

$$\sigma(M,\theta,\alpha\beta\gamma) = \tfrac{1}{2} \sum_{K} \frac{2K+1}{8\pi^2} P_K(\cos\beta)\, I_{00}^{(K)}(M,\theta) \tag{3.5}$$

$$+ \sum_{K}{}' \frac{2K+1}{8\pi^2} d_{qQ}^{(K)}(\beta) \cos(q\alpha + Q\gamma)\, Re(I_{qQ}^{(K)}(M,\theta))$$

$$+ \sum_{K}{}' \frac{2K+1}{8\pi^2} d_{qQ}^{(K)}(\beta) \sin(q\alpha + Q\gamma)\, Im(I_{qQ}^{(K)}(M,\theta)).$$

Here $\sum'$ contains all terms with $K > 0$, $q > 0$, $|Q| \geq 0$, and with $q = 0$, $Q > 0$. For unoriented molecules only terms with $K = 0$ contribute.

In order to obtain a symmetry relation for collisions with the optical isomer M' of M we perform a reflection of the total system (projectile and molecule M) in the scattering plane (which leaves the wave vectors of incoming and scattered projectiles unchanged). This reflection is equivalent to a rotation of the molecule with $\alpha \to -\alpha, \beta \to \beta, \gamma \to -\gamma$, followed by a reflection within the molecular xz–plane. This last operation transforms M into its enantiomer. Since the interaction is invariant under the total operation we have

$$\sigma(M, \theta, \alpha, \beta, \gamma) = \sigma(M', \theta, -\alpha, \beta, -\gamma). \tag{3.6}$$

Inserting eq. (3.6) into eq. (3.3), changing the integration variables, and using the symmetry properties of the rotation matrices [6], we obtain

$$I_{qQ}^{(K)}(M, \theta) = \int d\Omega \, D_{qQ}^{(K)*}(\alpha\beta\gamma) \, \sigma(M', \theta, \alpha\beta\gamma) \tag{3.7}$$

Defining steric factorse for the isomer M' similar to eq. (3.3) (interchanging M↔M') we obtain the symmetry relation

$$I_{qQ}^{(K)}(M, \theta) = I_{qQ}^{(K)*}(M', \theta). \tag{3.8}$$

Eq. (3.8) is our basic result. It relates the steric factors of M and M' to each other. For any orientation, M' is obtained by reflecting M in its xz–plane. The x– and z–axis of M and M' coincidence (fixed by $\alpha\beta\gamma$), and the position of each atom in M' is fixed. The corresponding differential cross section $\sigma(M', \theta, \alpha\beta\gamma)$ is obtained by inserting relation (3.8) into eq. (3.5). The first two terms remain unchanged, the last term changes its sign. The difference between $\sigma(M, \theta, \alpha\beta\gamma)$ and $\sigma(M', \theta, \alpha\beta\gamma)$ depends therefore on the numerical values of the imaginary parts of the steric factors.

If the molecule is non–chiral and possesses a symmetry plane (chosen as xz–plane) then the steric factors are real and the last term in eq. (3.5) vanishes.

It should be noted that these results hold for elastic, inelastic or reactive collisions if only one particle is observed in the final state, and if the sum over all unobserved variables is included in the cross sections occurring in the integrals in eq. (3.3).

2.2 NUMERICAL RESULTS

In order to obtain some estimates about the magnitude of the described effects we have performed numerical calculations for elastic electron collisions with H_2S_2 and $CHBrClF$. The main part is the calculation of the relevant partial–wave scattering amplitudes. The corresponding numerical procedure has been described in detail by I. M. Smith, D. G. Thompson, and K. Blum [3]. The relation between steric factors and partial wave amplitudes can be obtained by standard techniques (see eqs. (52) and (53) in A. Busalla and K. Blum [9]). Using results for the partial–wave amplitudes, obtained by I. M. Smith, D. G. Thompson, and K. Blum [3], we have calculated the relevant steric factors for M using eq. (3.3). The results for M' follow then immediately by applying the symmetry relation (3.8). The differential cross sections for M and M' can then be obtained by eq. (3.5). For H_2S_2 we have taken all dominant steric factors up to $K = 8$ into account, for $CHBrClF$ up to $K = 10$.

For H_2S_2 the molecular z–axis is parallel to the S–S–bond and the x–axis has been chosen parallel to the C_2–symmetry axis of the molecule (which divides the angle of $96,6°$ between the two S–H–bonds). From the C_2-symmetry follows the symmetry relation

$$I_{qQ}^{(K)}(M,\theta) = (-1)^K I_{q-Q}^{(K)}(M,\theta). \tag{3.9}$$

The geometry of the H_2S_2–molecule is shown in fig. 3.1.

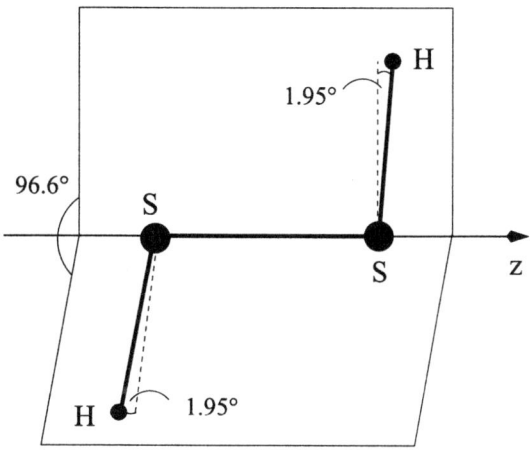

Figure 3.1 Geometry of H_2S_2

The numerical results depend considerably on the energy of the projectile, the scattering angle, and the molecular orientation. Some data are shown in figs. 3.2 and 3.3 for electron energy E=10eV and scattering angle $\theta = 70°$ which display remarkable differences between M and M'. For $\beta = 0°$ the S–S–bond is parallel to the incoming beam axis. The differential cross sections (DCS) are given as a function of γ which corresponds to a rotation around the S–S–bond.

M and M' have the same direction of the S–S–bond but the S–H–bonds are interchanged by a reflection in the xz–plane. For $\alpha = 0°$, $\beta = 90°$ the S–S–bond is perpendicular to the incoming beam axis, and lies in the scattering plane parallel to the x–axis. Again, we show results as a function of the angle γ. For $\alpha = \beta = 90°$ the S–S–bond would be perpendicular to the scattering plane. However, it follows from the C_2–symmetry about the molecular x–axis that in this case the difference between $\sigma(M, \theta, \alpha\beta\gamma)$ and $\sigma(M', \theta, \alpha\beta\gamma)$ vanishes.

Finally we note that for $\alpha = 0°$ the relation (see also eq. (3.6))

$$\sigma(M, \theta, 0°\beta, \gamma) = \sigma(M', \theta, 0°\beta, -\gamma)$$

98   COMPLETE SCATTERING EXPERIMENTS

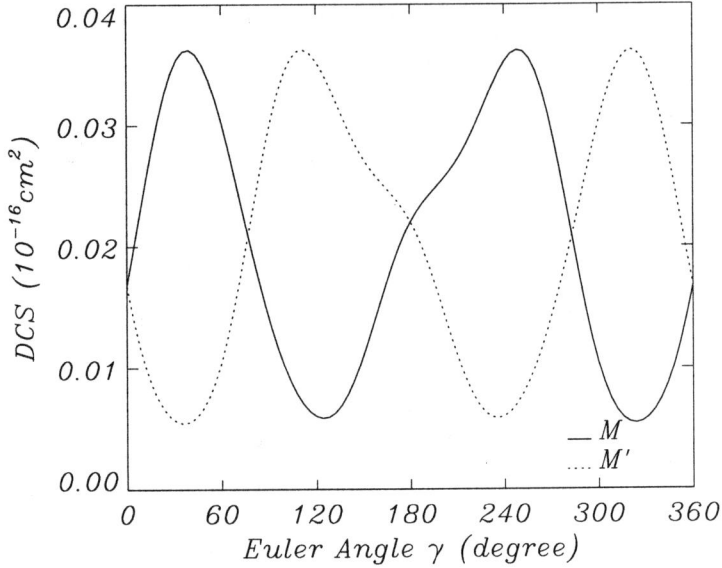

Figure 3.2 Differential cross section for H_2S_2: E= 10eV, $\alpha = \beta = 0°, \theta = 70°$

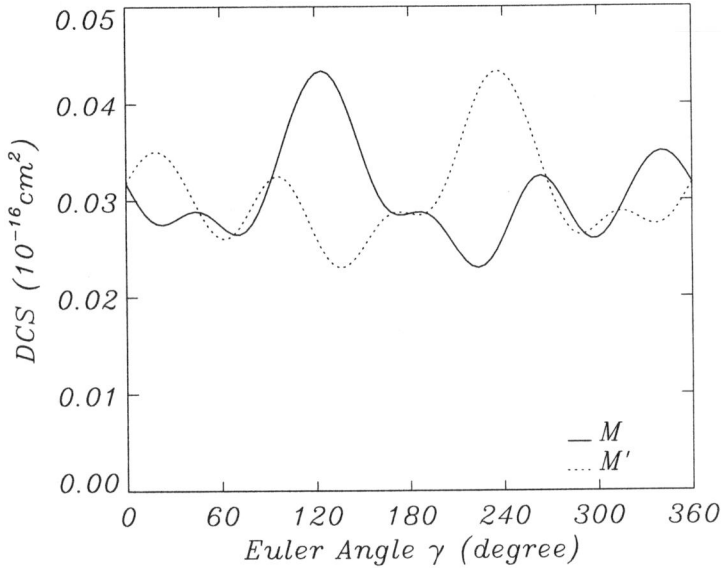

Figure 3.3 Differential cross section for H_2S_2: E= 10eV, $\alpha = 0°, \beta = 90°, \theta = 70°$

holds which is reflected in the figures.

The figures show considerable differences for particular orientations between $\sigma(M, \theta, \alpha\beta\gamma)$ and $\sigma(M', \theta, \alpha\beta\gamma)$. For example, fig. 3.2 shows that for $\gamma = 40°$

	r(au)		θ(degrees)
C-H	2.135	F-C-Cl	109.6
C-F	2.547	F-C-Br	109.2
C-Cl	3.313	Cl-C-Br	111.5
C-Br	3.642	H-C-X	108.8

Table 3.1 Structure of $CHBrClF$

nearly 75% of the electrons would be scattered by M, and about 25% by M'. The results indicate therefore high stereo selectivity for particular angles (if we define stereo selectivity as the *precision* with which the electrons select molecules of one specific spatial arrangement).

The geometry of the molecule $CHBrClF$ is summarised in table 3.1. However in our calculations, we have taken the X–C–X–bond angles to be tetrahedral (109.5°). The C–atom is chosen as origin of the molecular coordinate system. The C–Cl–bond is the molecular z–axis, and the molecular x–axis is chosen in such a way that the H–atom lies in the xz–plane (in positive x–direction). The Br–atom lies then above the xz–plane, the F–atom below. This corresponds to the R–configuration in the Ingold–Prelog–notation. The optical antipode M' is obtained by reflecting the molecule in the xz–plane, that is, Cl– and H–atom remain fixed, and F– and Br–atom are interchanged (S–configuration).

Some numerical results are shown in figs. 3.4 and 3.5 for $E = 5eV$ and scattering angle $\theta = 20°$. In fig. 3.4 we have $\alpha = \beta = 0°$, and for $\gamma = 0°$ the xz– and XZ–plane coincidence. Cl– and H–atom lie within the scattering plane as indicated in the inset of fig. 3.4. The molecule is then rotated around $z=Z$, and $\sigma(M, \theta, \alpha\beta\gamma)$ and $\sigma(M', \theta, \alpha\beta\gamma)$ are shown as a function of γ. After a rotation of 60° around z (C–Cl–bond) the Br–atom lies in the XZ–plane below the negative X–axis (indicated by $Br-$ in the figure). After a rotation of 120° the F–atom lies in the XZ–plane below the positive X–axis (indicated by $F+$) a.s.o. In fig. 3.5 we show results for the case where the molecule is rotated around the molecular x–axis about $\beta = 180°$ ($\alpha = 0°$). The Cl–atom lies on the negative $z=Z$–axis (incoming beam–axis), and the H–atom above the positive X–axis. $\sigma(M, \theta, \alpha\beta\gamma)$ and $\sigma(M', \theta, \alpha\beta\gamma)$ are again shown as a function of γ.

Comparing results for $\sigma(M, \theta, \alpha\beta\gamma)$ and $\sigma(M', \theta, \alpha\beta\gamma)$ it can be read off that stereo selectivity is in general large. For example, as shown by fig. 3.5, at $\gamma = 90°$ we have that $\sigma(M, \theta, \alpha\beta\gamma)$ is about three times larger than $\sigma(M', \theta, \alpha\beta\gamma)$.

100 COMPLETE SCATTERING EXPERIMENTS

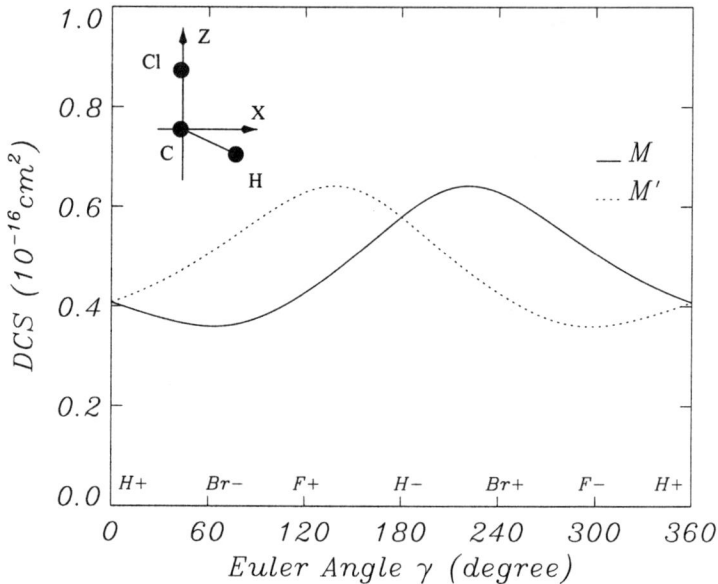

Figure 3.4 Differential cross section for $CHBrClF$: E= 5eV, $\alpha = \beta = 0°, \theta = 20°$

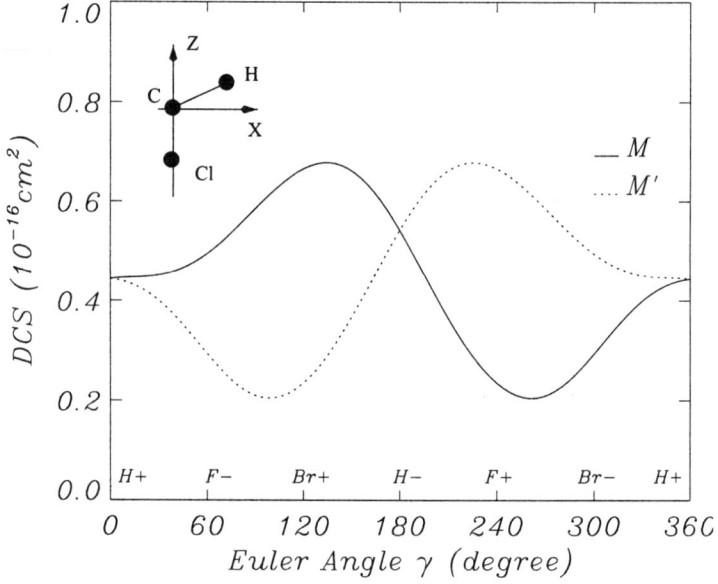

Figure 3.5 Figure 5: Differential cross section for $CHBrClF$: E= 5eV, $\alpha = 0°, \beta = 180°$, $\theta = 20°$

3. SPIN ASYMMETRIES

3.1 INTRODUCTION

In this section we will study the interaction between longitudinally polarised electrons and oriented chiral molecules. Let us denote by $\sigma(M, \theta, +, \alpha\beta\gamma) \equiv$

$\sigma(M+)$ the differential cross section for experiments where the initial electrons have "spin up" with respect to the incoming beam axis (right–handed electrons). Similarly, $\sigma(M-)$ refers to initial electrons with "spin down"(left–handed electrons). There are two asymmetries which we will study:

$$A = \frac{\sigma(M+) - \sigma(M-)}{\sigma(M+) + \sigma(M-)} \quad (3.10)$$

and

$$\eta = \frac{\sigma(M+) - \sigma(M'+)}{\sigma(M+) + \sigma(M'+)} . \quad (3.11)$$

A measures the difference in cross sections for collision with left– and right–handed electrons with the molecule M. η refers to the difference in the cross sections between molecule M and its isomer M' for collisions with right–handed electrons. For unoriented molecules we have $A = \eta$ because of symmetry requirements. For oriented molecules A and η may differ considerably which will be shown by our numerical results (see section 3.4). In particular, we will show that values of η up to 80 % can be obtained, whereas results for A are generally several orders of magnitude smaller.

There is a further difference between A and η. Consider collisions with non–chiral but oriented molecules, for example, collisions with oriented diatomics. A is different from zero if the molecular axis lies outside the scattering plane. On the other hand, η vanishes identically. In general, η is zero for a large class of non–chiral molecules namely those containing a symmetry plane since then the steric factors are real as discussed in section 2..

3.2 DISCUSSION OF THE ASYMMETRY A

We denote by $\sigma(M, \theta, m_1 m_0, \alpha\beta\gamma) \equiv \sigma(M, m_1 m_0)$ the differential cross section for collision with oriented molecules of definite handedness (M) where the initial electrons have spin component m_0, and the scattered electrons m_1 with respect to the incoming beam axis. We can then write eq. (3.10) in the form

$$\sigma(M)A = \sigma(M, ++) + \sigma(M, -+) - \sigma(M, +-) - \sigma(M, --) \quad (3.12)$$

where $\sigma(M) = \sigma(M, ++) + \sigma(M, -+) + \sigma(M, +-) + \sigma(M, --)$ is the unpolarised cross section (3.1). $\sigma(M, +-)$ denotes the cross section with $m_0 = -\frac{1}{2}$ and $m_1 = +\frac{1}{2}$ and similarly for the other spin combinations. Expanding the spin–dependent cross sections similar to eq. (3.1) we obtain

$$\sigma(M, m_1 m_0) = \sum_{KqQ} \frac{2K+1}{8\pi^2} I_{qQ}^{(K)}(M, \theta, m_1 m_0) D_{qQ}^{(K)*}(\alpha\beta\gamma) \quad (3.13)$$

where the spin–dependent steric factors are defined similar to eq. (3.3)

$$I_{qQ}^{(K)}(M,\theta,m_1 m_0) = \int d\Omega D_{qQ}^{(K)}(\alpha\beta\gamma)\sigma(M,\theta,m_1 m_0,\alpha\beta\gamma). \quad (3.14)$$

For the spin–dependent case the symmetry relations (3.4) and (3.8) are generalised to

$$I_{qQ}^{(K)}(M,\theta,m_1 m_0) = (-1)^{Q-q} I_{-q-Q}^{(K)*}(M,\theta,m_1 m_0) \quad (3.15)$$

and

$$I_{qQ}^{(K)}(M,\theta,m_1 m_0) = I_{qQ}^{(K)*}(M',\theta,-m_1,-m_0) \quad (3.16)$$

Using eqs. (3.12) and (3.13) we obtain

$$\sigma(M)A = \sum_{KqQ} \frac{2K+1}{8\pi^2} D_{qQ}^{(K)*}(\alpha\beta\gamma)$$

$$\times \left[I_{qQ}^{(K)}(M,\theta,++) + I_{qQ}^{(K)}(M,\theta,-+) - I_{qQ}^{(K)}(M,\theta,+-) - I_{qQ}^{(K)}(M,\theta,--) \right]$$
(3.17)

In all cases which we have investigated so far ($HBr, H_2S_2, CHBrClF$) we have found numerically that the spin–flip–terms in eq. (3.17) are very small. For simplicitye we will therefore ignore these terms in the following equations. We write eq. (3.17) in a more explicit form

$$\sigma(M)A = \sum_{K} \frac{2K+1}{8\pi^2} d_{00}^{(K)}(\beta) \left[I_{00}^{(K)}(M++) - I_{00}^{(K)}(M--) \right]$$

$$+2\sum{}' \frac{2K+1}{8\pi^2} d_{qQ}^{(K)}(\beta)\cos(\alpha q+\gamma Q) \left[Re(I_{qQ}^{(K)}(M++)) - Re(I_{qQ}^{(K)}(M--)) \right]$$

$$-2\sum{}' \frac{2K+1}{8\pi^2} d_{qQ}^{(K)}(\beta)\sin(\alpha q+\gamma Q) \left[Im(I_{qQ}^{(K)}(M++)) - Im(I_{qQ}^{(K)}(M--)) \right]$$
(3.18)

where $\sum'$ is defined below eq. (3.5).

If the initial electrons are unpolarised, and the longitudinal polarisation P_z of the final electrons is measured, the corresponding expression for σP_z is obtained by reversing the sign of the spin–flip–terms in eq. (3.17). Since these are very small we have approximately

$$A \approx P_z. \quad (3.19)$$

Let us substitute the symmetry relation (3.16) for the second term in the brackets of eq. (3.18). We obtain

$$\sigma(M)A = \sum_K \frac{2K+1}{8\pi^2} d_{00}^{(K)}(\beta) \left[I_{00}^{(K)}(M++) - I_{00}^{(K)}(M'++) \right]$$

$$+ 2 {\sum}' \frac{2K+1}{8\pi^2} d_{qQ}^{(K)}(\beta) \cos(\alpha q + \gamma Q) \left[Re(I_{qQ}^{(K)}(M++)) - Re(I_{qQ}^{(K)}(M'++)) \right]$$

$$- 2 {\sum}' \frac{2K+1}{8\pi^2} d_{qQ}^{(K)}(\beta) \sin(\alpha q + \gamma Q) \left[Im(I_{qQ}^{(K)}(M++)) + Im(I_{qQ}^{(K)}(M'++)) \right].$$

(3.20)

The first two terms contain the difference of the steric factors of M and M', but the last term is the *sum* of both contributions. It should be remembered that for M' a definite orientation has been chosen relative to M as discussed in section 2.1, and further illustrated in section 3.4. Taking this into account one might say that the third term in eq. (3.20) (and of course in eq. (3.18)) is *independent of the molecular handedness* and depends only on the molecular orientation, and might be called the "orientation term".

For diatomic molecules, the first two terms in eq. (3.18) and eq. (3.20) vanish and the spin asymmetry is entirely given by the last term. A would be zero for randomly oriented diatomics as follows also from symmetry requirements.

3.3   DISCUSSION OF THE ASYMMETRY η

Let us now consider the asymmetry (3.11). Using the abbreviation

$$Q = \sigma(M+) + \sigma(M'+) \quad (3.21)$$

and applying a similar procedure as in the derivation of eq. (3.18), we obtain

$$Q\eta = \sum_K \frac{2K+1}{8\pi^2} d_{00}^{(K)}(\beta) \left[I_{00}^{(K)}(M++) - I_{00}^{(K)}(M'++) \right]$$

$$+ 2 {\sum}' \frac{2K+1}{8\pi^2} d_{qQ}^{(K)}(\beta) \cos(\alpha q + \gamma Q) \left[Re(I_{qQ}^{(K)}(M++)) - Re(I_{qQ}^{(K)}(M'++)) \right]$$

$$+ 2 {\sum}' \frac{2K+1}{8\pi^2} d_{qQ}^{(K)}(\beta) \sin(\alpha q + \gamma Q) \left[Im(I_{qQ}^{(K)}(M++)) - Im(I_{qQ}^{(K)}(M'++)) \right],$$

(3.22)

where we have again neglected the very small spin–flip–terms. The corresponding equation for Q can be obtained with the help of eqs. (3.16) and (3.18).

Using the symmetry relation (3.16) we can rewrite eq. (3.22)

$$Q\eta = \sum_K \frac{2K+1}{8\pi^2} d_{00}^{(K)}(\beta) \left[I_{00}^{(K)}(M++) - I_{00}^{(K)}(M--) \right]$$
$$+ 2\sum' \frac{2K+1}{8\pi^2} d_{qQ}^{(K)}(\beta)\cos(\alpha q + \gamma Q) \left[Re(I_{qQ}^{(K)}(M++)) - Re(I_{qQ}^{(K)}(M--)) \right]$$
$$+ 2\sum' \frac{2K+1}{8\pi^2} d_{qQ}^{(K)}(\beta)\sin(\alpha q + \gamma Q) \left[Im(I_{qQ}^{(K)}(M++)) + Im(I_{qQ}^{(K)}(M--)) \right].$$
(3.23)

Eq. (3.23) reveals an interesting fact. The first two terms depend on the *differences* between the steric factors with opposite spin components. These terms are identical to the first two terms in the equation for A (eq. (3.18)). These terms depend on spin–dependent interactions and are generally small. The third term in eq. (3.23) contains the *sum* of the imaginary parts of the steric factors with opposite spin and is therefore independent of the spin components (contrary to the third term in eq. (3.18)).

This resulte is of great importance. In the case of randomly oriented molecules only the term with $K = 0$ in eq. (3.23) contributes (which is equal to the corresponding term in eq. (3.18)). This term is only significant if the spin–dependent interactions are not too small. In particular, it is usually required that the molecules contain heavy atoms so that the spin–orbit coupling is sufficiently strong. A similar conclusion can be drawn for the second terms in eq. (3.18) and (3.23). In contrast, the third term in eq. (3.23) is spin–independent and can be expected to be significant even for light–molecules which do not exhibit large spin–orbit couplings.

One can therefore expect that the third term in eq. (3.23) gives the dominant contribution to η. This is indeed the case as follows from our numerical calculations. We have then $Q\eta \approx \sigma(M) - \sigma(M')$ and η is closely related to the discussion in section 2.2.

It should be noted that these conclusions hold also if the spin–flip terms are included in eqs. (3.18) and (3.23).

3.4 NUMERICAL RESULTS AND DISCUSSION

We have calculated the asymmetries A and η for elastic electron collision with H_2S_2 and $CHBrClF$, using the numerical procedure described in section 2.

We have used eq. (3.18) for A, and eq. (3.23) for η. Figs. 3.6 to 3.7 present results for A for the two molecules at various orientations. The values of A are about one to two orders of magnitude larger than for randomly oriented ensembles [3].

In fig. 3.6 the S–S–bond lies parallel to the incoming beam axis ($\alpha = \beta = 0°$), and in fig. 3.8 perpendicular to it ($\alpha = 0°, \beta = 90°$). In both cases the

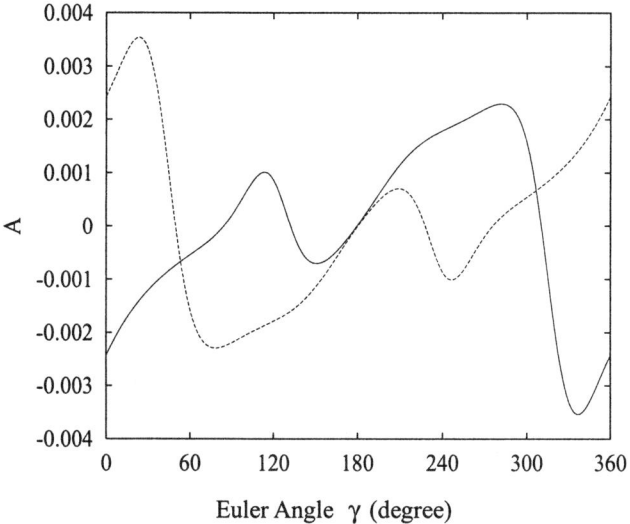

Figure 3.6 Asymmetry A for H_2S_2: E= 10eV, $\alpha = 0°$, $\beta = 0°$, $\theta = 70°$; molecule M (‖) and enantiomer M' (– –)

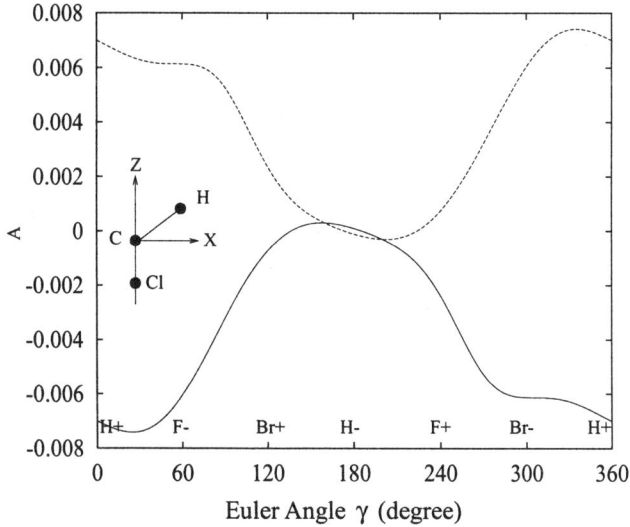

Figure 3.7 Asymmetry A for $CHBrClF$: E= 5eV, $\alpha = 0°$, $\beta = 180°$, $\theta = 20°$; molecule M (‖) and enantiomer M' (– –)

molecule is rotated around the S–S–bound about γ, and A is shown in both cases as a function of γ. In fig. 3.9 the C–Cl–bond is parallel to the incoming beam axis ($\alpha = \beta = 0°$), and the molecule is rotated about γ around this axis.

106   COMPLETE SCATTERING EXPERIMENTS

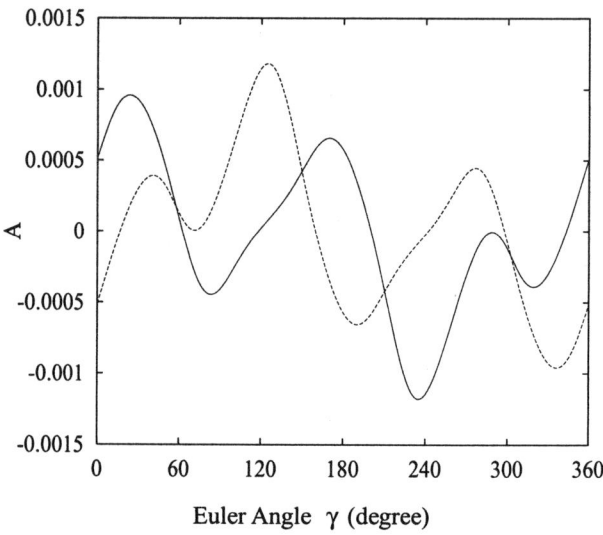

Figure 3.8 Asymmetry A for H_2S_2: E= 10eV, $\alpha = 0°$, $\beta = 90°$, $\theta = 70°$; molecule M (∥) and enantiomer M' (– –)

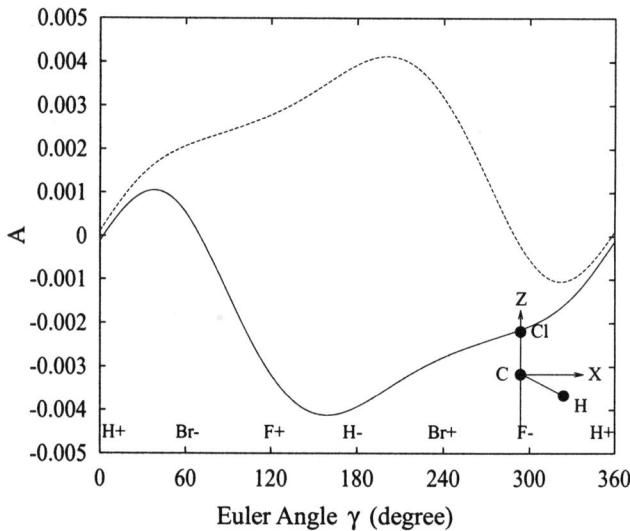

Figure 3.9 Asymmetry A for $CHBrClF$: E= 5eV, $\alpha = 0°$, $\beta = 0°$, $\theta = 20°$; molecule M (∥) and enantiomer M' (– –)

In order to visualise the rotation we have indicated various atomic positions in the figure as in figs. 3.4 and 3.5. The molecular orientation in fig. 3.7 is obtained by rotating M around the molecular x–axis about $180°$. The C–Cl–bound is then antiparallel to the incoming beam axis ($\alpha = 0°, \beta = 180°$), and results are again shown as a function of γ.

It is interesting that in figs. 3.9 and 3.7 the asymmetry for molecule M is mainly negative, that is, left-handed electrons are scattered preferentially. Comparison of both figures show that there is a large orientational effect. Results for the asymmetry for the isomer M' are also included in the figures.

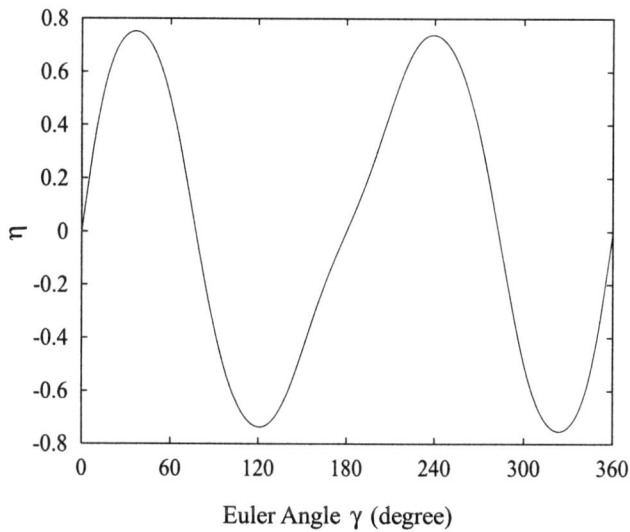

Figure 3.10 Asymmetry η for H_2S_2: E= 10eV, $\alpha = 0°, \beta = 0°, \theta = 70°$

Figs. 3.10 and 3.11 show results for η for H_2S_2, figs. 3.12 and 3.13 for $CHBrClF$ for the same orientations as in figs. 3.6 -3.7. The results are about two to three orders of magnitude larger than those obtained for A. This is due to the spin–independent term in eq. (3.23) which gives the dominant contribution to η. Consider $CHBrClF$ at a given orientation, for example, with the Br–atom closer to the scattering plane than the F–atom. In M' both atoms are interchanged. Since Br and F scatter the electrons with different probabilities, the cross sections will in general be different for M and M' even for unpolarised electrons. The collision is essentially a result of the Coulomb–interaction between target and projectile, and the spin–dependent forces will only give a small contribution, that is, in a good approximation we have

$$\eta \approx \frac{\sigma(M) - \sigma(M')}{\sigma(M) + \sigma(M')} \quad (3.24)$$

where $\sigma(M)$ and $\sigma(M')$ denote the cross sections for unpolarised electrons. The results for η should therefore be compared with those of section 2. The curve shown in fig. 3.12 has very nearly a pure sinus form. This is due to

108   COMPLETE SCATTERING EXPERIMENTS

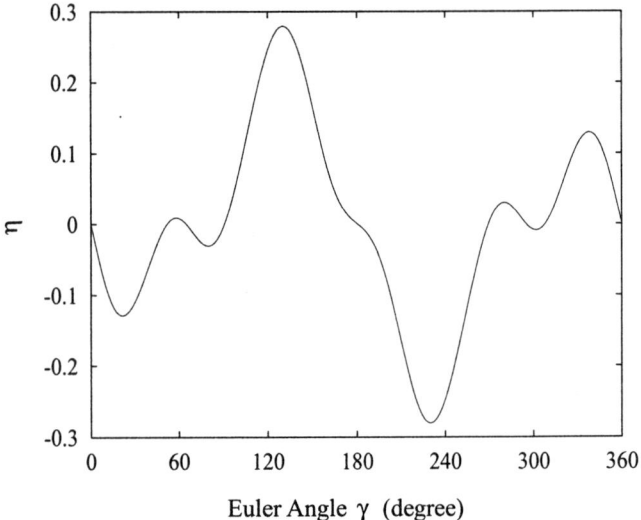

Figure 3.11   Asymmetry η for H_2S_2: E= 10eV, $\alpha = 0°$, $\beta = 90°$, $\theta = 70°$

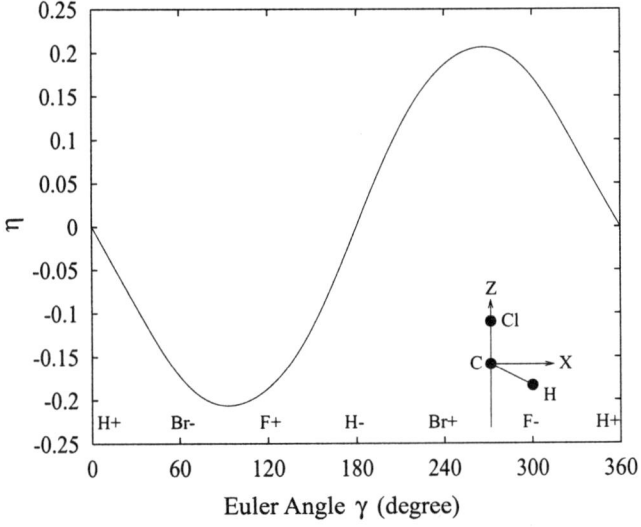

Figure 3.12   Asymmetry η for $CHBrClF$: E= 5eV, $\alpha = 0°$, $\beta = 0°$, $\theta = 20°$

the fact that the steric factors with $Q = 1$ have large imaginary parts at the chosen scattering energy and angle, and dominate the last term in eqs. (3.22) and (3.23).

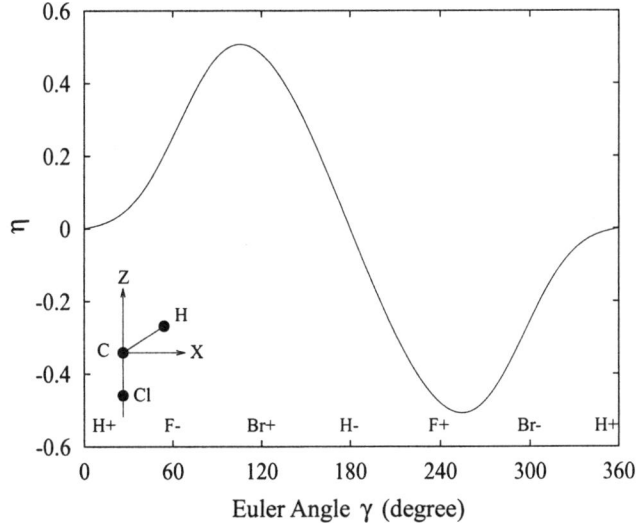

Figure 3.13 Asymmetry η for $CHBrClF$: E= 5eV, $\alpha = 0°, \beta = 180°, \theta = 20°$

References

[1] S. Mayer and J. Kessler, Phys. Rev. Lett. **74**, 4803 (1995).

[2] S. Mayer, C. Nolting, and J. Kessler, J. Phys. B: At. Mol. Opt. Phys. **29**, 3497 (1996).

[3] I. M. Smith, D. G. Thompson, and K. Blum, J. Phys. B: At. Mol. Opt. Phys. **31**, 4029 (1998).

[4] K. Blum and D. G. Thompson, Adv.At.Mol.Opt.Phys. **38**, 39 (1997).

[5] A. Busalla, K. Blum, T. Beyer, and B. M. Nestmann, J. Phys. B: At. Mol. Opt. Phys. **32**, 791 (1999).

[6] R. N. Zare, *Angular Momentum* (Wiley, New York, 1988).

[7] H. J. Loesch, Annu. Rev. Phys. Chem. **46**, 255 (1996).

[8] A. Busalla and K. Blum, J. Phys. Chem. A **101**, 7476 (1997).

[9] A. Busalla and K. Blum, in *Novel Aspects of Electron-Molecule Scattering*, edited by K. H. Becker (World Scientific Publishing, New Jersey, 1998).

Chapter 4

TOWARDS A COMPLETE EXPERIMENT FOR AUGER DECAY

A. N. Grum–Grzhimailo
Fakultät für Physik, Universität Freiburg, D-79104 Freiburg, Germany
Institute of Nuclear Physics, Moscow State University, Moscow 119899, Russia

A. Dorn
Fakultät für Physik, Universität Freiburg, D-79104 Freiburg, Germany

W. Mehlhorn
Fakultät für Physik, Universität Freiburg, D-79104 Freiburg, Germany

Abstract   A new method for performing a complete experiment in the case of the Auger decay is considered. The method assumes the LSJ-coupling approximation and is based on measurements of the relative intensities and the angular distribution of the Auger electrons. The method can be used for Auger transitions from states with nonvanishing spin S and orbital angular momentum L if the fine structure levels J_f of the final ionic state are resolved. As example, for the Auger decay $Na^+(2s2p^64p\ ^3P) \rightarrow Na^{2+}(2s^22p^5\ ^2P_{3/2},^2P_{1/2}) + e_{Auger}$ the absolute ratio of decay amplitudes and their relative phase are determined.

1. INTRODUCTION

A complete experiment refers to the experimental determination of the quantum-mechanical amplitudes characterizing the process. Although the first complete experiment for photoionization was reported as early as 1980 by Heinzmann [1], the first "almost" complete experiments for Auger (or autoionizing) decay were performed only very recently by West *et al* [2], Ueda *et al* [3] and Grum-Grzhimailo *et al* [4]. We use the term "almost" because the absolute values of the amplitudes were not determined.

The Auger process will be considered as part of a two-step process:

Step 1: Excitation of the Auger state by photon or particle impact

$$\gamma(e^-) + A \longrightarrow A^+(L, S, J) + e^-(2e^-) \qquad (4.1)$$

Step 2: Decay of the Auger state

$$A^+(L, S, J) \longrightarrow A^{++}(L_f, S_f, J_f) + e_A^-(lj). \qquad (4.2)$$

The assumption of a two-step process implies also that we neglect a direct amplitude for double ionization (i.e., there is no Fano-type interference [5]) and the post-collision interaction [6] between the Auger electron and the ejected and scattered electrons of process (4.1). The complete experiment on the Auger decay relates to step 2 and the aim of the experiment is to determine the absolute values (moduli) and relative phases of the decay amplitudes. The total number N of decay amplitudes depends on the total angular momenta J and J_f of the initial and final states of the Auger decay [7]:

$$N = 2J_f + 1 \quad \text{if} \quad J_f < J \quad \text{and} \quad N = 2J + 1 \quad \text{if} \quad J_f \geq J. \qquad (4.3)$$

For example, the Auger decay $M_4(^2D_{3/2})$ - $N_{2,3}N_{2,3}(^1D_2)$ with $J = 3/2$ and $J_f = 2$ is characterized by $N = 4$ decay amplitudes lj: $s_{1/2}$, $d_{3/2}$, $d_{5/2}$ and $g_{7/2}$. The complete Auger experiment with N decay amplitudes has to determine a total number of $2N - 1$ moduli and relative phases. Kabachnik and Sazhina [7] have theoretically shown that modern experiments on Auger decay (i. e., angular anisotropy and spin-polarization of Auger electrons) in principle can provide enough independent experimental quantities for a complete description of the Auger process. But in praxis part of the necessary experiments are hardly feasible and/or part of the quantities to be measured are very small. Then the experimental errors prevent to extract reliable values of the moduli and relative phases of decay amplitudes. As result of this a complete experiment for Auger decay has not been performed until today by measuring the angular anisotropy and spin-polarization of Auger electrons.

Instead, other kinds of complete experiments for Auger (autoionizing) decay ("almost" complete experiments) were suggested and carried out. West *et al* [2] and Ueda *et al* [3] measured the angular correlation between the autoionization electron and the subsequent fluorescence photon. Grum-Grzhimailo *et al* [4] proposed and performed a new approach for an "almost" complete experiment by the measurements of the intensity ratio and the angular distribution of Auger electrons. In the following section we will treat an example of a complete experiment by measuring the angular anisotropy and spin-polarization of Auger electrons. Then, in section 3. a new approach for an "almost" complete experiment as proposed by Grum-Grzhimailo *et al* [4] is discussed. In section 4. we present the performance of this "almost" complete experiment and compare the experimental results with theoretical values. Finally, in section 5. we discuss the conditions and the applicability of the new approach.

2. EXAMPLE FOR A POSSIBLE COMPLETE EXPERIMENT

We assume that the initial Auger state $A^+(J)$ decays into several final states $A^{++}(J_{f_k})$ (see Fig. 4.1) and we wish to perform a complete experiment on

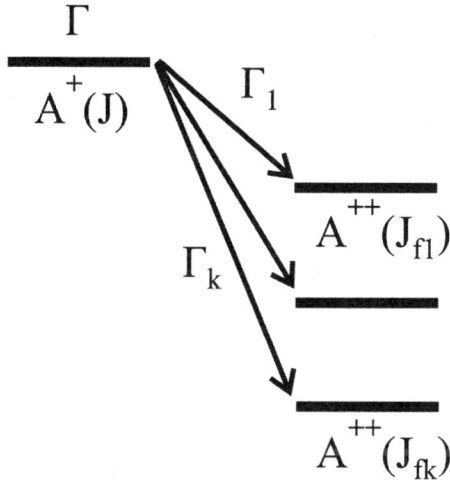

Figure 4.1 Diagram for Auger transitions $A^+(J) \to A^{++}(J_{f_k}) + e_A^-$. The total decay probability and the partial decay probabilities are given by the widths Γ and Γ_k, respectively.

the Auger decay to the final state J_{f_1}. Furthermore, we assume that the Auger decay is characterized by only two decay amplitudes, $V_1 = |V_1|\exp(i\Delta_1)$ and $V_2 = |V_2|\exp(i\Delta_2)$. Then the complete experiment has to determine the two moduli $|V_1|$ and $|V_2|$ and the relative phase $\Delta_{12} = \Delta_1 - \Delta_2$. What independent experimental quantities are available?

1. From the ratio of integral intensities W_k of transitions $J \to J_{f_k}$ relative to the integral intensity W_1 of transition $J \to J_{f_1}$

$$W_k/W_1 = \Gamma_k/\Gamma_1 \qquad (4.4)$$

and the total decay width Γ of the initial Auger state $A^+(J)$ with

$$\Gamma = \sum_k \Gamma_k \qquad (4.5)$$

the partial decay width due to the transition $J \to J_{f_1}$ can be determined experimentally via

$$\Gamma_1 = \Gamma W_1 / \sum_k W_k. \qquad (4.6)$$

Here, Γ_k is the partial decay width due to the transition $J \to J_{f_k}$. The partial width Γ_1 is related to the transition amplitudes V_1 and V_2 by

$$\Gamma_1 = 2\pi \sum_{n=1,2} a_n |V_n|^2, \qquad (4.7)$$

where the coefficients a_n are known and depend on the particular configurations involved in the Auger transition and coupling schemes.

2. The angular intensity $W_1(\vartheta, \varphi)$ of the Auger electrons from transition $J \to J_{f_1}$ measured by a spin-unsensitive detector and when the polarization state of the residual ion $A^{++}(J_{f_1})$ is not detected is given by [4,8]

$$W_1(\vartheta, \varphi) = \frac{W_1}{4\pi} \left[+\alpha_2 \sum_q \mathcal{A}_{2q}(J) \sqrt{\frac{4\pi}{5}} Y_{2q}(\vartheta, \varphi) \right.$$

$$\left. +\alpha_4 \sum_q \mathcal{A}_{4q}(J) \sqrt{\frac{4\pi}{9}} Y_{4q}(\vartheta, \varphi) + ... \right]. \qquad (4.8)$$

Here W_1 is the integral intensity of the Auger transition $J \to J_{f_1}$, $Y_{kq}(\vartheta, \varphi)$ are the spherical harmonics and $\mathcal{A}_{kq}(J)$ are the reduced statistical tensors (state multipoles) which describe the polarization of the decaying Auger state and depend only on the excitation process (step 1, see (4.1)). The decay parameters α_k ($k = 2, 4, ...; k_{max} \leq 2J$) depend only on the dynamics of the Auger decay and are expressed as functions of $|V_1|$, $|V_2|$ and $\cos \Delta_{12}$. In praxis only the decay parameter α_2 is measureable, the term with α_4 is generally too small if present in Eq.(4.8).

3. A third experimental quantity to obtain information on the decay amplitudes is the spin-polarization of Auger electrons. The angular intensity of Auger electrons measured with a spin-sensitive detector is given in a general form analogous to (4.8) by

$$W_1(\vartheta, \varphi, \psi) = \frac{W_1}{4\pi} \left[1 + \eta_1 \sum_q \mathcal{A}_{1q}(J) f_{1q}(\vartheta, \varphi, \psi) \right.$$

$$\left. +\eta_2 \sum_q \mathcal{A}_{2q}(J) f_{2q}(\vartheta, \varphi, \psi) + ... \right], \qquad (4.9)$$

where an additional angle ψ characterizes the direction of the axis of the electron spin analyser. The coefficients η_k ($k = 1, 2, ...; k_{max} \leq 2J$) are again functions of $|V_1|$, $|V_2|$ and Δ_{12}. The angular functions $f_{kq}(\vartheta, \varphi, \psi)$ are fully determined by the geometry of the experiment and the effectiveness of the spin-sensitive detector. Note, that in contrast to Eq.(4.8) terms with both, even and odd, ranks k are present in Eq.(4.9). Therefore, spin-resolved measurements are basically more informative than measurements of the ordinary angular distribution (4.8). Unfortunately, experiments with a spin-resolved detector suffer from low counting rates. Furthermore, the coefficient η_2, standing for the dynamical spin-polarization [9], is small for normal Auger transitions [10] and is therefore hardly measurerable with high enough accuracy. To measure η_1, which stands for the spin-transfer mechanism [9], one has to introduce orientation, i.e. the first rank tensor $\mathcal{A}_{1q}(J)$, into the Auger state. Presently the only practical method to achieve this is to produce the vacancy by a circular polarized photon beam. This method requires high brilliant sources of circular polarized VUV or X-ray radiation [11]. Recently, a new method for producing Auger states with large orientation which leads also to large spin-polarization of the Auger electrons was suggested by means of laser-excited polarized atoms [12], but it has not yet been realized.

A general problem for the measurements of the decay coefficients α_k and η_k is also to disentangle them from the statistical tensors $\mathcal{A}_{kq}(J)$ characterizing the excitation step. Independent measurements are generally needed to solve this problem. In conclusion, for the full determination of two Auger amplitudes V_1 and V_2 (two moduli and one relative phase) we need to measure the relative integral intensities W_k/W_1, the total width Γ of the initial Auger state and the decay parameters α_2 and η_1. Strictly speaking, one more measurement is needed to avoid an ambiguity in the relative phase Δ_{12}, since the physical quantities depend either on $\cos \Delta_{12}$ or on $\sin \Delta_{12}$. Up to now such a complete experiment has not been performed.

3. PROPOSAL FOR AN "ALMOST" COMPLETE EXPERIMENT

Again we assume that the Auger decay is characterized by the two amplitudes V_1 and V_2. But instead of determining the two absolute values $|V_1|$ and $|V_2|$ we are interested here in the determination of only the ratio $r = |V_1|/|V_2|$ and of $\cos \Delta_{12}$. Therefore we use the term "almost" complete experiment. We further assume that the LSJ-coupling approximation is valid. This implies that the spin-orbit interaction causes only a fine-structure splitting of the atomic levels, but does not change the atomic wavefunctions, including the wavefunction of the Auger electron. The Auger decay partial waves are then described by l instead of lj, where l and j are the orbital and total angular momenta, respectively. This approximation is accurate enough for Auger transitions in not too heavy atoms. Introducing the LSJ-coupling approximation, in general, reduces the number of decay amplitudes. As an example we consider the decay of one fine-structure level J of multiplet LS into the two final levels J_f and $J_{f'}$ of the final multiplet $L_f S_f$ (see Fig. 4.2). The Auger decay is characterized by

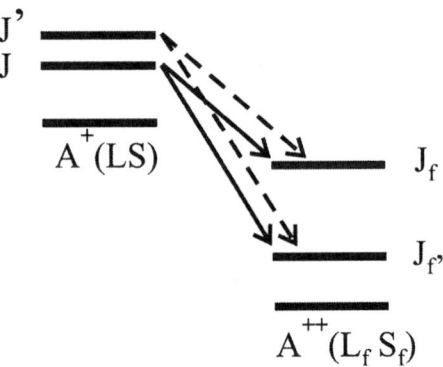

Figure 4.2 Diagram for Auger transitions assuming LSJ coupling. Here we consider the pair of transitions $J \to J_f$ and $J \to J_{f'}$ (solid lines) or $J' \to J_f$ and $J' \to J_{f'}$ (dashed lines).

the two decay amplitudes V_l and $V_{l'}$, where the amplitudes are now given by $V_l = |V_l| \exp(i\Delta_l)$ and are independent of J, J_f and j. For the determination of the two quantities $r = |V_l|/|V_{l'}|$ and $\cos \Delta_{ll'}$ we need two independent experimental quantities. Here we propose the measurements of the ratio of integral intensities of transitions $J \to J_f$ and $J \to J_{f'}$ (see Fig. 4.2) and the angular distribution of Auger electrons of transition $J \to J_f$ with spin-unsensitive detector.

The ratio R of integral intensities of transitions $J \to J_f$ and $J \to J_{f'}$ is given in analogy to Eq. (4.7) by [4]

$$R = \frac{W(J, J_f)}{W(J, J_{f'})} = \frac{\sum_l Z_{ll0}(J, J_f) |V_l|^2}{\sum_l Z_{ll0}(J, J_{f'}) |V_l|^2}. \tag{4.10}$$

Here, due to the LSJ-coupling approximation the coefficients $Z_{ll0}(J, J_f)$ are pure algebraic. The general expression for the coefficient $Z_{ll'k}$ is of the form

$$Z_{ll'k}(J, J_f) = (-1)^k \hat{J}_f^2 \hat{S}^2 \hat{J} \hat{l} \hat{l}' (l0, l'0 | k0) \sum_x (-1)^{J+x} \hat{x}^2$$

$$\times \begin{Bmatrix} L_f & S & x \\ \frac{1}{2} & J_f & S_f \end{Bmatrix}^2 \begin{Bmatrix} L_f & S & x \\ J & l & L \end{Bmatrix} \begin{Bmatrix} J & l & x \\ l' & J & k \end{Bmatrix} \begin{Bmatrix} l' & J & x \\ S & L_f & L \end{Bmatrix}. \tag{4.11}$$

We used standard notations for the 6j-symbols and Clebsch-Gordan coefficients and abbreviated $\hat{J} \equiv \sqrt{2J+1}$. Since the integral intensities $W(J, J_f)$ depend only on the squared moduli $|V_l|^2$, the ratio R of integral intensities of the fine-structure components of Auger lines carries information on the ratio of moduli, $r = |V_l|/|V_{l'}|$.

The angular distribution of Auger electrons of transition $J \to J_f$ measured without spin-detection is given by Eq. (4.8). Here we are interested only in the first term of the anisotropy with decay parameter α_2. This parameter is given in the LSJ-coupling approximation by [4]

$$\alpha_2(J, J_f) = \frac{2 \sum_{ll'} Z_{ll'k}(J, J_f) \, Re(V_l V_{l'}^*)}{\sum_l Z_{ll0}(J, J_f) \, |V_l|^2}, \qquad (4.12)$$

and depends therefore on $r = |V_l|/|V_{l'}|$ and $\cos \Delta_{ll'}$. As can be seen, from Eqs. (4.10) and (4.12) the quantities $|V_l|/|V_{l'}|$ and $\cos \Delta_{ll'}$ of the "almost" complete experiment can be evaluated from the experimental values R and α_2.

Information on the ratio r of moduli can be extracted from the ratio R of integral intensities only if the quantum numbers J and J_f are not factorizing from the sums in the nominator and denominator of Eq. (4.10). From the explicit form (4.11) for the coefficients $Z_{ll0}(J, J_f)$ it follows that factorization occurs in each of the following cases:

$$L_f = 0, \; S_f = 0, \; L = 0, \; S = 0. \qquad (4.13)$$

Each of these conditions give therefore restrictions on the use of the proposed method to perform an "almost" complete experiment. On the other side, there is an additional advantage of the proposed method. Besides the measurement of the ratio of integral intensities of transitions $J \to J_f$ and $J \to J_{f'}$ any other intensity ratio, e. g., of transitions $J' \to J_f$ and $J' \to J_{f'}$ (dashed lines in Fig. 4.2), can be measured and used for cross checking the first result.

4. PERFORMANCE OF AN "ALMOST" COMPLETE EXPERIMENT

We have performed an "almost" complete experiment on the following Auger decay (see Fig. 4.3) [4]:

$$Na^{+*}(2s2p^6 4p\, ^3P_J) \longrightarrow Na^{++}(2s^2 2p^5\, ^2P_{1/2},\, ^2P_{3/2}) + e_A^-(\varepsilon s, \varepsilon d). \qquad (4.14)$$

The aim of this investigation was to obtain information on the ratio of moduli, $r = |V_s|/|V_d|$, and on $\cos(\Delta_s - \Delta_d) = \cos \Delta_{sd}$. The Auger state $Na^{+*}(2s2p^6 4p\, ^3P_J)$ was produced via 2s-ionization of laser-excited sodium atoms by fast electron impact [13–15] (see Fig. 4.3):

$$e^-(1.5 \text{ keV}) + Na^+(3p_{3/2}) \longrightarrow Na^{+*}(2s2p^6 3p, 4p\, ^3P_J) + 2e^-. \qquad (4.15)$$

118   COMPLETE SCATTERING EXPERIMENTS

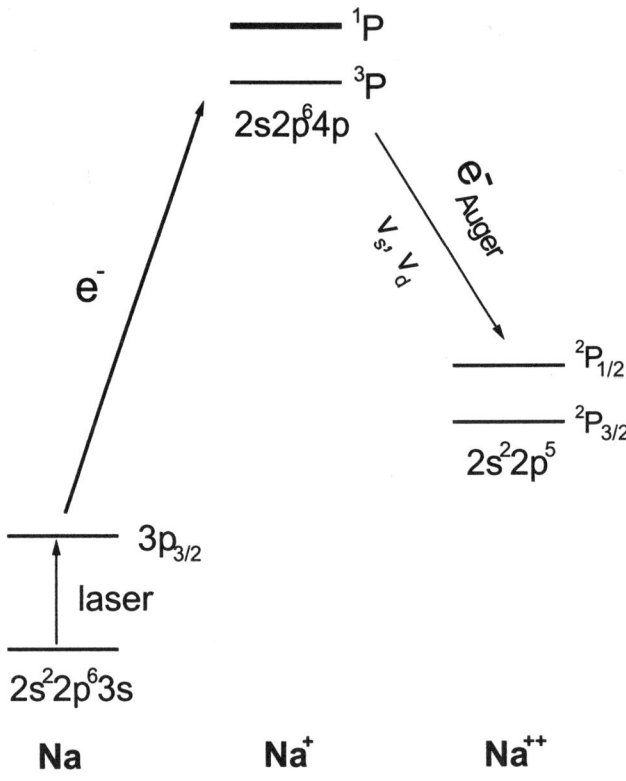

Figure 4.3 Level scheme for the excitation of the Auger states $Na^{++}(2s2p^64p\ ^3P,\ ^1P)$ by electron impact of laser-excited $Na^*(2s^22p^63p_{3/2})$ atoms and the Auger decay to $Na^{++}(2s^22p^5\ ^2P_{1/2}, ^2P_{3/2})$. $Na^{++}(2s2p^64p\ ^3P,\ ^1P)$ is excited via $3p \to 4p$ shake-up during the $2s$-ionization process.

Besides the $2s$ ionization without excitation of the laser-excited $3p_{3/2}$ electron a large fraction (about 31%) of $2s$ ionization is accompanied by a $3p \to 4p$ shake-up process [13].

The experimental procedure is briefly discussed in the following, for more details see references [13] and [14]. The laser beam (along the y-axis) and the electron beam (along the z-axis) cross the atomic beam (x-axis) (see Fig. 4.4). The Auger electrons are measured by means of a 75° sector-shaped cylindrical mirror analyser (CMA) which can be rotated in the yz plane in the angular range $\vartheta = 50....150°$. The detector is either a channeltron or, at smaller target densities, a position sensitive detector (PSD). The laser beam is linearly polarized with the polarization axis parallel to the axis of the electron beam. Then the experiment has rotational symmetry around the z-axis (direction of electron beam) and the angular distribution (4.8) depends only on the angle ϑ

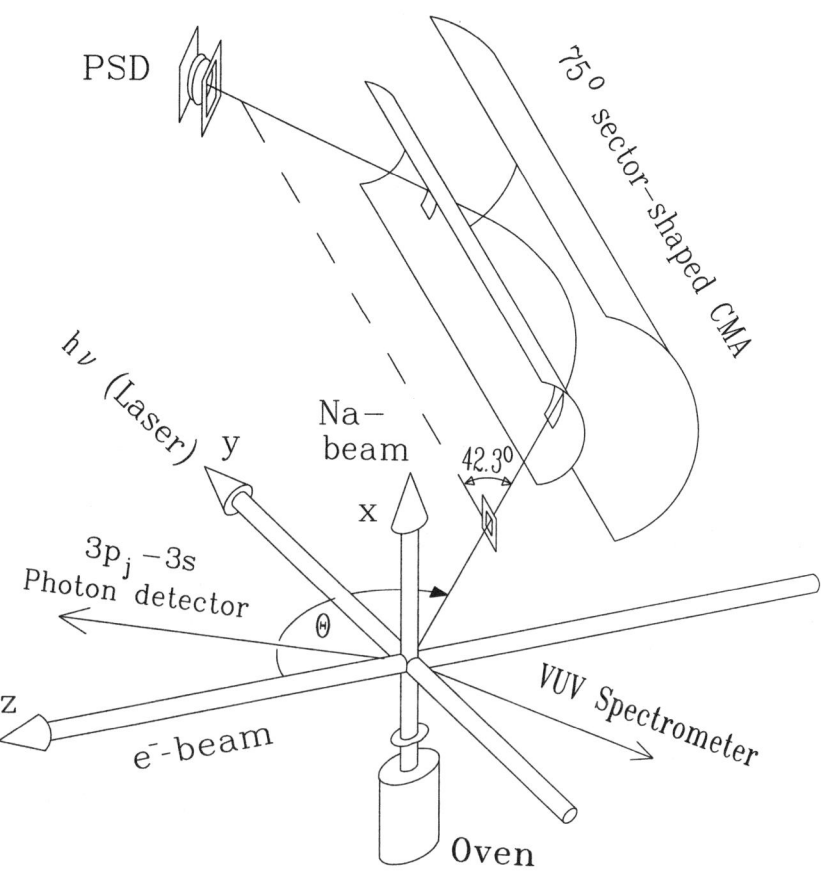

Figure 4.4 Experimental setup (from ref. [14], Fig. 1).

relative to the electron beam direction. For the $3s \to 3p_{3/2}$ excitation of the sodium atoms we use an argon ion laser pumped linear CW dye laser operating at $\lambda = 589.4$nm with 1MHz bandwidth. In order to achieve a maximum relative density of the $Na^*(3p_{3/2})$ atoms the laser is operated in the two-mode version [16, 17]: both hyperfine transitions $3s$, $F = 1 \to 3p_{3/2}$, $F = 2$ and $3s$, $F = 2 \to 3p_{3/2}$, $F = 3$ are excited. With a laser power of 100mW at beam diameter of 2mm the relative density $n_{rel}(3p\ ^2P_{3/2}) = 0.45$ has been reached. The alignment of the excited $Na^*(3p\ ^2P_{3/2})$ state can be determined via the degree of linear polarization of the fluorescence radiation $3p_{3/2} \to 3s$. The alignment decreases by radiation trapping for vapour densities higher than about 10^{11}

atoms/cm³ [18]. For a target density of about $5 \cdot 10^{11}$ atoms/cm³ the alignment was measured to be $\mathcal{A}_{20} = -0.50$ [19].

We performed two experiments:

- Measurement of the ratio of integral intensities, $R = \frac{W(^3P \rightarrow {}^2P_{1/2})}{W(^3P \rightarrow {}^2P_{3/2})}$, and

- measurement of the angular distribution $W(\vartheta)$ of Auger electrons of the transition $2s2p^6 4p\ ^3P \rightarrow 2s^2 2p^5\ ^2P_{3/2} + e_A^-$ and determination of the decay parameter α_2.

Experiment 1:

The target density in this experiment was high (about $3 \cdot 10^{12}$ atoms/cm³) and the alignment of the initial $3p_{3/2}$ state induced by the optical laser pumping was destroyed due to radiation trapping. Ionization from the $2s$ shell by fast electrons does not bring additional polarization into the Auger state [12]. Therefore, all tensors $\mathcal{A}_{kq}(J)$ in (4.8) vanish and the Auger electrons are emitted isotropically. Thus, the measurement of integral intensities can be performed at an arbitrary fixed angle, in the experiment the magic angle $\vartheta = 125,7°$ was used. In Fig. 4.5 the electron spectrum of Auger electrons due to the decay of $Na^+(2s2p^6 4p\ ^1P, {}^3P) \rightarrow Na^{++}(2s^2 2p^5\ ^2P_{1/2}, {}^2P_{3/2}) + e_A^-$ is shown. Here we need only the two intensities of Auger lines $^3P \rightarrow {}^2P_{1/2}$ and $^3P \rightarrow {}^2P_{3/2}$. The value of the ratio $R_{exp} = W(^3P \rightarrow {}^2P_{1/2})/W(^3P \rightarrow {}^2P_{3/2})$ has been measured [15] to be $R_{exp} = 1 : 6.6(3)$. Although in the present case the fine-structure splitting of the initial ^3P_J components is much smaller than the width of the Auger states $2s2p^6 4p\ ^3P_J$ ($\Gamma = 9.4 meV$ [20]), i. e. the fine-structure components completely overlap, contributions from different J sum incoherently into the integral intensity [21]. The branching ratio of the Auger decay to different fine-structure components J_f of the final ion $Na^{++}(2p^5\ ^2P_{J_f})$ takes then the form [4]

$$R = \frac{W(J_f)}{W(J_{f'})} = \frac{\sum_J P(J)\left(Z_s(J, J_f)|V_s|^2 + Z_d(J, J_f)|V_d|^2\right)}{\sum_J P(J)\left(Z_s(J, J_{f'})|V_s|^2 + Z_d(J, J_{f'})|V_d|^2\right)}, \quad (4.16)$$

where $Z_s(J, J_f)$ and $Z_d(J, J_f)$ are exactly known constants. In the case of resolved fine-structure J of the initial state the excitation probability $P(J)$ cancels from (4.16) and the whole treatment contains only the decay step. In order to proceed further in the present case we have to know the relative excitation probabilities $P(J)$ of the different fine-sructure states. These can be calculated assuming the one-particle model for the excitation process. The $P(J)$ depend on the total angular momentum j_0 of the initial fine-structure level of laser-excited state. For $j_0 = 3/2$ one obtains $P(0) : P(1) : P(2) = 0 : 1 : 5$ and the

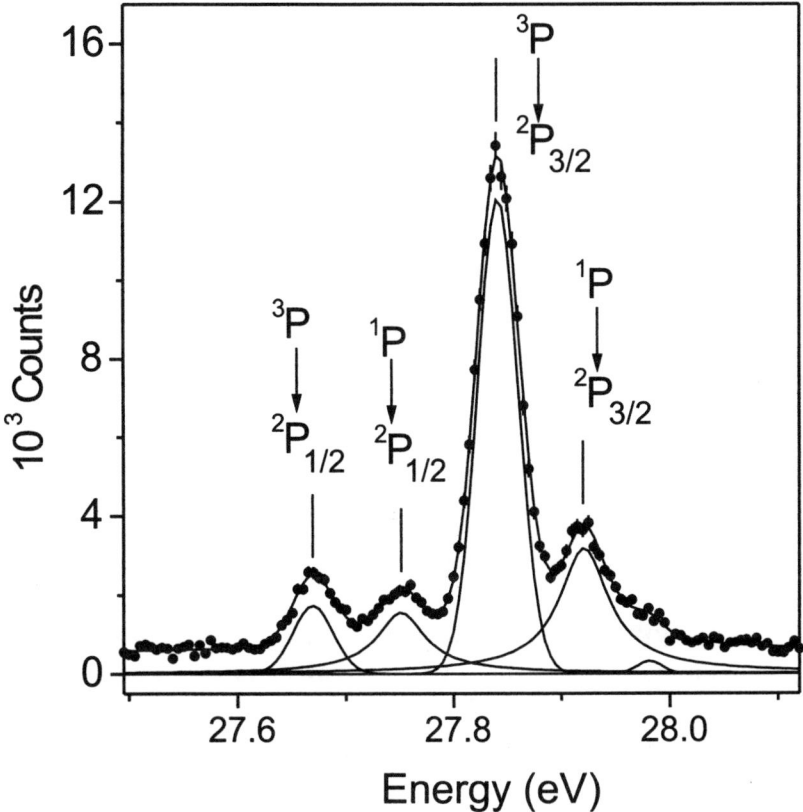

Figure 4.5 Electron spectrum due to the decay $Na^+(2s2p^64p\ \ ^3P, ^1P) \rightarrow Na^{++}(2s^22p^5\ ^2P_{1/2}, ^2P_{3/2}) + e_A^-$ measured at $\vartheta = 125.3°$ and for high-density Na target (from ref. 15, Fig. 3b).

final form of Eq. (4.16) is given by [15]

$$R = \frac{W(J_f = 1/2)}{W(J_f = 3/2)} = \frac{4+r^2}{5+8r^2}, \qquad (4.17)$$

where $r = |V_s|/|V_d|$. From the experimental value $R_{exp} = 1 : 6.6(3)$ we evaluate $r^2 = 15.3^{+5.2}_{-3.4}$. So the probability for emission of an s partial wave during the Auger decay is 15 times larger than the emission of a d partial wave.

Experiment 2:

In the second experiment the angular distribution of Auger electrons of the transition

$$Na^{+*}(2s2p^64p\ ^3P) \longrightarrow Na^{++}(2s^22p^5\ ^2P_{3/2}) + e_A^- \qquad (4.18)$$

was measured. The angular intensity distribution is given by

$$W(\vartheta) = \frac{W_0}{4\pi}(1 + \mathcal{A}_{20}\,\alpha_2\,P_2(\cos\vartheta)), \qquad (4.19)$$

where the decay parameter $\alpha_2(^3P\to{}^2P_{3/2})$ is a function of $r = |V_s|/|V_d|$ and $\cos\Delta_{sd}$ [4]:

$$\alpha_2(^3P\to{}^2P_{3/2}) = \frac{8\sqrt{2}\,r\cos\Delta_{sd} - 1}{8r^2 + 5}. \qquad (4.20)$$

In order to obtain an anisotropic part of the angular distribution (4.19) an alignment $\mathcal{A}_{20}$ has to be introduced into the initial Auger state ^3P. This is done via the laser excitation $3s \to 3p_{3/2}$ of Na atoms with linear polarized light and low enough target density [19]; the alignment $\mathcal{A}_{20}(^3P_{3/2})$ was determined via the linear polarization of fluorescence radiation $3p_{3/2} \to 3s$ to be $\mathcal{A}_{20}(^3P_{3/2}) = -0.50$ [19]. The ionization of a $2s$ electron by impact of fast electrons does not bring in an additional alignment [12], i.e., $\mathcal{A}_{20}(2s2p^64p\ ^3P) = \mathcal{A}_{20}(2s^22p^63p\ ^2P_{3/2})$. The intensity distribution $W(\vartheta)$ of Auger transition (4.18) relative to the isotropic intensity of transition $Na^{+*}(2p^53s(^1P)3p\ ^2S_{1/2}) \to Na^+(2p^6\ ^1S_0) + e_A^-$ is plotted in Fig. 4.6. The solid line is a fit function according to Eq.(4.19), a total anisotropy $\mathcal{A}_{20}(^3P)\alpha_2(^3P\to{}^2P_{3/2}) = +0.15(3)$ was obtained. Using the alignment value $\mathcal{A}_{20}(^3P) = -0.50$ yields the decay parameter $\alpha_2(^3P\to{}^2P_{3/2}) = -0.30(6)$. Together with the above listed experimental value of r_{exp}^2 the relative phase Δ_{sd} can be evaluated via Eq.(4.20) to be $\Delta_{sd} = 147°^{+33°}_{-21°}$ or $213°^{+21°}_{-33°}$. The first value agrees well with the calculated relative phase $(\Delta_{sd})_{th} = 143°$ [20, 22], therefore the second value can be ruled out.

In table 4.1 we compare the experimental values of $r^2 = |V_s|^2/|V_d|^2$ and Δ_{sd} with theoretical results obtained for different approximations of the wavefunctions [20,22]. One can see from table 4.1 that the experimetal values agree well with the theoretical results obtained with wavefunctions which include both the core correlation and the core polarization. Especially the correlation of the core wavefunction increases strongly the ratio of squared amplitudes $|V_s|^2/|V_d|^2$ and brings experiment and theory in good agreement. On the other side it is interesting to note that the various approximations of wavefunctions have only very little effect on the relative phase. From this it seems that the experimental determination of the relative phase will much less stringently test its theoretical value than the ratio r of moduli of amplitudes (or the absolute values of amplitudes in case of a complete experiment). The reason for this weak dependence on the wavefunction approximation is that the relative phase is a property integrated from the inner part of the potential of the atom to infinity whereas the moduli depend directly on the quality of wavefunctions beeing active in the Auger decay.

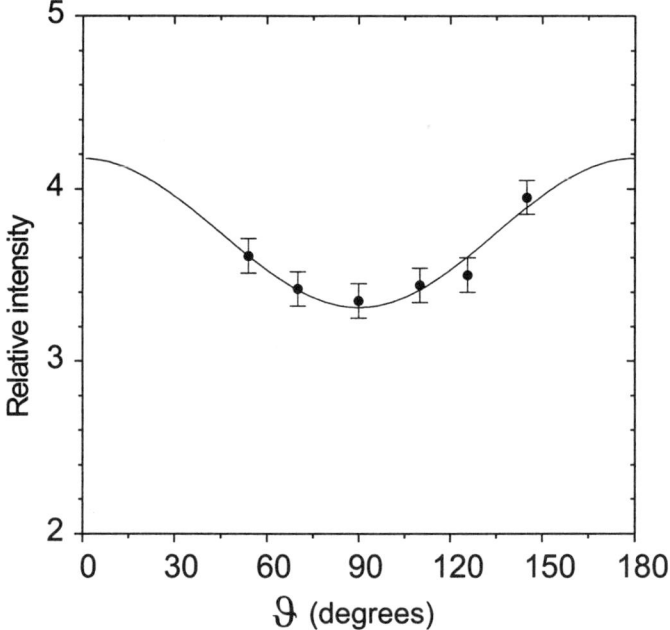

Figure 4.6 Intensity of transition $Na^{+*}(2s2p^64p\,^3P) \to Na^{++}(2s^22p^5\,^2P_{3/2})+e_A^-$ relative to the isotropic transition $Na^*(2p^53s(^1P)3p\,^2S) \to Na^+(2p^6\,^1S)+e_A^-$ as function of angle ϑ relative to the direction of the electric vector of exciting laser beam (which is parallel to the direction of ionizing electron beam). The target density is much lower than in the case of Fig. 4.5; the alignment of laser-excited atoms is $\mathcal{A}_{20}(^2P_{3/2}) = -0.50$. The solid line is the fit function $W(\vartheta)$ to the experimental values (from ref. 4, Fig. 3).

The present "almost" complete experiment can be extended to a real complete experiment by measuring the total width $\Gamma(^3P)$ of the decaying Auger state $2s2p^64p\,^3P$. The total width is given by

$$\Gamma(^3P) = 2\pi \sum_{J_f} \left[Z_{000}(J,J_f)|V_s|^2 + Z_{220}(J,J_f)|V_d|^2 \right] \qquad (4.21)$$

which yields

$$\Gamma(^3P) = \frac{2\pi}{3} \left(|V_s|^2 + |V_d|^2 \right). \qquad (4.22)$$

$\Gamma(^3P)$ was calculated [20] to be 9.4meV, which is much smaller than the energy resolution of the electron analyser of about 35 to 40meV in the present experiment. Therefore, the determination of the total width Γ will be hardly feasible unless the resolution of the analyser is much improved.

Table 4.1 Experimental and theoretical values of the squared ratio of amplitudes, $r^2 = |V_s|^2/|V_d|^2$, and of the relative phase Δ_{sd} of Auger transition $Na^{+*}(2s2p^64p\ {}^3P) \rightarrow Na^{++}(2s^22p^5\ {}^2P) + e_A^-(s,d)$.

	Experiment	Theory [20,22] (a)	(b)	(c)
$\|V_s\|^2/\|V_d\|^2$	$15.3^{+5.2}_{-3.4}$	16.0	12.6	4.9
Δ_{sd}	$147°^{+33°}_{-21°}$	143°	144°	147°

(a): Calculation with both correlated core and core polarization
(b): Differs from case (a) by excluding core polarization
(c): Differs from case (a) by using the Hartree-Fock core

5. CONDITIONS AND APPLICABILITY OF THE PROPOSED METHOD

A complete experiment is complete only within the framework of the theoretical model which is used to describe it. As outlined in the introduction the basic theoretical model for the Auger decay is that the decay can be treated as step 2 of a two-step process. For the proposed method of an "almost" complete experiment we furthermore assumed (see section 3.) that LSJ coupling is valid for the initial and the final state of the Auger decay. Then several additional conditions should be fulfilled in order that the proposed method can be applied.

Condition 1: In section 3. it has been shown, that the ratio of intensities of transitions $J \rightarrow J_f$ and $J \rightarrow J_{f'}$ is non-statistically and depends on the ratio of moduli $|V_l|$ if the following conditions are simultaneously fulfilled: $L \neq 0$, $S \neq 0$, $L_f \neq 0$, $S_f \neq 0$. Condition 1 is therefore a rigorous condition.

Condition 2: In order to measure the intensities of transitions $J \rightarrow J_f$ and $J \rightarrow J_{f'}$ the fine-structure splitting $\Delta E(J_f, J_{f'})$ should be larger than the total width $\Gamma(J)$ of decaying state but still small enough that LSJ coupling is a good approximation.

Condition 3: In order to resolve the fine-structure $\Delta E(J_f, J_{f'})$ of transitions $J \rightarrow J_f$ and $J \rightarrow J_{f'}$ the resolution of the electron energy analyser should be high enough.

Now we consider the general applicability of the proposed method. In the case of diagram Auger transitions of initially closed-shell atoms the initial Auger state has to be produced by ionization in an inner shell with $l \neq 0$ (condition

1), and the final doubly charged state after the Auger decay must have $L_f \neq 0$ and $S_f \neq 0$, e.g. it could be a ^3P state. Therefore the proposed method could be applicable to the following Auger transitions:

$$\text{Ar } L_{2,3}\text{-}M_1M_{2,3}(^3P_J), \text{Kr } M_{4,5}\text{-}N_1N_{2,3}(^3P_J), \text{Xe } N_{4,5}\text{-}O_1O_{2,3}(^3P_J). \quad (4.23)$$

Condition 2 is well fulfilled for the Ar case of transitions (4.23). In view of the recent results on the complete experiment of the Xe 4d photoionization [23], where it has been shown that instead of the three amplitudes R_{lj} and the two relative phases $\Delta_{lj,l'j'}$ in the relativistic approach the two amplitudes R_l and the relative phase $\Delta_{ll'}$ in the LS coupling are sufficient to describe the photoionization process, one can argue that the LSJ coupling approach is also sufficient for the above listed Auger transitions in Kr and Xe. Condition 3 is easily fulfilled for the Kr and Xe transitions: high-resolution Auger spectroscopy clearly separated the Auger lines according to their final states J_f [24]. In the case of Ar the resolution was not high enough to resolve the $L_{2,3}$-$M_1M_{2,3}(^3P_J)$ lines, the fine-structure intensities were obtained by a line fitting procedure [25].

In the case of resonant Auger transitions following the photoexcitation by synchrotron radiation one faces the same restrictions as for the non-resonant Auger transitions. Therefore the proposed method is not applicable for resonant Auger transitions excited from the $^1S_0$ ground state of closed-shell atoms, but it is applicable in case of initially open-shell atoms. By utilizing the resonant Raman Auger effect the experimental width of Auger lines is mainly given by the bandpass ΔE_γ of exciting photons (for $\Delta E_\gamma \ll \Gamma$) and the resolution of the spectrometer. Measurements made with very high photon and electron energy resolutions allowed to resolve almost all J_f fine-structure components in resonant Raman Auger transitions of noble gas atoms [26]. Therefore, the proposed method should be well suited for resonant Raman Auger transitions in initially open-shell atoms provided the two-step model is valid for the resonant Raman Auger effect.

6. CONCLUSIONS

We have presented a proposal for an "almost" complete experiment on Auger decay and have performed such an experiment for the Auger decay $Na^{+*}(2s2p^64p\ ^3P) \rightarrow Na^{++}(2s^22p^5\ ^2P_{1/2},^2P_{3/2}) + e_A^-(s,d)$. We determined the ratio of squared moduli, $r^2 = |V_s|^2/|V_d|^2$, and the relative phase Δ_{sd} and compared them with calculated values. We have given the conditions which should be fulfilled that the proposed method can be applied either to normal or to resonant Auger transitions.

Acknowledgments

The authors thank O. I. Zatsarinny for sending his results on the relative phase Δ_{sd}. This work is part of a joint program supported by the Deutsche Forschungsgemeinschaft (DFG) and the Russian Foundation for Basic Research (grant 96-02-00204), it is also supported by the DFG via SFB 276. ANG greatfully acknowledges the hospitality of the Albert-Ludwigs-Universität, Freiburg.

References

[1] U. Heinzmann, J. Phys. B: At. Mol. Opt. Phys. **13**, 4353 (1980).

[2] J. B. West, K. J. Ross, and H. J. Beyer, J. Phys. B: At. Mol. Opt. Phys. **31**, L647 (1998).

[3] K. Ueda et al., J. Phys. B: At. Mol. Opt. Phys. **31**, 4801 (1998).

[4] A. N. Grum-Grzhimailo, A. Dorn, and W.Mehlhorn, Comm. At. Mol. Phys. Comm. Mod. Phys. D **1**, 29 (1999).

[5] U. Fano, Phys. Rev. **124**, 1866 (1961).

[6] S. A. Sheinerman, W. Kuhn, and W. Mehlhorn, J. Phys. B: At. Mol. Opt. Phys. **27**, 5681 (1994).

[7] N. M. Kabachnik and I. P. Sazhina, J. Phys. B: At. Mol. Opt. Phys. **23**, L353 (1990).

[8] E. G. Berezhko, N. M. Kabachnik, and V. V. Sizov, J. Phys. B: At. Mol. Opt. Phys. **11**, 1819 (1978).

[9] H. Klar, J. Phys. B: At. Mol. Opt. Phys. **13**, 4741 (1980).

[10] H. Merz and J. Semke, in *X-Ray and Inner Shell Processes*, American Inst. Phys., edited by T. A. Carlson, M. O. Krause, and S. T. Manson (AIP Conference Proceedings 215, New York, 1990), pp. 719–728.

[11] G. Snell et al., Phys. Rev. Lett. **76**, 3923 (1996).

[12] A. N. Grum-Grzhimailo and W. Mehlhorn, J. Phys. B: At. Mol. Opt. Phys. **30**, L9 (1997).

[13] A. Dorn et al., J. Phys. B: At. Mol. Opt. Phys. **28**, L225 (1995).

[14] A. Dorn, O. Zatsarinny, and W. Mehlhorn, J. Phys. B: At. Mol. Opt. Phys. **30**, 2975 (1997).

[15] A. N. Grum-Grzhimailo and A. Dorn, J. Phys. B: At. Mol. Opt. Phys. **28**, 3197 (1995).

[16] E. E. B. Campbell, H. Hülser, R. Witte, and I. V. Hertel, Z. Phys. D **16**, 21 (1990).

[17] P. Strohmeier, Opt. Commun. **79**, 187 (1990).

[18] A. Fischer and I. V. Hertel, Z. Phys. A **304**, 103 (1982).

[19] A. Dorn *et al.*, J. Phys. B: At. Mol. Opt. Phys. **27**, L529 (1994).

[20] O. I. Zatsarinny, J. Phys. B: At. Mol. Opt. Phys. **28**, 4759 (1995).

[21] K. Blum, *Density Matrix Theory and Applications* (Plenum Press, New York and London, 1981).

[22] O. I. Zatsarinny, 1998, private communication.

[23] G. Snell *et al.*, Phys. Rev. Lett. **82**, 2480 (1999).

[24] L. O. Werme, T. Bergmark, and K. Siegbahn, Phys. Scripta **6**, 141 (1972).

[25] D. Ridder, J. Dieringer, and N. Stolterfoht, J. Phys. B: At. Mol. Opt. Phys. **9**, L307 (1976).

[26] H. Aksela *et al.*, Phys. Rev. A **55**, 3532 (1997).

Chapter 5

THE TRANSFER OF ANGULAR MOMENTUM FOR ELECTRON–ATOM COLLISION PROCESSES INVOLVING INELASTICALLY EXCITED $S \rightarrow P$ AND $P \rightarrow S$ TRANSITIONS

M. Shurgalin
Harvard-Smithsonian Centre for Astrophysics, Cambridge, MA 02138, USA

A. J. Murray
The Schuster Laboratory, University of Manchester, Manchester, M13 9PL, UK

W. R. MacGillivray
The Laser Atomic Physics Laboratory, Griffith University, Brisbane, QLD 4111, Australia

M. C. Standage
The Laser Atomic Physics Laboratory, Griffith University, Brisbane, QLD 4111, Australia

Abstract Since the initial electron-photon coincidence and superelastic scattering experiments showed that substantial transfers of angular momentum between incident electrons and target atoms could occur, apart from stimulating much theoretical activity to develop better quantum mechanical scattering calculations, several classical/semi-classical models have also been developed. In this paper, new experimental results from electron inelastic scattering from the laser excited $3^2 P_{3/2}$ state of sodium has provided a further test of scattering theories, including a very recent convergent close coupling calculation.

One of the surprising results that came out of the early electron-photon coincidence and superelastic scattering investigations of electron-atom collision processes was the observation that a substantial transfer of angular momentum could occur between the incident electron and the target atom at small scattering

Complete Scattering Experiments, Edited by Becker and Crowe
Kluwer Academic/Plenum Publishers, New York 2001

angles. This outcome pointed to the deficiencies of existing theories including the first Born approximation which predicted that the transfer of angular momentum would be zero. Apart from stimulating much theoretical activity to develop new quantum mechanical treatments of the electron-atom scattering problem, a number of attempts were also made to introduce classical/semi-classical models to describe the processes involved.

Early attempts focussed on the "rolling ball" or "grazing incidence" pictures in which the incident electron pushes and distorts the atomic charge cloud, imparting angular momentum to it in this process. In 1980, Steph and Golden [1] introduced a semi-classical model in which the observed positive values for the transferred angular momentum at small positive scattering angles were attributed to the attractive long range atomic polarization potential, whereas the negative values observed at large scattering angles were attributed to the repulsive potential of the target atom's electrons. In spite of the attractiveness of this model, Madison et al 1986 [2] showed from theoretical considerations based on a distorted wave treatment that neither the atomic polarization nor the electron-electron repulsion determine the behavior of the transferred angular momentum as a function of scattering angle. Based on an analysis of the contributions of different partial waves to the transferred angular momentum, these authors concluded that a quantum mechanical interference phenomenon principally determined the behavior.

In 1986, Andersen and Hertel [3] further explored ways of qualitatively describing electron-atom collision processes. These authors considered the behavior of the transferred angular momentum for electron and positron scattering. They drew on earlier theoretical work by Madison and Winters 1981 [4] who showed how a Born series expansion of the T matrix led to an expression for the transferred angular momentum that depended on, amongst other factors, the sign of the charge of the incident projectile. The argument is as follows. The scattering amplitude for exciting an atom from state $|a\rangle$ to state $|b\rangle$ is given in terms of a Born series expansion as

$$f_{ba} = \langle b|V|a\rangle + \langle b|VG^+V|a\rangle + \ldots \quad (5.1)$$

where the wave functions are asymptotic states, V is the Coulombic interaction between the incident electron and the target atom, and G^+ is the free particle Green's function. Noting that the potential V depends on the charge q of the incident particle, equation 5.1 can be rewritten as

$$f_{ba} = qT_{ba}^1 + q^2T_{ba}^2 + \ldots \quad (5.2)$$

Taking the quantization axis along the direction of the momentum transfer, the $s \to p$ transition can be described in terms of $p\sigma$ and $p\pi^+$ orbitals. In the first Born approximation, only the $p\sigma$ orbital is excited and equation 5.2

becomes,
$$f_\sigma = qT_\sigma^1 + q^2 T_\sigma^2 + \cdots$$
and
$$f_\pi = qT_\pi^1 + q^2 T_\pi^2 + \cdots \quad (5.3)$$

The transferred angular momentum is given by
$$\begin{aligned} L_\perp &= -2Im(f_\sigma f_\pi^*) \\ &= -2q^3 Im(T_\sigma^1) Re(T_\pi^{2*}) - 2q^4 Im(T_\sigma^2 T_\pi^{2*}) + \cdots \end{aligned} \quad (5.4)$$

As is well known, in terms of this Born series expansion treatment, to first order, $L_\perp$ is zero. The first term in equation 5 depends on both the first and second order Born terms and on the incident particle charge to the third power, and the second term depends on second order terms and the fourth power of the charge. Accordingly, the higher order Born terms play a very significant role in $L_\perp$ for all scattering angles. The typically observed behavior of $L_\perp$ as a function of electron scattering angle for a number of target atoms such as helium, sodium, and rubidium is for a cross-over point to be observed at in the middle range of scattering angles from 0° to 180°. This behavior can be interpreted in terms of equation 5 as the second term growing sufficiently large as to equal the first term at the cross-over angle. Equation 5 predicts that for positron scattering, unlike electron scattering, the first term will reverse sign, whereas the second term will have the same sign, with the result that it would be expected that for positron scattering, $L_\perp$ will be negative for all angles. This prediction was confirmed by theoretical calculations performed by Madison and Winters 1981 [4].

Andersen and Hertel went on to draw a very interesting analogy between the behavior of $L_\perp$ for electron and positron scattering and $s \rightarrow p$ excitation and $s \rightarrow p$ de-excitation. For zero angle scattering, the transferred linear momentum, treated as a vector, reverses sign comparing the $s \rightarrow p$ excitation and $s \rightarrow p$ de-excitation cases. Since T_σ^1 changes sign when the transferred linear momentum reverses sign, while the second-order terms are unchanged, these authors suggested that at least at smaller scattering angles, the behavior of $L_\perp$ for $s \rightarrow p$ excitation and $s \rightarrow p$ de-excitation might mirror the behavior predicted for electron versus positron scattering. Thus, $L_\perp$ was predicted to be negative at all scattering angles for $s \rightarrow p$ de-excitation. Of course, given the approximations involved, this prediction could only be expected to hold for smaller angle scattering. Furthermore, they proposed an experiment which was based on a variant of the superelastic scattering method using transitions in sodium that could test this suggestion.

In this paper, we discuss results from the first experiment undertaken to compare all the atomic collision parameters, including $L_\perp$, for $s \rightarrow p$ excitation and $s \to p$ de-excitation processes, thus providing an experimental test of the suggestion made by Andersen and Hertel. Excitation processes for $s \to p$ transitions have been investigated using electron-photon coincidence and superelastic scattering techniques since the mid-1970's following the pioneering work of Kleinpoppen and associates and Hertel and associates. A comprehensive listing of papers associated with the work of these two groups is given in Andersen et al 1988 [5]. In these experiments, alignment and orientation parameters, which in this paper will be called atomic collision parameters (ACP) are determined. Assuming L-S coupling holds, for electron excitation of an $s \to p$ transition, four ACP's are required, the angular momentum transferred to an atom from the incident electron which is perpendicular to the scattering plane, $L_\perp$, the tilt of the atomic charge cloud with respect to the direction of the incident electron, γ, the degree of anisotropy of the atomic charge cloud, P_l, and the degree of polarization, $P+$. For superelastic scattering experiments, the differential scattering rate of electrons superelastically scattered in the $p \to s$ de-excitation process, measured as a function of laser polarization provides measurements of all four ACP's. Such superelastic scattering processes are the time-inverse of the excitation of an $s \to p$ transition with the principle of micro-reversibility being invoked to interpret the superelastic scattering data.

A technique similar to superelastic scattering can be used to measure ACP's for transitions from an excited p to a higher lying s state. Such a transition is the time-inverse of the de-excitation of an $s \to p$ transition. In this case, the inelastically scattered electrons inducing the $p \to s$ transition are detected as a function of laser polarization. These measurements require resolution of the $p \to s$ transition in the electron detection channel to obtain unambiguous measurements of the ACP's. The first attempt to obtain ACP's for transitions between excited s and p states was performed by Hermann et al 1977 [6]. These authors measured alignment parameters for the electron-impact induced $3^2 p_{3/2} \to 4^2 s_{1/2}$ transition of sodium at 6, 10 and $20 eV$ incident electron energy. No measurements of the $L_\perp$ parameter were reported and the data was only obtained for a very limited range of scattering angles which does not allow comprehensive comparisons to be made with data obtained from the superelastic $3^2 p_{3/2} \to 3^2 s_{1/2}$ transition.

In Figures 5.1 to 5.3, we present for completeness sets of all the spin-averaged ACP's obtained from the inelastic excitation of the $3p \to 4s$ transition of sodium for an incident electron energy of $22 eV$ and over the range of electron scattering angles $-10°$ to $+30°$. The data is compared with data obtained using the same apparatus for the corresponding parameters for superelastic scattering measurements for the well-known $3p \to 3s$ transition of sodium. The measurements are compared with a Convergent Close Coupling (CCC) and Second Order Distorted Wave Born (DWB2) calculation and the first Born approximation. Details of the experimental methods used by the authors together with

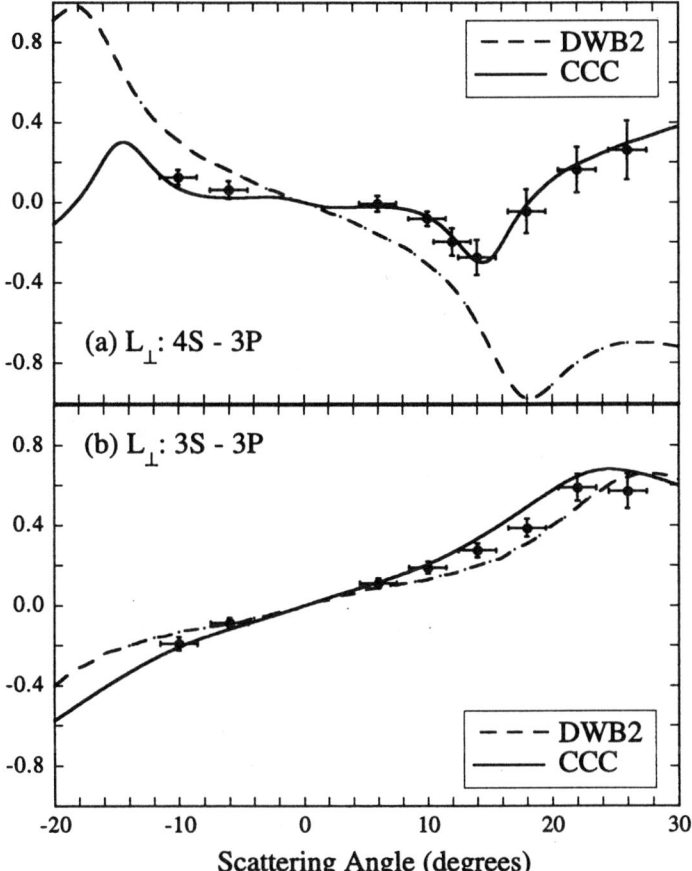

Figure 5.1 The $L_\perp$ parameter for (a) the $4s \to 3p$ and (b) the $3s \to 3p$ transitions as a function of scattering angle. The solid lines show the CCC calculation, and the dashed lines show the DWB2 calculation.

a discussion of the data and comparison with theory for all the ACP's can be found in references [7, 8]. In this paper, further discussion is restricted to the ACP's that are relevant to the transfer of momentum, both angular and linear, in collision processes.

It will be noted that the behavior of the $L_\perp$ parameter is very different for the $4s \to 3p$ data as compared with the $3s \to 3p$ data. For the $3s \to 3p$ transition, $L_\perp$ is positive for positive scattering angles and increases with scattering out to around 24°, reaching a maximum of 0.7, at which point it starts to fall. This behavior is in accord with the predictions of Andersen and Hertel, and the CCC and BWB2 calculations are in good agreement with the data. For the $4s \to 3p$ data, $L_\perp$ is almost zero for scattering angles below 8°, then

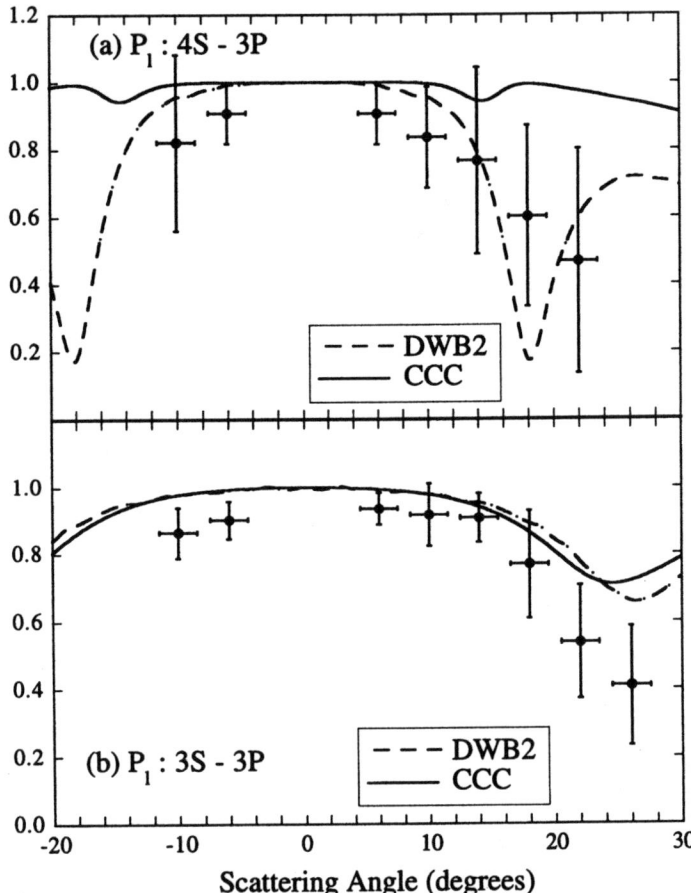

Figure 5.2 The P_l parameter for (a) the $4s \rightarrow 3p$ and (b) the $3s \rightarrow 3p$ transitions as a function of scattering angle. The solid lines show the CCC calculation, and the dashed lines show the DWB2 calculation.

decreases to a negative value of -0.25 at a scattering angle of $15°$. $L_\perp$ then turns around and becomes positive with increasing values of scattering angle. For the $4s \rightarrow 3p$ case, the CCC calculation is in excellent agreement with the experimental data, whereas the DWB2 calculation, although qualitatively predicting some of the trends in the data correctly, gives much larger values. For the $3s \rightarrow 3p$ transition, the CCC and DWB2 calculations are in reasonable accord with each other and with the experimental data.

For small angles, it can be argued that the first Born approximation, which predicts zero for the $L_\perp$ parameter at all scattering angles, gives a much better prediction for the $4s \rightarrow 3p$ transition than for the $3s \rightarrow 3p$ transition.

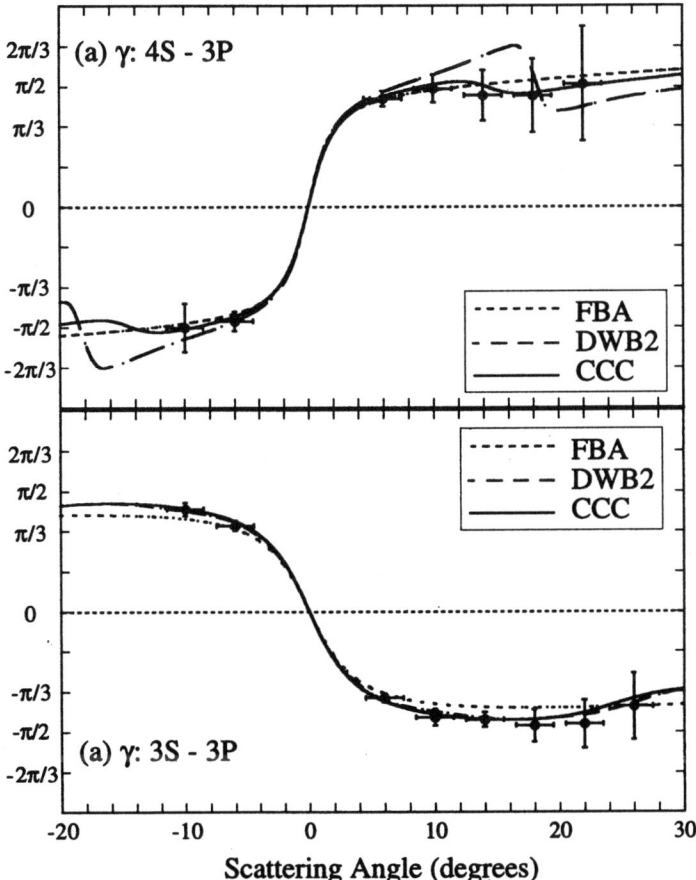

Figure 5.3 The γ parameter for (a) the $4s \to 3p$ and (b) the $3s \to 3p$ transitions as a function of scattering angle. The solid lines show the CCC calculation, the full dashed lines show the DWB2 calculation and the smaller dashed lines show the first Born results.

It is also of interest to look at the behavior of the alignment angle, γ, for both cases. In the first Born approximation, the alignment angle gives the direction of the transferred linear momentum relative to the direction of the incident electron. A comparison of the data obtained for the alignment angle and the theoretical calculations shows over the angular range investigated that for the $3s \to 3p$ transition, both the DWB2 and CCC calculations are in good agreement with the experimental data. For the $4s \to 3p$ transition, the CCC calculation is in good agreement with the data, whereas the DWB2 calculation shows some departure from the CCC calculation in the region of 15° scattering angle. Over the angular range 0° − 30°, the first Born approximation differs only slightly from the data for the $3s \to 3p$ transition, and is in good agreement

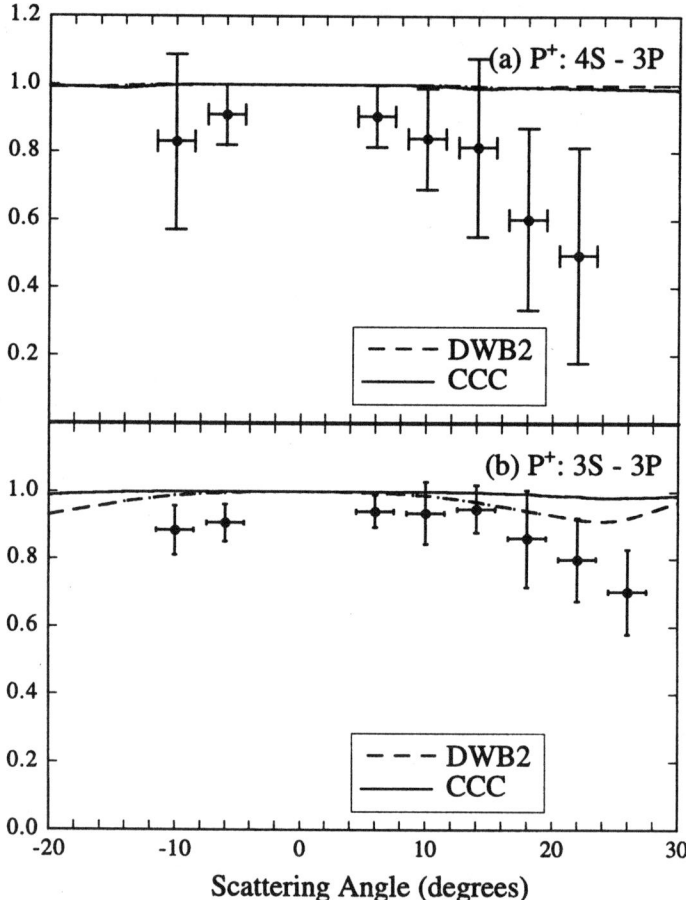

Figure 5.4 The P^+ parameter for (a) the $4s \to 3p$ and (b) the $3s \to 3p$ transitions as a function of scattering angle. The solid lines show the CCC calculation, and the dashed lines show the DWB2 calculation.

with the data for the $4s \to 3p$ transition. In both cases, the simple kinematical description provided by the first Born approximation gives a good account of the linear momentum transferred between the incident electron and the target atom over a significant range of scattering angles. This is to be contrasted with the case for the angular momentum transferred, where only for the $4s \to 3p$ transition, and only at small scattering angles, does the first Born approximation give a good indication of the behavior of $L^\perp$.

We have seen that the qualitative prediction of Andersen and Hertel, which has very successfully accounted for the behavior of $L^\perp$ as a function of scattering angle for $s \to p$ transitions involving excitation does not provide a satisfactory account for the one case so far investigated for $s \to p$ transitions involving de-excitation. Out to 10° angle scattering, $L_\perp$ is only slightly negative. In the range 10° to 16°, $L_\perp$ reaches its maximum negative value and then becomes strongly positive. It can be said that $L_\perp$ behaves in accord with the prediction for the $4s \to 3p$ transition in the sense that this parameter is negative out to 16°, although it does not behave as simply the negative of the $L_\perp$ behavior for the $3s \to 3p$ transition which is strongly positive over the same angular range. In terms of equation 5, the behavior of $L_\perp$ for the $3s \to 3p$ transition is explained in terms of the dominance of the first term with its first and second order Born terms over the higher order terms in the expansion. However, for the $4s \to 3p$ transition, it appears that in terms of equation 5, at small scattering angles, either the first term is not dominant, or all terms are essentially zero.

In the authors' view, several conclusions can be drawn from the research reported here. Continuing care needs to be exercised concerning the range of applicability of various theoretical approximations and models for electron-atom scattering processes. The inelastic scattering experiment reported here has provided a very stringent test of theory, in that it has shown that even a sophisticated theory such as the DWB2 is unable to quantitatively reproduce the $L_\perp$ data, although it is very interesting to note that the CCC theory provides good agreement. This outcome points to the need to undertake further experimental and theoretical investigations of collision processes involving higher lying transitions in atoms to further probe the range of validity of existing scattering theories.

References

[1] N. C. Steph and D. E. Golden, Phys. Rev. A **21**, 1848 (1980).
[2] G. C. D. H. Madison and D. C. Cartwright, J. Phys. B: At. Mol. Opt. Phys. **19**, 3361 (1986).
[3] N. Andersen and I. V. Hertel, Comments At. Mol. Phys. **19**, 1 (1986).
[4] D. H. Madison and K. H. Winters, Phys. Rev. Lett. **47**, 1885 (1981).
[5] N. Andersen, J. W. Gallagher, and I. V. Hertel, Phys. Rep. **165**, 1 (1988).
[6] H. W. Hermann et al., J. Phys. B: At. Mol. Opt. Phys. **10**, 251 (1977).
[7] M. Shurgalin et al., Phys. Rev. Lett. **81**, 4604 (1998).
[8] M. Shurgalin et al., J. Phys. B: At. Mol. Opt. Phys. (1999), in press.

II
PHOTON IMPACT

Chapter 6

COMPLETE PHOTOIONISATION EXPERIMENTS USING POLARISED ATOMS

G. Prümper
Fritz–Haber–Institut der Max–Planck–Gesellschaft, Faradayweg 4–6, D-14195 Berlin

B. Zimmermann
Fritz–Haber–Institut der Max–Planck–Gesellschaft, Faradayweg 4–6, D-14195 Berlin

U. Becker
Fritz–Haber–Institut der Max–Planck–Gesellschaft, Faradayweg 4–6, D-14195 Berlin

H. Kleinpoppen
Unit of Atomic and Molecular Physics, University of Stirling, Stirling, Scotland

Abstract We performed quantum mechanically complete photoionisation experiments using polarised thallium atoms. We determined the ratio of the matrix elements R_{l_0-1} and R_{l_0+1} and the asymptotic phase difference Δ as a function of photon energy for the Tl 5d- and 6p-shell. The results are in good agreement with Random Phase Approximation with Exchange (RPAE) -calculations and reveal the strong influence of continuum mixing for the Tl 6p photoelectron line. Additionally we checked the consistency of our partial wave analysis with measurements of the dynamical spin polarisation of the Tl-5d multiplet.

1. INTRODUCTION

The experiment is based on an idea of Professor Kleinpoppen, who suggested to use angle resolved photoelectron spectroscopy on polarised atoms for a complete analysis [1–3] of the photoionisation process [4,5], in the same way as it has been done before using spin sensitive detection and unpolarised rare gas atoms [6].

The measurements for thallium are part of a larger collaborative effort by Uwe Becker and Hans Kleinpoppen on photoionisation of polarised atoms and were performed by Burkhard Langer and Georg Prümper at BESSY I. The theoretical support in form of Hartee-Fock (HF) and Random Phase Approximation with Exchange (RPAE)-calculations came from Nikolai Cherepkov and Björn Zimmermann. Spin resolved measurements were performed in cooperation with the Group of Professor U. Heinzmann from the University of Bielefeld (Germany), in particular Bernd Schmidtke and Markus Drescher.

1.1 APPROXIMATIONS

The final state after photoionisation

$$\text{Atom}_{\gamma_0, J_0} + h\nu \rightarrow \underbrace{\text{Ion}^+_{\gamma', J'} + e^-_{\epsilon, l, j}}_{J = J' + j}$$

is a linear combination of continuum states with different quantum numbers for the angular moments.

$$|\text{final}\rangle = \sum_{\substack{\gamma', J', l, j, \\ J, M_J, M'_J, m_j}} d_{\gamma', J', M'_J, l, j, m_j, J, M_J} \cdot |\text{Ion}^+_{\gamma', J', M'_J}\rangle \times |e^-_{\epsilon, l, J, j, m_j}\rangle.$$

The measurement of the kinetic energy of the photoelectron determines the quantum numbers γ' and J' that describe the ion but in general all combinations of l, J, and j contribute to the final state.

$$|\text{final}\rangle = \sum_{\substack{l, j, \\ J, M_J, M'_J, m_j}} d_{l, j, m_j, J, M_J, M'_{J'}} |\text{Ion}^+_{\gamma', J', M'_J}\rangle \times |e^-_{\epsilon, l, J, j, m_j}\rangle.$$

All non-dipole contributions are neglected, because they either vanish in the detector plane shown in figure 6.3 or are smaller by several orders of magnitude. The number of independent parameters in the photoionisation is reduced by the LS-approximation [7, 8], that includes the angular momentum algebra for all angular momentum quantum numbers in the angular and spin parts of the wave function. In the LS-approximation the matrix elements and phases describing the state of the photoelectron depend only on the kinetic energy and the orbital angular momentum $l = l_0 \pm 1$ of the photoelectron where l_0 denotes the angular momentum of the shell where the electron is removed from. This assumption reduces the number of independent parameters to three as only two partial waves occur that are different in their radial parts. The coefficients $d_{l,j,m_j,J,M_J,M_{J'}}$ are known functions of the two radial integrals R_{l_0-1} R_{l_0+1} and total asymptotic phase difference Δ of the two partial waves [9]. For the p-ionisation we need to determine the matrix elements R_s, R_d and the phase difference Δ of the two partial waves. For the ionisation of a d-shell we need the matrix elements R_p and R_f and the phase difference between the f-and d-wave.

2. THE OBSERVABLES IN PHOTOIONISATION

β: use unpolarised atoms
and rotate spectrometer

$$I \propto (1 + \frac{\beta}{4}[1 + 3P_1 \cos(2\Theta)])$$

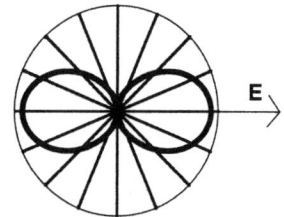

asymmetry: measure at
45 degrees the variation of
the intensity for different
directions of atomic polarisation

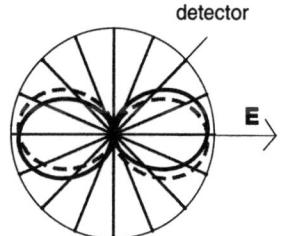

$$\text{asymmetry} = \frac{(I_{\uparrow\uparrow} - I_{\uparrow\downarrow})}{(I_{\uparrow\uparrow} + I_{\uparrow\downarrow})}$$

Figure 6.1 Measurements of β and the LMDAD-asymmetry. P_1 denotes the degree of linear polarisation of the radiation.

Generally angle integrated photoelectron detection, ion detection or absorption measurements do only depend on the matrix elements and not on the phase difference. Angle- and spin- resolved electron detection can yield quantities that additionally depend on the phase difference. The problem of absolute determination of the matrix elements by measuring the absolute cross section can be treated as a separate problem as all quantities in this experiment can be expressed as functions of the ratio $\gamma = R_{l_0+1}/R_{l_0-1}$ and Δ. The photoelectron angular distribution of unpolarised atoms yields the anisotropy parameter β. In polarized atoms the symmetry is lowered from spherical to axial one, and polarized atoms are characterized by an additional vector in the direction of polarization. In the experiment one observes a tilt of the angular distribution around this axis. (Compare figure 6.3). The maximum difference between the two electron count rates i. e. the Linear Magnetic Dichroism in the Angular Distribution (LMDAD) is observed for the detector placed at $\theta = 45°$ with respect to the linear polarisation of the ionising radiation. The anisotropy parameter β and the LMDAD-asymmetry calculated from the two count rates (see figure 6.1) include γ, $\cos(\Delta)$ and $\sin(\Delta)$ [5, 7]. So it is possible to determine γ and Δ from them. In case of ambiguous solutions theoretical assumptions or a further measurement is necessary to determine the correct solution. The

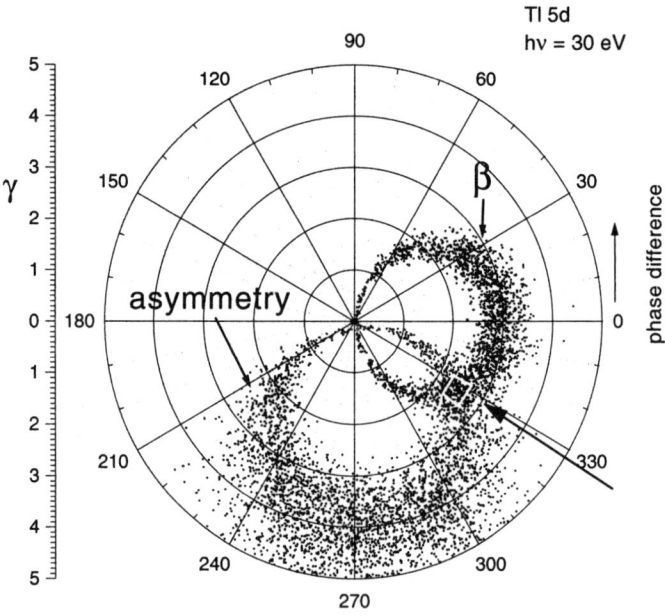

Figure 6.2 The determination of γ and Δ from β and the LMDAD asymmetry: The scatter plots represent the information given by the two measured quantities. The white frame at the crossing area of the two diffuse rings is the result of the combined information from β and the asymmetry. The second possible solution at the center with $\gamma < 1$ is ignored.

figure 6.2 illustrates how the ratio of matrix elements and the phase difference are determined from a combination of these two measurements.

2.1 EXPERIMENTAL SET-UP

The experiment was originally designed for atomic hydrogen and was then used for atomic oxygen [10–13] and later on modified for metal vapor measurements [14]. A vacuum chamber contains all components: The oven for the evaporation of the thallium, a skimmer to form the atomic beam, a hexapole magnet, a small rotatable electron spectrometer and the coils for the guid-

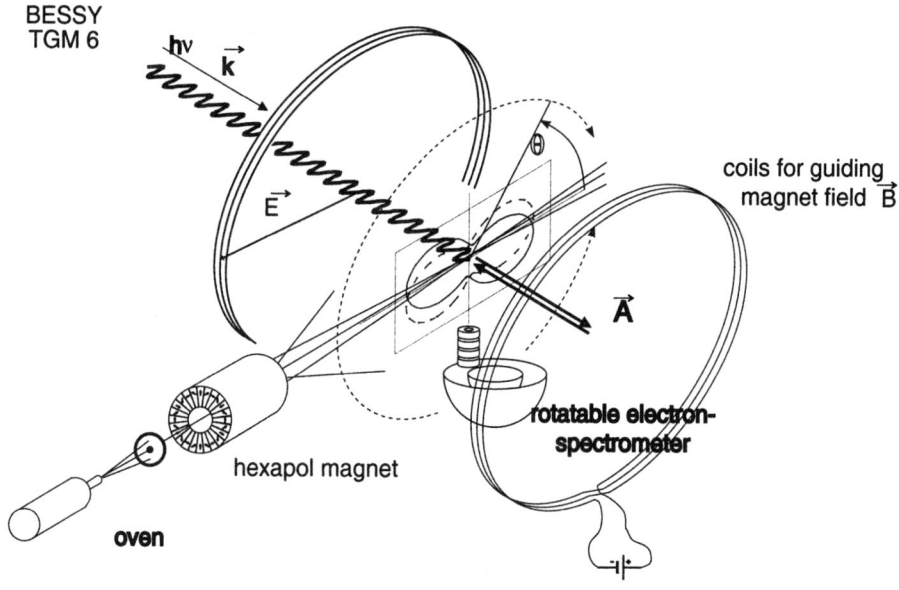

Figure 6.3 Apparatus for LMDAD measurement of polarised thallium atoms

ing magnet field. We used VUV radiation from the BESSY I U1 Undulator. Its most important properties are a flux of 10^{13} to 10^{14} photons/s, the degree of linear polarisation $P_1 > 0.95$ for photon energies from 30 to 70 eV and a monochromator band pass of 200 meV. After appropriate collimation, an atomic beam passes a hexapole magnet that focuses the atoms with positive magnetic quantum numbers M_J while those with negative M_J are defocused. The atoms leaving the strong field of the hexapole magnet approach adiabatically the weak field region. The vacuum chamber contains two concentric μ-metal cylinders in order to shield the magnetic earth field of 44 μT. The remaining magnetic field of 1 μT is further compensated by three orthogonal pairs of coils (in figure 6.3 only one pair is drawn). In order to define the orientation of the atomic polarisation and to have sufficient polarisation of the atoms at the crossing point of the atomic beam with the ionising synchrotron radiation, a guiding magnetic field of 2 μT is required, which is small enough to avoid a serious influence on the electron spectrometer. This influence was measured using unpolarisable targets like rare gases. This field parallel or antiparallel to the synchrotron beam axis is also provided by the three pairs of coils. A rotatable electrostatic hemispherical electron spectrometer is used for the detection of the photoelectrons; its mean radius is 50 mm and its mean energy resolution at 10 eV pass energy was measured to 150 meV. The guiding magnetic fields B for the polarisation direction of the atoms were parallel or antiparallel to the propagation vector $\vec{k}$ of

146   COMPLETE SCATTERING EXPERIMENTS

the incoming light. To avoid long term instabilities, the two photoelectron spectra where taken almost simultaneously, by inverting the magnetic field every second. For a precise knowledge of the degree of atomic polarisation a Monte Carlo simulation of the atomic beam is necessary. The depolarisation effects due to hyperfine interaction must be included, too. In this experiment the degree of atomic polarisation was 0.33 ± 0.02. The experiment for the measurement of the dynamical spin polarisation was performed in the same vacuum chamber. Instead of the apparatus generating the atomic beam of polarised atoms a small oven was placed next to the light beam. The Mott-scattering foil is also positioned at 45 degrees with respect to the plane of polarization of the ionising radiation. Spectra were measured with a Mott-Time of Flight spectrometer in

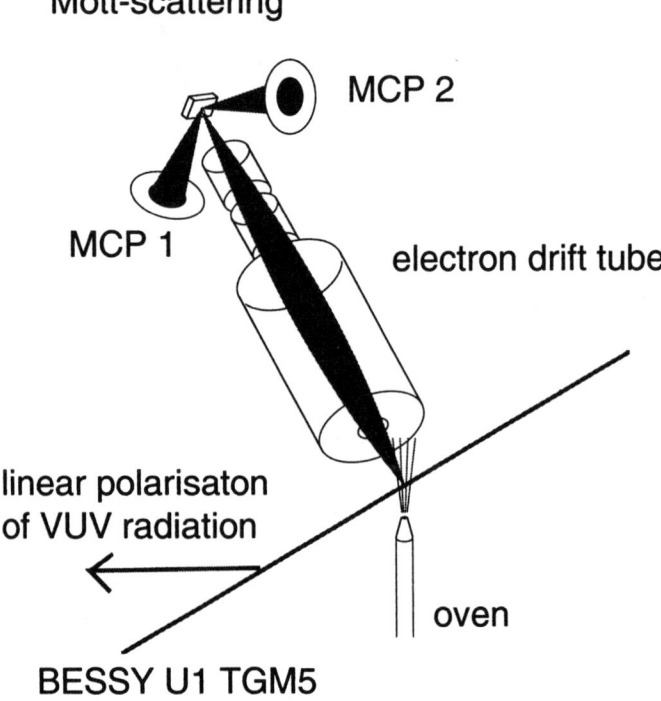

Figure 6.4   measurement of dynamical spin polarisaton

the BESSY I single bunch mode. The spin detection direction is defined by the positions of the MCPs. The dynamical spin polarisation is a quantity that is very similar to LMDAD. The spin detection axis corresponds to the axis of atomic polarisation in the LMDAD measurement and both quantities can be expressed in a very similar way as functions of γ and Δ. Both are proportional to $\sin(\Delta)$.

2.2 THE THALLIUM 5D SPECTRUM

Thallium has a $5d^{10}\ 6s^2\ 6p\ (^2P_{1/2})$ ground state. There is no significant population of the $^2P_{3/2}$ fine structure level for the vapour temperature of $800°C$. In the VUV region the 5d-, 6s- and 6p-subshells can be ionised. In the 5d- and 6p-ionisation non vanishing LMDAD was observed. The 6p electron carries

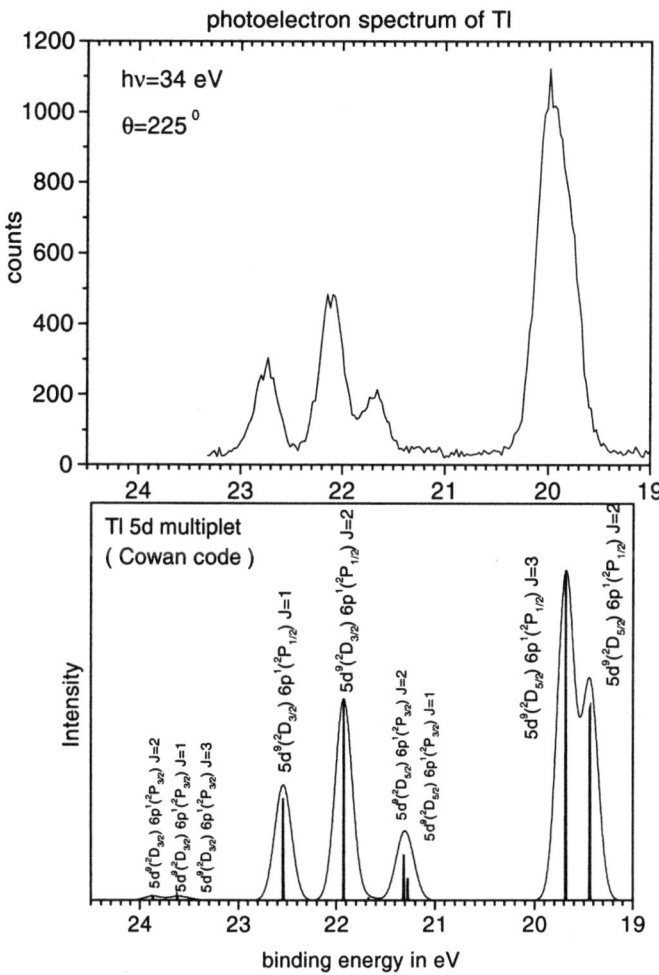

Figure 6.5 Tl 5d-photoelectron spectrum and the Cowan code results.

the polarisation of the atom. If it is removed during ionisation the ion remains in the $^1S_0$ state. For the ionisation of the 5d-subshell the situation is more complicated. It is a good approximation to describe the ion in jj-coupling. A $5d_{3/2}^{-1}$- or a $5d_{5/2}^{-1}$-hole can be created that couple with the 6p-electron to two different values of total angular momentum J. The observed intensities and positions of the main lines agree well with the Cowan Code [15] HF-calculations. The deviations are smaller than 10% for the branching ratios and less than 500 meV for the positions. $6p^1$ $^2P_{3/2}$ satellite lines are also present in the spectrum. This is a deviation from single-particle approximation, because these states can not be excited by dipole transitions in pure jj-coupling. They occur due to the mixing of Tl^+ $5d^9$ $(^2D_{3/2,5/2})$ $6p^1(^2P_{1/2})$-states with Tl^+ $5d^9$ $(^2D_{3/2,5/2})$ $6p^1(^2P_{3/2})$-states. A more detailed analysis of the Cowan Code calculation for the intermediate coupling of the ionic states shows that the LMDAD-asymmetry is not significantly affected by the intermediate coupling and pure jj-coupling model can be applied despite the presence of the satellite lines.

The Tl-$5d$ LMDAD-measurement

Figure 6.6 shows the RPAE-calculation for the $5d$-multiplet. The binding energies have been corrected to the optical data [16]. The bars in the lower part of the figure indicate the intensities that would have been measured with an atomic polarisation of 100%. The curves above include the atomic polarisation of 0.33 and the limited resolution of the spectrometer. The comparison with the measured data yields a good agreement with theory. In the middle part of figure 6.6 the difference of the two experimental spectra and the difference of the theoretical curves is compared. The theoretical curves are not scaled to fit the experimental difference but to the same total area in the multiplet.

Results of the partial wave analysis

The asymmetry of the $5d^9$ $(^2D_{3/2})$ $6p^1$ $(^2P_{1/2})$ $J = 1$ line has the highest value in the $5d$-multiplet and thus the highest statistical significance. The asymmetry of this line has been used to determine the matrix elements and phases of the partial waves. This line changes the sign of the asymmetry in the photon energy range used while the asymmetry of the $6p$-line varies only in magnitude (compare figures 6.7 and 6.8. Figure 6.7 shows the measured data for the $5d$-multiplet and the ratio of the matrix elements and the phase difference of the f- and p-wave. The RPAE-method was originally developed for closed shell atoms [17]. Vesnicheva et al [18] extended this method to the case of one unfilled subshell. The HF- and the RPAE-calculation do not refer to one special line of the $5d$-multiplet. Only the factors for the asymmetry are chosen for the $5d^9$ $(^2D_{3/2})$ $6p^1$ $(^2P_{1/2})$ $J = 1$ line. The intrinsic phase shift, i.e. the total phase shift without the Coulombic contribution matches the quantum defect difference at threshold as predicted by Seaton's theorem [19]. Experiment and theory agree very well. Only one of the solutions for the pair γ and Δ has been plotted in the figure. The solution with $\gamma < 1$ is ignored. For the $6p$-line the HF-and the RPAE-calculation yield completely different results,

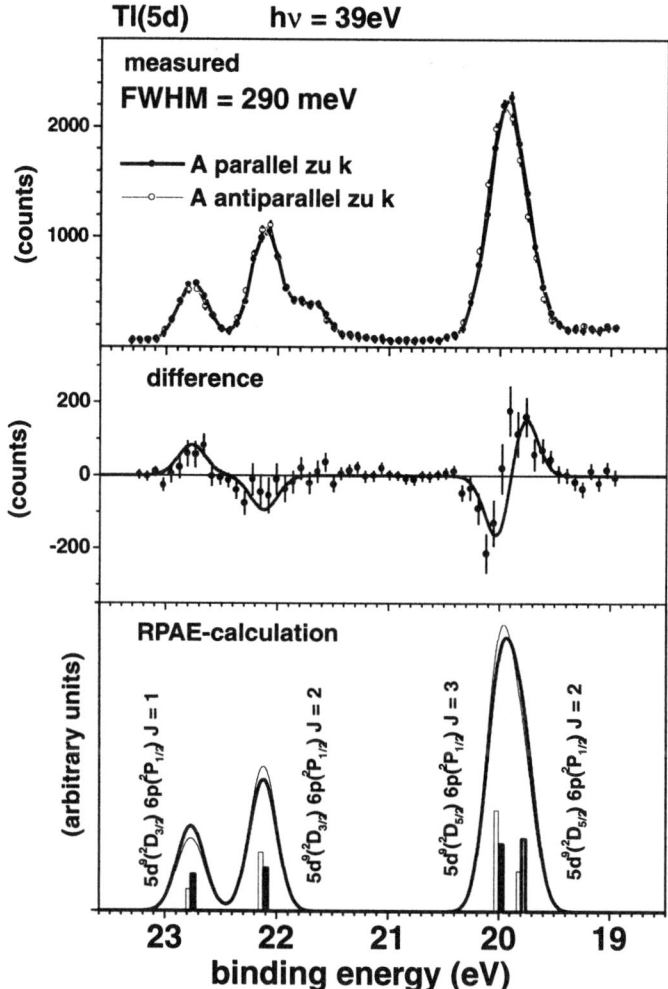

Figure 6.6 Measurement of the LMDAD asymmetry.

because in this case the interchannel coupling in the continuum can not be neglected. A trivial indication for the failure of the HF-method is the strength of the 6p-line. The HF-calculation suggests that the 6p-cross section should be lower by 2 or 3 orders of magnitude than the cross section of the 5d-multiplet and could never be measured in our experiment. For kinetic energies above 25 eV the values of Δ differs in the HF- and RPAE-calculation about $\pi/2$. We think that the coupling of the 5d- and 6p-continuum channels are responsible for that. The values measured for the LMDAD-asymmetry of the 6p-line agrees better with the RPAE-calculation than with the HF-calculation which yields the wrong sign for the asymmetry. Even though the experimental value for β sug-

150 COMPLETE SCATTERING EXPERIMENTS

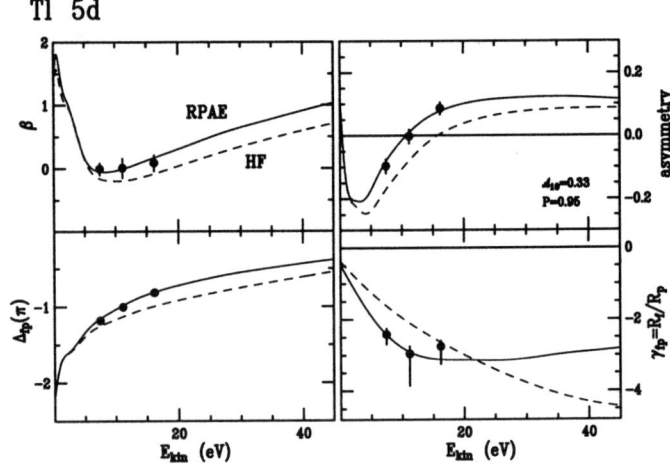

Figure 6.7 Results of the partial wave analysis for the Tl 5d-line

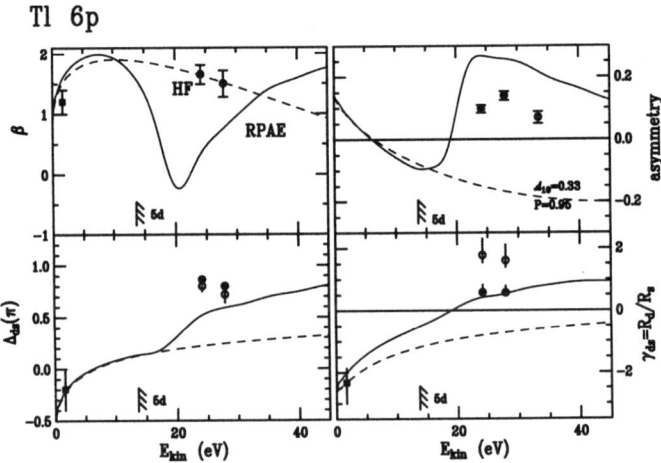

Figure 6.8 Results of the partial wave analysis for the Tl 6p-line. The data points close to threshold are the non-resonant data from [20]

gest that the HF-calculation is better, the combined information from β and the LMDAD-asymmetry in terms of γ and Δ clearly favor the RPAE-calculation. The phase shift changes very rapidly at the 5d-threshold. This situation is further complicated by resonances in this energy region, which are not included in the calculations and might cause the deviations between experiment and calculation. Both solutions of the partial wave analysis are shown in figure 6.8. Both favor the RPAE-results. So it is very likely that matrix element of the

d-wave really crosses zero at this low energy and not at 100 eV as predicted by the HF-calculation.

2.3 RESULTS FOR THE DYNAMICAL SPIN POLARISATION

The completeness of this experiment relies on the validity of LS-coupling. To verify this at least one redundant measurement is necessary. We measured the dynamical spin polarisation of the whole $5d$-multiplet. Figure 6.9 shows the calculation for the $5d$-multiplet. Matrix elements and phase differences are the same that are used to describe the LMDAD measurements. The bars in the lower part of the figure indicate the intensities that would have been measured with a perfect Mott-detector. The limited spin selectivity of the Mott-scattering is described by the Shermann function. It reduces the measurable effects just like the limited polarisation of the target atoms reduces the measured asymmetry. The curves above the bars include the Sherman function and the limited resolution of the spectrometer. The comparison with the measured data yields a good agreement with theory. The theoretical difference curve is not scaled to fit the experimental difference but to the same total area in the multiplet.

2.4 DIFFERENCES BETWEEN THE LMDAD-ASYMMETRY AND DYNAMICAL SPIN POLARISATION

Despite the close connection of LMDAD-asymmetry and dynamical spin polarisation they obey completely different symmetry rules within the multiplet. This becomes obvious when the figures 6.6 and 6.9 are compared. The calculation of the size of the spin polarisation or the asymmetry is very cumbersome, but the rules which lines show the same or opposite effect are quite simple. In the spectator model the creation of the $^2D_{3/2}$-hole is independent from the polarisation in the $6p$-electron but the measurement of the kinetic energy which is a measurement of J means a detection of the $^2D_{3/2}$-hole polarisation when the direction of the polarisation of the $6p$ electron is fixed. So the two $^2D_{3/2}$-components must have opposite asymmetry. As the $^2D_{3/2}$-hole are created without a preferred orientation along the k-vector of the light, no dichroism is left after averaging over the two $^2D_{3/2}$-components of the multiplet. The same is true for the $^2D_{5/2}$-hole. The situation is completely opposite for the detection of the electron spin for unpolarised atoms. In this case the measurement of J is no selection of the direction of the hole. The spin polarisation of the two $^2D_{3/2}$-components of the multiplet have the same value. The same is true for the $^2D_{5/2}$-lines of the multiplet.

152 COMPLETE SCATTERING EXPERIMENTS

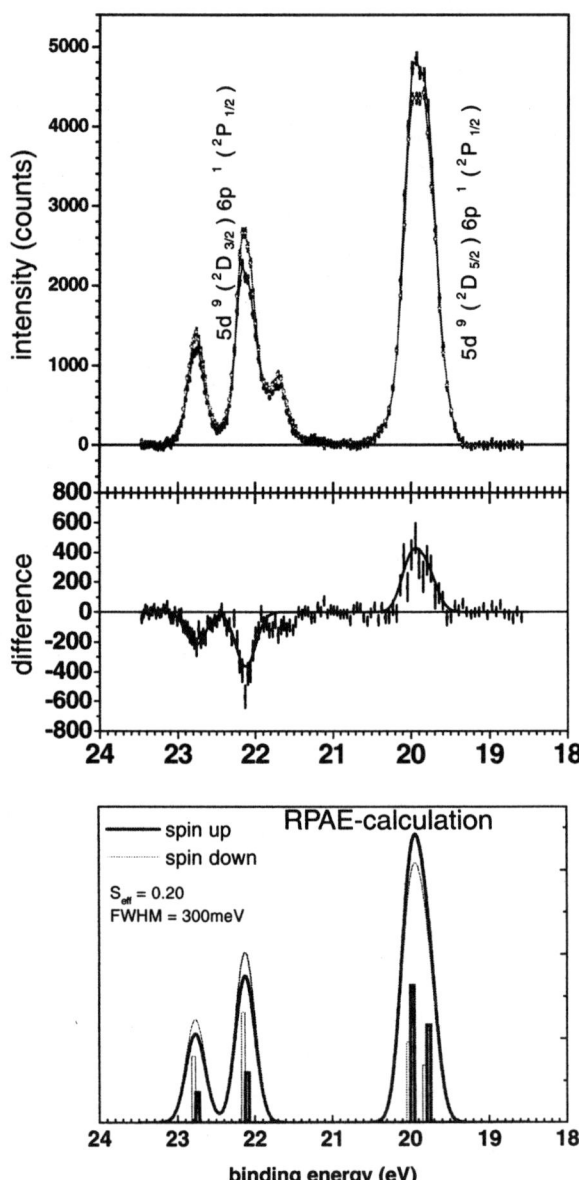

Figure 6.9 Measurement of the dynamical spin polarisation.

2.5 SUMMARY

A partial wave analysis has been performed successfully for the Tl $5d$- and $6p$-ionisation. This is a quantum mechanically complete description provided the LS-approximation is valid. The experiment is in good agreement with the RPAE-calculation for the $5d$-shell. This description provides a good prognosis for the dynamical spin polarisation. For the $6p$-shell only a qualitative discussion was possible as the continuum coupling and autoionising resonances complicate the situation. However a strong deviation from the HF-calculation was observed in the LMDAD.

Acknowledgments

This work was supported in part by the Deutsche Forschungsgemeinschaft. One of us (H. K.) acknowledges support by the BESSY European Union TMR Program. The authors are most greatful to B. Langer, Professor U. Heinzmann and his group, and N. A. Cherepkov for their help with the measurements and the calculation.

References

[1] U. Fano, Rev. Mod. Phys. **29**, 74 (1957).
[2] B. Bederson, Comments At. Mol. Phys. **1**, 41 (1969).
[3] B. Bederson, Comments At. Mol. Phys. **2**, 65 (1969).
[4] H. Klar and H. Kleinpoppen, J. Phys. B: At. Mol. Opt. Phys. **15**, 933 (1982).
[5] N. A. Cherepkov, V. V. Kuznetsov, and V. A. Verbitskii, J. Phys. B: At. Mol. Opt. Phys. **28**, 1221 (1995).
[6] U. Heinzmann and N. A. Cherepkov, in *VUV- and soft X-ray photoionization*, edited by U. Becker and D. A. Shirley (Plenum Press, New York, 1996), p. 521.
[7] J. Cooper and R. N. Zare, J. Chem. Phys. **48**, 942 (1968).
[8] J. W. Cooper and R. N. Zare, in *Lectures in Theoretical Physics*, edited by S. Geltman, K. T. Mahanthappa, and W. E. Britting (Gordon & Breach, Science Publ., New York, 1969), Vol. IIC, p. 317.
[9] K. Blum, *Density Matrix Theory and Applications* (Plenum Press, New York and London, 1981).
[10] O. Plotzke *et al.*, Phys. Rev. Lett. **77**, 2642 (1996).
[11] O. Plotzke, Ph.D. thesis, Technische Universität Berlin, 1997.
[12] G. Prümper *et al.*, J. Phys. B: At. Mol. Opt. Phys. **30**, L683 (1997).
[13] G. Prümper *et al.*, Europhysics Letters **38**, 19 (1997).
[14] G. Prümper *et al.*, Phys. Rev. Lett. (2000), submitted.

[15] R. D. Cowan, *The Theory of Atomic Structure and Spectra* (University of California Press, Berkeley, Los Angeles, London, 1981).

[16] C. E. Moore, *Atomic Energy Levels* (Natl. Bur. Stand. (U.S. GPO), Circ. 467, Washington, DC, 1949).

[17] M. Y. Amusia and N. A. Cherepkov, Case Studies in Atomic Physics **5**, 47 (1975).

[18] G. A. Vesnicheva, G. M. Malyshev, V. F. Orlov, and N. A. Cherepkov, Sov. Phys. Tech. Phys **31**, 402 (1986).

[19] M. J. Seaton and G. Peach, Proc. Phys. Soc. London **79**, 1296 (1962).

[20] N. Müller, N. Böwering, A. Svensson, and U. Heinzmann, J. Phys. B: At. Mol. Opt. Phys. **23**, 2267 (1990).

Chapter 7

ON THE CONTRIBUTION OF PHOTOELECTRON AUGER ELECTRON COINCIDENCE SPECTROMETRY TO COMPLETE PHOTOIONIZATION STUDIES

N. Scherer
Fakultät für Physik, Universität Freiburg, D-79104 Freiburg, Germany

S. J. Schaphorst
Fakultät für Physik, Universität Freiburg, D-79104 Freiburg, Germany
Now at: Delta Information Systems, 300 Welsh Road, Horsham PA 19044-2273

V. Schmidt
Fakultät für Physik, Universität Freiburg, D-79104 Freiburg, Germany

Abstract Angular correlation patterns of $4d_{5/2}$ photoelectrons and N_5-$O_{2,3}O_{2,3}$ $^1S_0$ Auger electrons in xenon are investigated with respect to their sensitivity to the involved photoionization matrix elements and relative phases. We propose a measurement of two complementary patterns for which experimental matrix elements and relative phases can be extracted unambiguously within and possibly beyond the LSJ coupling description (three-parameter model).

1.   INTRODUCTION

Within a theoretical model, an experiment that provides the maximum information about the underlying process uniquely characterizes a pure state and is called a "complete" (Fano 1957 [1]) or "perfect" (Bederson 1969 [2], Kessler 1981 [3]) experiment. Hans Kleinpoppen and coworkers were the first to report complete experiments in atomic physics, in the experimental determination of complex excitation amplitudes in electron-helium collisions by means of angular correlation measurements using a coincidence method (Eminyan et al 1973 [4]). This study initiated an enormous and extremely exciting field of research for many different processes of interactions with atoms. (In addition, the topic of the first study is still of current interest, see for example Yalim et al 1997 [5].) It is an honour for us to be included in the diverse range of studies contributing to the "Hans Kleinpoppen Symposium on Complete Scattering Experiments". We will evaluate a particular photoelectron Auger electron coincidence experiment with respect to its contribution to and elucidation of complete photoionization studies.

The first complete experiments in photoionization were performed by Heinzmann [6, 7] by measuring the spin polarization of $5p_{3/2}$ and $5p_{1/2}$ photoelectrons in xenon, emitted by unpolarized and circularly polarized VUV radiation. Within the dipole approximation photoionization in a closed-shell atom or in a one-electron subshell can have at most three complex transition amplitudes because of the selection rule $\Delta J = \pm 1$ or 0. Therefore, the three spin polarization parameters, ξ, η, ζ, together with known values for the partial photoionization cross section σ and the angular distribution parameter β of the photoelectrons give a sufficient number of independent observables for a complete determination of the three complex photoionization matrix elements: namely, three real values of the matrix elements and two relative phases (an absolute phase is irrelevant for observables). Again, this experiment prompted a plethora of complete photoionization studies, including the exploration of different approaches designed to avoid the cumbersome detection of spin polarization (for a recent review see Becker and Langer 1998 [8]).

Complete photoionization experiments are of interest for several reasons. The comparison of matrix elements and relative phases extracted from complete experiment with theoretical predictions provides the most sensitive and therefore ultimate test of the underlying theory. In particular, direct insight can be gained into the influences of electron-electron interactions in the photoionization process. An example for 2p photoionization in magnesium is discussed by Schmidt 1997 [9]. Furthermore, as the name implies, the matrix elements and relative phases determined in a particular complete experiment allow the derivation of numerical values for the observables of any other experiment. One application is described by Lörch et al 1998 [10], in which the known matrix elements and their relative phase for 2p photoionization in magnesium are used to predict the shape of photoelectron Auger electron angular correlation patterns for arbitrary polarization of the incident light, and thereby determine the polarization of the incident light. A few studies have pointed out, however, that one must be aware of trivial dependencies on angular functions in which case the angular correlation patterns provide no information on the transition matrix elements (see Végh and Becker 1992 [11], Kämmerling and Schmidt 1993 [12], Schmidt 1994 [13]). Finally, the critical analysis performed by Becker and Langer 1998 [8] for available data of complete photoionization experiments gives strong evidence that a nonrelativistic Cooper-Zare model (LSJ coupling case) is valid for a large number of cases, even for the system selected here.

In the present communication we concentrate on a complete study of $4d_{5/2}$ photoionization in xenon, based on a coincidence measurement between $4d_{5/2}$ photoelectrons and subsequent N_5-$O_{2,3}O_{2,3}$ $^1S_0$ Auger electrons. This experiment showcases the potential of such studies (Kämmerling and Schmidt 1991 and 1993 [12, 14, 15]), but also demonstrates the difficulties in the extraction of the desired matrix elements and phases from the observables (Kämmerling and Schmidt 1993 [12], Schaphorst et al 1997 [16]). Here we will concentrate on the results of a detailed investigation into the sensitivity of the coincident angular correlation patterns to small changes in the matrix elements and phases involved. First we discuss general information on the selected process. Then we recapitulate the results of our previous experiment and point out the need of more sensitive experiments. Finally, with this information we propose an experiment in which two complementary angular correlation patterns with special angles for the fixed electron emission are measured in order to unambiguously determine the relevant matrix elements and relative phases.

2.   SELECTED PROCESS

$4d_{5/2}$ photoionization in xenon with subsequent N_5-$O_{2,3}O_{2,3}$ $^1S_0$ Auger decay represents a special case of two-electron emission because direct double photoionization is negligible and the indirect sequential process can be described well by a two-step model if the energies of the two electrons are sufficiently different (and effects of post-collision interaction can be neglected). Both points lead to a considerable simplification in the theoretical description, because the observables depend then on dynamical parameters which separate for photoionization and Auger decay, respectively. In addition, for the Auger transition selected here the Auger parameters are simply the Auger yield ω_A (important for absolute intensities, $\omega_A = 0.053 \pm 0.005$) and known coefficients from angular momentum couplings. Therefore, the relevant matrix elements for $4d_{5/2}$ photoionization and subsequent N_5-$O_{2,3}O_{2,3}$ $^1S_0$ Auger transition in xenon are those of the photoionization process only, and a coincidence experiment between $4d_{5/2}$ photoelectrons and N_5-$O_{2,3}O_{2,3}$ $^1S_0$ Auger electrons can provide information on these elements. As explained above, three complex dipole amplitudes are necessary,

$$D_+ = d_+ e^{i\Delta_+}, \quad D_0 = d_0 e^{i\Delta_0}, \quad D_- = d_- e^{i\Delta_-}, \qquad (7.1)$$

to describe the photoionization channels $4d_{5/2} \to \epsilon f_{7/2}$, $4d_{5/2} \to \epsilon f_{5/2}$, and $4d_{5/2} \to \epsilon p_{3/2}$, respectively. Since only relative phases ($\Delta_{ij} = \Delta_i - \Delta_j$) are of relevance for the observables, five parameters provide a complete description.

In the LSJ coupling limit the constraints

$$\Delta_{0+} = 0 \quad \text{and} \quad d_0/d_+ = 1/\sqrt{20} \qquad (7.2)$$

are imposed, and a three-parameter model with two matrix elements and one relative phase is obtained.

To evaluate the dependencies of the observed two-electron emission on the transition matrix elements, a suitable parametrization formula and a convenient graphical representation are needed. Here we concentrate on the angular correlation between the two emitted electrons measured in coincidence. For reasons of simplicity and without loss of generality, we restrict further discussion to complete linear polarization of the incident light and to the observation of the electrons in a plane perpendicular to the photon beam direction. In this case the angular correlation $P(\theta_1, \theta_2)$ depends only on the angle settings θ_1 and θ_2 of the electron spectrometers, where these angles are measured relative to the electric field vector. Usually, one electron spectrometer (index 1) is kept at a fixed position, whilst the angle of the other (index 2) is varied. Angular correlation patterns are recorded by plotting the coincident intensity of the Auger and photoelectrons in a polar diagram. Two distinguishable and complementary patterns are obtained by detecting either the photoelectron or the Auger electron with the spectrometer in fixed direction. Since absolute intensities are not measured, only the shapes of these patterns are of interest.

3. RESULTS FROM A PARTICULAR CORRELATION PATTERN

In a recent experiment at 132.2 eV photon energy the correlation pattern for $4d_{5/2}$ photoionization in xenon with N_5-$O_{2,3}O_{2,3}$ $^1S_0$ Auger decay has been measured keeping the angle of Auger electron emission fixed, ($\theta_1 = \theta_A = 149°$; Schaphorst et al 1997 [16]). Difficulties in extracting the desired matrix elements from the observables were discussed and finally led to three conclusions which are partly reproduced here and are described in the upper part of Table 7.1:

(i) The "best-fit" results (set #4 in the table) with the smallest χ^2 value are not compatible with the theoretical assertion of $\Delta_{0+} = 0$.

(ii) The experimental data are equally well reproduced for a "recommended" data set with Δ_{0+} fixed close to zero (set #18), albeit leading to a slightly higher χ^2 value. This data set, in full agreement with negligible spin-orbit effects in the photoelectron continuum, also gives a d_0/d_+ ratio close to $1/\sqrt{20}$. Therefore, this data set is compatible with the three-parameter model.

(iii) The "recommended" data set is rather close to "unrelaxed RPA" results (set #2*, Cherepkov 1979 and 1998 [17]), but differs significantly from the predictions of relaxed RRPA calculations (set #3*, Johnson and Cheng 1992 [18]) which also fulfills the LSJ coupling limit relations and, therefore, shall be termed here "relaxed RPA" results.

Table 7.1 Matrix element ratios and relative phases for $4d_{5/2}$ photoionization in xenon. The five-parameter description contains three real values d_+, d_0, d_- and two relative phases Δ_{ij} (in rad; two phases fix the third which is listed for convenience). Since only the shape of the angular correlation pattern is of interest here, the d_+ value is used for normalization, i.e., ratios d_0/d_+ and d_-/d_+ are listed. Different sets of relative matrix elements and phases are shown in the different columns. The #-numbers refer to those in Table 2 of Schaphorst et al 1997 [16], except for #2* ("unrelaxed RPA" results from Cherepkov 1979 [17]), and for #3* ("relaxed RPA" results; average of length and velocity results which were mistakenly attributed in Table 2 of Schaphorst et al [16] as relaxed and unrelaxed RRPA results); #4 is the "best-fit" result, #18 the "recommended" data set. The dev_i values represent deviations from a three-parameter model as defined by eq. (7.4) and described in the main text.

	reference data set	unrelaxed RPA	relaxed RPA	best-fit result	recommended data set
	#0	#2*	#3*	#4	#18
d_0/d_+	0.224	0.224	0.224	0.115	0.21
d_-/d_+	0.360	0.367	0.295	0.299	0.36
Δ_{0-}	2.6	2.38$^{(a)}$	2.27	3.32	2.6
Δ_{+-}	2.6	2.38$^{(a)}$	2.27	3.06	2.6
Δ_{0+}	0.0	0.0	0.0	0.27	0.0
dev_1	0.0	0.0	0.0	-0.109	-0.014
dev_2	0.0	0.007	-0.065	-0.061	0.0
dev_3	0.0	0.0	0.0	0.27	0.0
dev_4	0.0	-0.22	-0.33	0.72	0.0

$^{(a)}$ The phase difference calculated by Cherepkov is -7.048 rad which is equivalent to -0.76 rad; it is augmented here by π in order to adapt to the different equations used to describe the dependences of observables on the matrix elements and their relative phases, in the present case to the equation for the angular distribution parameter β (compare eq.(32) of Cherepkov (1979) [17] with eq.(II.19) of Cooper and Zare [19]).

The selected angular correlation pattern is given by (Schaphorst *et al* 1997 [16])

$$P(\theta_P, \theta_A = 149°) = A_0 + A_2 \cos 2\theta_P + A_4 \cos 4\theta_P + A_6 \cos 6\theta_P$$
$$+ B_2 \sin 2\theta_P + B_4 \sin 4\theta_P + B_6 \sin 6\theta_P \quad (7.3)$$

where the A- and B-coefficients depend in particular forms on the matrix elements and their relative phases. The complex dependence of the pattern on the six individual angle-dependent terms prevents a simple explanation for the effects of the transition matrix elements and phases on the shape of the observed pattern. In other words it can be difficult to ascertain the sensitivity of such angular correlation patterns to the dynamical parameters involved, and in the work of Schaphorst *et al* 1997 [16] no error bars could be assigned to the extracted matrix elements and relative phases. This unsatisfactory result has prompted us to search for angular correlation patterns which are more sensitive to these parameters, thus allowing a better evaluation of the results from different theoretical models.

4. PROPOSED STUDY

By performing many calculations of angular correlation patterns with different angles for the fixed electron (photoelectron or Auger electron) and by allowing systematic variations of the matrix elements and phases, we finally found that two complementary patterns with a fixed angle $\theta_1 = 90°$ are promising candidates for a detailed investigation. Changes in the matrix elements and/or phases have the largest effect on the shape of the angular correlation patterns at this θ_1 value. To elucidate this property and to show more clearly the influences of the parameters involved, we introduce a "reference" data set, #0 in Table 7.1 (close to the "recommended" data set), which fulfills the conditions of the three-parameter model exactly. Then we introduce in the five-parameter description deviations from the three-parameter model using the relations

$$dev_1 = d_0/d_+ - 0.224, \quad dev_2 = d_-/d_+ - 0.36,$$
$$dev_3 = \Delta_{0+}, \quad dev_4 = \Delta_{0-} - 2.6 (rad), \quad (7.4)$$

and we expand the A- and B-coefficients of the angular correlation function up to first order in these dev_i. This linearization approximation leads to parametrizations which can be discussed with respect to the effects of changes in the dev_i on the shape of the angular correlation pattern. With this method we hope to be able to rule out certain sets of matrix elements and phases by comparing the expected patterns with experimental data. Answers to four questions are of particular interest:

(A) Can the unrealistic "best-fit" result be ruled out?

(B) Can the disagreement between the "recommended" data set and the "relaxed RPA" results be confirmed?

(C) Can the slight difference between the "recommended" data set and the "unrelaxed RPA" results be experimentally detected?

(D) Can deviations from the three-parameter model be observed?

The simplest angular correlation pattern is presented for $\theta_P = 90°$ for which

$$P(\theta_P = 90°, \theta_A) = b_0 + b_2 \cos 2\theta_A + b_4 \cos 4\theta_A \qquad (7.5)$$

with

$$\begin{aligned}
b_0 &= d_+^2 \; 0.273 \;(\;\;1.0 \;\;\; -0.553 \; dev_1 - 2.283 \; dev_2 + 1.012 \; dev_3 - 0.969 \; dev_4)\\
b_2 &= d_+^2 \; 0.273 \;(\;\;\;\;\;\;\;\;\;\;\;\;\; 1.382 \; dev_1 \;\;\;\;\;\;\;\;\;\;\;\;\;\;\; -0.107 \; dev_3 \;\;\;\;\;\;\;\;\;\;\;\;\;\;)\\
b_4 &= d_+^2 \; 0.273 \;(-0.384 \; -1.382 \; dev_1 + 0.878 \; dev_2 - 0.266 \; dev_3 + 0.373 \; dev_4).
\end{aligned}$$
(7.5b)

The analysis shows that the shape (not the magnitude) of the pattern does not

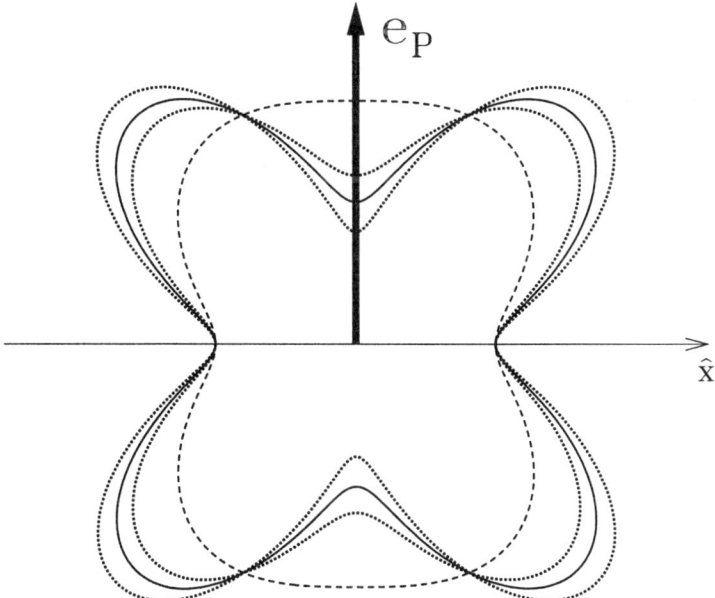

Figure 7.1 Angular correlation patterns for $4d_{5/2}$ photoionization in xenon and subsequent N_5-$O_{2,3}O_{2,3}$ $^1S_0$ Auger decay, for 132.2 eV linearly polarized photons (field vector along $\hat{x}$), observed in the perpendicular plane geometry, keeping the direction of the photoelectron (P) fixed. All patterns are predicted within the linearization method described in the main text, using the following data sets from Table 7.1: solid curve: "reference" set (includes "unrelaxed RPA", "relaxed RPA" and "recommended" data sets within drawing uncertainty); dashed curve: "best-fit" results; inner dotted curve: "reference" set, but d_0/d_+ decreased by 20%; outer dotted curve: "reference" set, but d_0/d_+ increased by 20%.

162   COMPLETE SCATTERING EXPERIMENTS

change if the condition

$$dev_1 = 0.077 \, dev_3 , \qquad (7.6)$$

is fulfilled, because for this case $b_2 = 0$ and $b_4 = -0.384 \, b_0$. Using the dev_i listed in the lower part of Table 7.1 for the matrix elements and relative phases of interest, one can verify that this condition holds exactly for the "unrelaxed RPA" and the "relaxed RPA" results, but not for the other data sets. How strongly the angular correlation patterns will change for these other data sets can only be answered by comparing the calculated shapes. The results are shown in Figure 7.1 where the solid line is the "reference" pattern. Within drawing uncertainty, this pattern agrees with the one from the "recommended" data set; however, the dashed curve, representing the previous "best-fit" data, displays a rather distinct shape arising from differences in both the d_0/d_+ ratio and the phases. As a result, the answer to question (A) is clearly "yes", but (B) and (C) must be answered with "no".

In addition, observable changes might occur for data sets beyond the three-parameter model as shown in Figure 7.1 for the matrix element ratio d_0/d_+ assumed to be 20% larger or smaller (dotted curves). These dotted curves would be slightly modified further, if deviations in Δ_{0+}, i.e., $dev_3 \neq 0$ are taken into account. Hence, the answer to (D) is "yes" only in a four-parameter model in which Δ_{0+} is still restricted to zero, but d_0/d_+ can deviate from the value of the LSJ coupling limit.

Now we concentrate on the complementary pattern in which the Auger electron is fixed at an angle $\theta_A = 90°$ and for which we derive

$$P(\theta_P, \theta_A = 90°) = a_0 + a_2 \cos 2\theta_P + a_4 \cos 4\theta_P + a_6 \cos 6\theta_P \qquad (7.7)$$

with

$$\begin{aligned}
a_0 &= d_+^2 \, 1.030 \, (\ 1.0 \ \ +0.424 \, dev_1 + 0.319 \, dev_2 + 0.035 \, dev_3 - 0.010 \, dev_4) \\
a_2 &= d_+^2 \, 1.030 \, (\ 0.199 + 0.201 \, dev_1 + 0.911 \, dev_2 - 0.223 \, dev_3 + 0.195 \, dev_4) \\
a_4 &= d_+^2 \, 1.030 \, (-0.280 \, -0.642 \, dev_1 + 0.228 \, dev_2 - 0.035 \, dev_3 + 0.049 \, dev_4) \\
a_6 &= d_+^2 \, 1.030 \, (\ 0.347 + 0.444 \, dev_1 \hspace{5.5cm}) .
\end{aligned}$$
(7.7b)

For this case the analysis shows that the shape of the patterns does not change if dev_1 and dev_3 obey the following relations

$$dev_1 = 1.63 \, (0.423 \, dev_4 - dev_2) , \quad dev_3 = 2.64 \, (0.423 \, dev_4 + dev_2) . \qquad (7.8)$$

Inspecting the dev_i in Table 7.1 one can see that none of the data sets fulfills both requirements (a trivial exception is the "reference" data set), but different degrees of violation exist: strong violations of both conditions for the "best-fit" results, a strong violation of the dev_3 condition for the "relaxed RPA" data, and decreasing degrees of violations for the other sets. Hence, for the angular correlation patterns based on the previous "best-fit" results and based on the "relaxed RPA" data one can expect strong deviations from the "reference" pattern, and for the other data sets the corresponding calculations must show particular sensitivities. The results are plotted in Figure 7.2 where the solid curve is again the "reference" pattern (note that this pattern has more structure than that shown in Figure 7.1, because this case has an additional term dependence on $\cos 6\theta_P$ which can be traced to the higher partial wave $\ell = 3$ of the photoelectron).

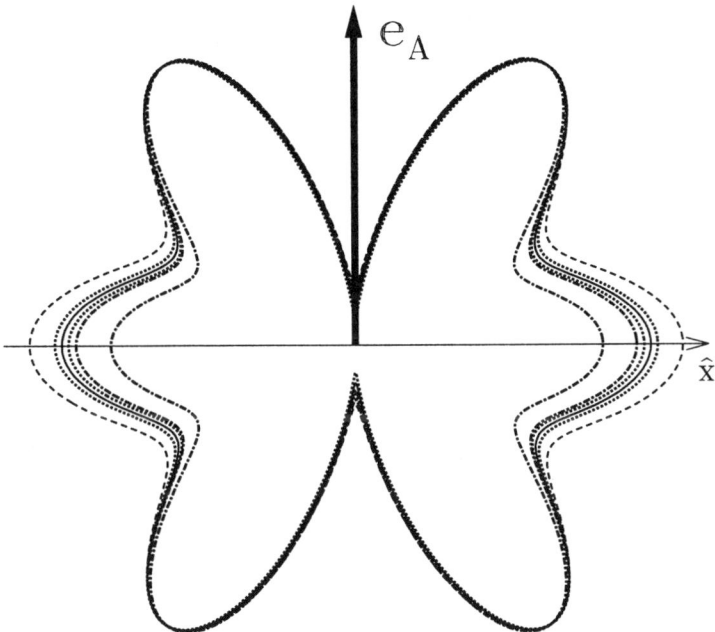

Figure 7.2 Angular correlation patterns for $4d_{5/2}$ photoionization in xenon and subsequent N_5-$O_{2,3}O_{2,3}$ $^1S_0$ Auger decay, for 132.2 eV linearly polarized photons (field vector along $\hat{x}$), observed in the perpendicular plane geometry, keeping the direction of the Auger electron (A) fixed. All patterns are predicted within the linearization method described in the main text, using the following data sets from Table 7.1: solid curve: "reference" set (includes "recommended" data set within drawing uncertainty); dashed curve: "best-fit" data set; inner dashed-dotted curve: "relaxed RPA" result; outer dashed-dotted curve: "unrelaxed RPA" result; inner dotted curve: "reference" set, but d_0/d_+ increased by 20%; outer dotted curve: "reference" set, but d_0/d_+ decreased by 20%.

Clearly one can see in Figure 7.2 the rather distinct shapes for the previous "best-fit" data (dashed curve) and the "relaxed RPA" data (inner dashed-dotted curve), as compared to the "reference" pattern. Therefore, questions (A) and (B) can be answered with a definite "yes".

To differentiate between the solid curve, the outer dashed-dotted curve, and the two dotted curves, experimental data would need sufficiently high accuracy since these patterns are seen to lie close to one another. The outer dashed-dotted pattern in Figure 7.2 is based on "unrelaxed RPA" results. Its closeness to the "reference" pattern already reflects the rather good agreement of the two data sets. In this context it should be noted that there still exists an inconsistency between experiment and theory for the phase-dependent angular distribution parameter of non-coincident photoelectrons: $\beta(relaxed\ RPA) = 1.25$, $\beta(unrelaxed\ RPA) = 1.41$, but $\beta(experiment) = 1.54 \pm 0.02$. The small difference in the pattern using the "unrelaxed RPA" data to the "reference" pattern can only be attributed unambiguously to the relative phase Δ_{0-}, if the three-parameter model is applied, because otherwise this effect is hidden by the influences discussed in the following for deviations beyond the three-parameter model. Therefore, no simple answer can be given to question (C). The two dotted curves demonstrate such deviations beyond the three-parameter model, namely $\pm 20\%$ deviations in d_0/d_+. In addition there exist similar changes in the shape of these pattern s if deviations in Δ_{0+} are present. This is in contrast to the complementary pattern discussed above where the influence of Δ_{0+} was rather small. In the present case a variation of $\Delta_{0+} = \pm 0.2$ rad leads to modifications of equal order of magnitude as those caused by the 20% differences in d_0/d_+ and, depending on the sign of Δ_{0+}, these contributions shift the dotted curves towards or away from the solid curve. Therefore, combined with the outcome for d_0/d_+ from the pattern of Figure 7.1 it might be possible to answer question (D) with "yes".

Summarizing the results for the complementary angular correlation patterns for $\theta_A = 90°$ and $\theta_P = 90°$ with respect to questions (A) to (D):

1. The measurements should clearly distinguish the "reference" data set (which is almost identical with the "recommended" data set) from the previously obtained unrealistic "best-fit" result.

2. The measurements should clearly distinguish the "reference" data set from the "relaxed RPA" result.

3. Only within the non-relativistic three-parameter model and for sufficiently accurate experimental data it will be possible to distinguish the "reference" data set from the "relaxed RPA" results.

4. For sufficiently accurate data it should be possible to distinguish between the "reference" data set and significant deviations in d_0/d_+, i.e., to decide whether a four-parameter model is necessary. In special cases it might be even possible to test the five-parameter model which allows deviations of Δ_{0+} from zero.

5. CONCLUSIONS

Our study demonstrates that difficulties in the extraction of matrix elements and relative phases from experimentally observed angular correlation patterns of coincident photoelectrons and Auger electrons can be overcome by properly selecting the geometry of the experimental setup. Since difficulties in the extraction of matrix elements and phases have also been reported for experiments based on the powerful method of spin-resolved photoelectron and Auger electron spectrometry (Snell 1997 [20], Snell *et al* 1999 [21]), the technique of electron-electron coincidence spectrometry should provide an important contribution to complete experiments. This implies, in particular, the possibility to measure effects beyond the LSJ coupling description (three-parameter model), even though such measurements require an extremely high accuracy for the experimental data.

Acknowledgments

It is our pleasure to thank Dr. N. A. Cherepkov and Dr. B. Zimmermann for communicating to us the unrelaxed random-phase-approximation results.

References

[1] U. Fano, Rev. Mod. Phys. **29**, 74 (1957).

[2] B. Bederson, Comments At. Mol. Phys. **1**, 41 (1969).

[3] J. Kessler, Comments At. Mol. Phys. **10**, 47 (1981).

[4] M. Eminyan, K. B. MacAdam, J. Slevin, and H. Kleinpoppen, Phys. Rev. Lett. **31**, 576 (1973).

[5] H. A. Yalim, D. Cvejanovic, and A. Crowe, Phys. Rev. Lett. **79**, 2951 (1997).

[6] U. Heinzmann, J. Phys. B: At. Mol. Opt. Phys. **13**, 4353 (1980).

[7] U. Heinzmann, J. Phys. B: At. Mol. Opt. Phys. **13**, 4367 (1980).

[8] U. Becker and B. Langer, Phys. Script. **T78**, 13 (1998).

[9] V. Schmidt, *Electron Spectrometry of Atoms using Synchrotron Radiation*, Cambridge Monographs on Atomic, Molecular and Chemical Physics (Cambridge University Press, Cambridge, 1997).

[10] H. Lörch, N. Scherer, and V. Schmidt, in *6th EPS Conference on Atomic and Molecular Physics*, 6th EPS Conference on Atomic and Molecular Physics, edited by V. Biancalana and E. Mariotti (Siena, 1998), pp. 8–23.

[11] L. Végh and R. L. Becker, Phys. Rev. A **46**, 2445 (1992).

[12] B. Kämmerling and V. Schmidt, J. Phys. B: At. Mol. Opt. Phys. **26**, 1141 (1993).

[13] V. Schmidt, Nucl. Instr. Meth. **87**, 241 (1994).

[14] B. Kämmerling and V. Schmidt, Phys. Rev. Lett. **67**, 1848 (1991).

[15] B. Kämmerling and V. Schmidt, Phys. Rev. Lett. **69**, 1145 (1992).

[16] S. J. Schaphorst et al., J. Phys. B: At. Mol. Opt. Phys. **30**, 4003 (1997).

[17] N. A. Cherepkov, J. Phys. B: At. Mol. Opt. Phys. **12**, 1279 (1979), present data communicated by B. Zimmermann.

[18] W. R. Johnson and K. T. Cheng, Phys. Rev. A **46**, 2952 (1992).

[19] J. W. Cooper and R. N. Zare, in *Lectures in Theoretical Physics*, edited by S. Geltman, K. T. Mahanthappa, and W. E. Britting (Gordon & Breach, Science Publ., New York, 1969), Vol. IIC, p. 317.

[20] G. Snell, Ph.D. thesis, University Bielefeld, 1997.

[21] G. Snell et al., Phys. Rev. Lett. **82**, 2480 (1999).

Chapter 8

COMPLETE EXPERIMENTS IN MOLECULAR PHOTOIONIZATION

N. A. Cherepkov
State University of Aerospace Instrumentation, 190000 St. Petersburg, Russia

1. INTRODUCTION

Photoionization is the simplest process which allows a detailed investigation of atomic and molecular structure. The explanation of atomic photoabsorption spectra played a key role in the foundation of quantum mechanics. Nevertheless, for a long time there was a great distance between a very detailed information with which theorists are dealing like matrix elements of the dipole operator and phase shifts of different partial waves of a photoelectron wave function, and a very general information like the total or partial photoionization cross section for randomly oriented molecules which was usually measured by experimentalists. Only during the last two decades, owing to the application of synchrotron radiation sources in conjunction with photoelectron spectroscopy methods, it became possible to determine directly from experimental data the matrix elements and phase shifts calculated by theorists and in this way to establish a close connection between theory and experiment. An experiment (or usually a set of several experiments) from which one can extract all matrix elements and phase shifts necessary for a theoretical description of the process, is called a complete (or perfect) experiment [1–3]. Since every theoretical description is performed within some approximation, for example, in the electric-dipole approximation and neglecting the hyperfine interactions in the case of the photoionization process, the concept of the complete experiment is valid within the same approximation, too. Below we will always imply that the electric-dipole approximation is valid.

2. ON THE POSSIBILITIES OF COMPLETE EXPERIMENTS WITH ATOMS AND MOLECULES

The photoionization of atoms is simpler for interpretation than the photoionization of molecules, therefore the first complete experiments have been performed with atoms. At low photon energies (below $1 KeV$) when the electric-dipole approximation is valid, the dipole selection rules restricts the number of the photoelectron partial waves by only two with $l_1 = l \pm 1$ where l is the orbital angular momentum of an initial state. As a result, in the simplest case of closed shell atoms the photoionization process is described theoretically by at most five dynamical parameters (provided spin-orbit interaction in continuous spectrum is taken into account), namely, by three dipole matrix elements corresponding to the transitions from an initial nlj subshell ($j = l \pm 1/2$) to the final $\varepsilon(l-1)(j-1), \varepsilon(l \pm 1)j$, and $\varepsilon(l+1)(j+1)$ states, and two phase shift differences between these three partial waves [4]. We will imply in the following that the complex dipole matrix element is presented in a polar form as a product of a module and an exponent with the total (Coulomb plus short range) phase shift, therefore the dipole matrix element is always real and positive. In order to perform a complete experiment one needs to measure five different parameters from which five theoretical values can be determined. For the subshells with $j = 1/2$ only two transitions are allowed by the dipole selection rules, and the process is fully described by three dynamical parameters. The number of parameters is reduced to three also in the case when spin-orbit interaction in continuous spectrum is neglected and the final photoelectron state is characterized solely by the orbital angular momentum l.

As it follows from theory [4] the angular distribution of photoelectrons with defined spin orientation ejected from unpolarized atoms by circularly polarized light is characterized by five parameters, namely, by the partial photoionization cross section σ_{nlj}, the angular asymmetry parameter β, and three spin polarization parameters describing projections of photoelectron spin onto three mutually perpendicular directions. So, the measurements of spin polarization of photoelectrons as a function of an ejection angle, combined with an absolute measurement of the partial photoionization cross section, can constitute a complete experiment in the electric dipole approximation. The first complete experiments in photoionization [5, 6] have been performed with the rare gas atoms having all shells closed in the initial state.

In the case of photoionization of open shell atoms when the initial total angular momentum of the atom J_i is different from zero, the final state of the system ion+photoelectron can have three different values of the total angular momentum J_f, $J_f = J_i - 1, J_i, J_i + 1$, which is an additional characteristic of the final state. As a result, the number of allowed dipole transitions is becoming greater than three, the number of parameters greater than five, and measurements of spin polarization of photoelectrons ejected from unpolarized atoms is becoming not sufficient for the complete experiment. On the other hand, contrary to the closed shell atoms, the open shell atoms with nonzero total angular momentum can be polarized in the initial state. In the pioneering work of Klar and Klein-poppen [7] it was shown that the angular distribution of photoelectrons ejected from polarized atoms is characterized by more than five parameters, and one can perform the complete experiment with polarized atoms even without measuring the spin polarization of photoelectrons. Instead of the angular distribution itself, one can measure the difference between two angular distributions obtained, for example, with left and right circularly polarized light, or with two mutually perpendicular linear polarizations of light [8]. These values are called Circular or Linear Dichroism in the Angular Distribution of photoelectrons (CDAD or LDAD) respectively. One can define also the difference between two angular distributions obtained with the opposite polarizations of atoms, which is called Magnetic Dichroism in the Angular Distribution (MDAD) [8].

The standard way to obtain polarized (aligned or oriented) atoms is to excite them by laser light, with subsequent ionization by a second laser or by a synchrotron radiation. Though many experiments have been performed already with polarized excited atoms, usually the simplest cases have been chosen (the photoionization of excited alkali or alkali-earth elements) when three parameters were sufficient for the complete characterization of the process (see for example [9–13] and references therein). The only example where the strong dependence of the dipole matrix elements and phase shifts on the total angular momentum of the system ion+photoelectron has been established, is the photoionization of aligned Ar atoms in the excited $3p^54p(^3D_3)$ state [13]. In that particular case 11 dynamical parameters are necessary for a complete characterization of the process, 6 dipole matrix elements and 5 phase shift differences. Though they have not been determined directly from the measured data, the theoretical analysis clearly demonstrated that one could not reproduce the observed results with a smaller number of parameters.

In the case of molecules, due to the violation of spherical symmetry, the orbital angular momentum l is not a good quantum number, and the ground state wave function could not be characterized by a single value of l but rather presented as an infinite expansion in partial waves. As a consequence, the dipole selection rules could not restrict the infinite expansion in partial waves of the continuous spectrum wave function. Therefore formally the photoionization process is described by an infinite number of dipole matrix elements corresponding to different transitions, and the complete experiment is not feasible. Fortunately, the partial wave expansions for both bound and continuous spectrum wave functions are converging rather rapidly, and to a good approximation one can restrict the corresponding summations by a limited number of terms (though the number of theoretical parameters characterizing the photoionization of molecules will be in every case larger than 5). In this more restricted sense the complete experiments are feasible also with molecules, and the general discussion of complete experiments is presented below.

3. PRINCIPLES OF COMPLETE EXPERIMENTS WITH MOLECULES

Consider the general symmetry properties of the angular distributions of photoelectrons ejected from atoms or molecules. These properties are defined by the number of vectors characterizing the process, and in many cases are the same both for atoms and molecules, therefore we will not distinguish between these two cases when possible. If we have an ensemble of unpolarized atoms or randomly oriented and unpolarized molecules, the angular distribution of photoelectrons is characterized by two vectors, the photoelectron momentum **p** (which is implied to be a unit vector), and the photon polarization vector **e** for linearly polarized light, or the unit vector in the direction of the photon beam **q** for circularly polarized and unpolarized light, and depends on just one angle between these two vectors. As it follows from symmetry arguments [14], the angular distribution is expressed as a sum of a constant term and a term proportional to the second Legendre polynomial depending on this angle. It is characterized by two parameters, the partial photoionization cross section $\sigma_{nl}^J(\omega)$ where J is the total angular momentum of the initial state and ω is a photon energy, and the angular asymmetry parameter β^J. For the case of circularly polarized and unpolarized light it has the well known form

$$I^J(\mathbf{p}) = \frac{\sigma_{nl}^J(\omega)}{4\pi}\left[1 - \frac{\beta}{2}P_2(\mathbf{p}\cdot\mathbf{q})\right] \tag{8.1}$$

Suppose now that in addition to the direction of photoelectron momentum, the direction of photoelectron spin is also measured, that is the third vector is added to the characterization of the process, the vector of photoelectron spin. From the symmetry arguments it again follows [4] that three new terms appear in the angular distribution of photoelectrons. They characterize the projections of spin onto three mutually perpendicular directions. The total number of parameters describing such a process is equal to five, both for atoms and molecules. And if in the case of atoms it can be sufficient for a complete experiment, in the case of molecules it will be most probably not sufficient, though one can use these measurements in combination with some other experiments to gain an additional information. But even in the case of photoionization of closed shell atoms it was recently discovered [15] that four parameters (the angular asymmetry parameter β and three spin-polarization parameters) are not mutually independent, there is one analytical equation connecting them. As a consequence, one can extract at maximum only four dynamical parameters from the measurements of the angular distribution and spin polarization of photoelectrons, that is the complete experiment of that kind with closed shell atoms is impossible. One must look for additional information, for example, from fluorescence or Auger decay processes provided they are allowed.

Since the spin polarization measurements with the conventional Mott detector are connected with the loss of at least three orders of magnitude in intensity [16], they are rarely used in practice, and we will discuss later only the experiments which do not require the spin polarization measurements. Under this restriction the only way to add a new vector for characterization of the photoionization process is to consider an ensemble of atoms or molecules which is non isotropic in the initial state. One possibility is to polarize them before ionization. It can be done, for example, by optical pumping, or by investigating the two-step process in which the first light beam excites the target, that is prepares the non isotropic initial state, and the second light beam ionizes it [9–13,17–19]. In particular, one can prepare an ensemble of aligned or oriented molecules by excitation of randomly oriented molecules by a polarized narrow band laser beam to a state with a well defined rotational quantum number. That produces a nonstatistical distribution of magnetic sublevels with different projections M_J of the total angular momentum J in the laboratory frame [17–19]. This excited state will be aligned by linearly polarized or unpolarized light, and oriented by circularly polarized light. The ionization of this state by a second light beam allows to measure the angular distributions of photoelectrons which contain a sufficiently large number of parameters. The process is characterized by three vectors, the third vector being the unit vector in the direction of molecular (or atomic) polarization $\mathbf{n}$. One of these three vectors is taken to fix the coordinate system, while two others can have an arbitrary direction. In most cases the photon coordinate frame is taken as a laboratory one, then the angular distribution is presented as a double expansion in spherical harmonics of two directions, $\mathbf{p}$ and $\mathbf{n}$. To derive it in the most general case, we need to introduce the density matrices for both the initial state and the photon beam. The result is [8]

$$I^J(\mathbf{p},\mathbf{n}) = \sigma^J(\omega)\sqrt{3(2J+1)}$$

$$\sum_{k,L,x} C^J_{kLx} \sum_{q,M,\xi} \begin{Bmatrix} k & L & x \\ q & M & \xi \end{Bmatrix} \rho^\gamma_{kq}\rho^\mathbf{n}_{x0} Y^*_{LM}(\hat{\mathbf{p}}) Y^*_{x\xi}(\mathbf{n}) \qquad (8.2)$$

where the coefficients C^J_{kLx} contain the products of dipole matrix elements and cosine or sine functions of the phase shift differences, $\rho^\mathbf{n}_{x0}$ are the state multipoles characterizing the polarization of atoms or molecules [20], and ρ^γ_{kq} are the photon state multipoles expressed through the Stokes parameters ξ_i by the equations

$$\rho^\gamma_{00} = \frac{1}{\sqrt{3}}, \quad \rho^\gamma_{10} = \frac{\xi_2}{\sqrt{2}}, \quad \rho^\gamma_{20} = \frac{1}{\sqrt{6}},$$
$$\rho^\gamma_{1\pm 1} = \rho^\gamma_{2\pm 1} = 0, \quad \rho^\gamma_{2\pm 2} = -\frac{1}{2}(\xi_3 \mp i\xi_1). \quad (8.3)$$

In the photon coordinate frame with Z-axis directed along the photon beam $\xi_3 = +1(-1)$ corresponds to the linear polarization along the $X(Y)$ axis, $\xi_1 = +1(-1)$ corresponds to the linear polarization along the direction making the angle $+45(-45)$ degrees with the X axis, and $\xi_2 = +1(-1)$ corresponds to the right (left) circular polarization, described by the spherical function $Y_{11}(Y_{1-1})$. Atoms or molecules are said to be oriented if $\rho^n_{10} \neq 0$ (when the populations of sublevels with projections M_J and $-M_J$ are not equal), and aligned if $\rho^n_{10} = 0, \rho^n_{20} \neq 0$ (when the sublevels with projections M_J and $-M_J$ are equally populated, but for different $|M_J|$ the populations are different).

The number of additional terms now can be greater than three and depends on the value of the total angular momentum J of the initial state, as well as on the way in which this state has been prepared. In principle, the larger is J, the more new terms can appear in the angular distribution. Apparently, for $J = 1/2$ there could be at most three new terms as it was in the case of detection of photoelectron spin from unpolarized atoms or molecules. For $J = 1$ there could be already as many as 8 new terms, and the total number of parameters can be as large as $9J+1$ [7]. The actual number of additional terms is connected with the number of nonzero state multipoles describing the polarization of the initial state. In the case of one–photon excitation of randomly oriented initial state the highest rank of the state multipole ρ^n_{x0} is equal to that of the exciting photon, that is equal to 2 (see eq. 8.3), so that all the summations in 8.2 are restricted by the conditions $k \leq 2, x \leq 2, L \leq 4$, and the total number of terms in 8.2 is relatively small. Another disadvantage of the two-step process is that one investigates not the ground but the excited state of the atom or molecule. An example of the complete experiment with diatomic molecule based on this idea is presented in the nest section.

For molecules there is another way to break the spherical symmetry of the initial state by investigating the photoionization of fixed–in–space molecules. From the point of view of the vector characterization, this case is analogous to the photoionization of polarized molecules with an infinitely large total angular momentum J, so that the summation over x in 8.2 extends till infinity, and the total number of terms in the angular distribution 8.2 is becoming infinite [21,22], as is the number of theoretical parameters. Therefore the photoionization of fixed–in–space molecules is the most direct and promising way of performing the complete experiment with molecules in the ground state. For practical purposes it is more convenient to write the angular distribution of photoelectrons for fixed–in–space molecules not in the photon frame as it was done in 8.2, but in the molecular frame with Z axis directed along the molecular axis. Then instead of 8.2 we obtain [21]

$$I^m_{\Lambda^i}(\mathbf{p},\mathbf{e}) = \sqrt{3}\sigma_{\Lambda^i}(\omega)(-1)^{1-m}\sum_{L,M}\sum_J^\infty\begin{Bmatrix}1 & 1 & J\\ -m & m & 0\end{Bmatrix}A^{J-M}_{LM}Y_{LM}(\hat{\mathbf{p}})Y_{J-M}(\hat{\mathbf{e}})$$

(8.4)

where Λ^i is the projection of orbital angular momentum of a final ionic state on the molecular axis, $m = 0$ for linearly polarized light and $m = \pm 1$ for circularly polarized light, $\mathbf{e}$ is the polarization vector for linearly polarized light which should be substituted by the unit vector $\mathbf{q}$ in the direction of the light beam for circularly polarized light, $\sigma_{\Lambda^i}(\omega)$ is a partial photoionization cross section for a given photon energy ω, and the parameters A^{J-M}_{LM} contain the dipole matrix elements and cosine or sine functions of the phase shift differences. In practice the infinite expansion over L in 8.4 can always be restricted to some finite number of terms.

There are basically two ways to investigate photoionization of fixed–in–space molecules. One is based on the use of molecules which are not rotating due to some external constrains like in the cases of molecules adsorbed on surfaces [23], or molecules in liquid crystals. A disadvantage of this method is that instead of free molecules, we have molecules deformed by the substrate field or by the interaction with surrounding molecules, respectively. The other way is based on the use of coincidence technique which allows to select the processes corresponding to a well defined direction of molecular axis in space from the photoionization of an ensemble of randomly oriented (rotating) molecules. It is done by performing an angular resolved detection of a photoelectron in coincidence with an atomic ion resulting from dissociation of a molecule, provided the dissociation time is much shorter than the period of molecular rotation [24–27]. The last condition restricts the possible photoionization processes by those which correspond to strongly repulsive final molecular ion states. This method is particularly convenient for diatomic or linear polyatomic molecules for which the direction of momentum of the atomic ion after dissociation coincides with the direction of molecular axis at the moment of photoionization. Two different geometries of experiment have been used up to now. In one geometry [24–26] the position of the parallel-plate ion-energy analyzer is fixed, while the electron detector (also a parallel-plate electron-energy analyzer) is rotated allowing to measure the angular distribution of photoelectrons ejected from fixed–in–space molecules. In the other geometry [27] the position of the electron time–of–flight detector is fixed, and a three-dimensional angular distribution of residual ions (extracted from the interaction region by a static electric field) is measured with a two-dimensional position-sensitive detector. And from a most advanced experiment with time-resolved detection of all photoelectrons and all ions by using two separate position-sensitive detectors for electrons and ions (extracted from the interaction region by a static electric field) [28,29] one can extract either three-dimensional angular distributions of photoelectrons for a fixed direction of residual ions, or three-dimensional angular distributions of residual ions for a fixed direction of photoelectrons. That kinds of experiment are considered in more detail in section 5.

4. PHOTOIONIZATION OF EXCITED ALIGNED MOLECULES

The first complete photoionization experiment with molecules have been performed in [18, 19] using the (1 + 1') Resonance-Enchanced Multiphoton Ionization (REMPI) technique when the first photon beam produces the resonantly excited polarized molecules which are then ionized by the second light beam. In particular, the authors investigated photoionization of NO

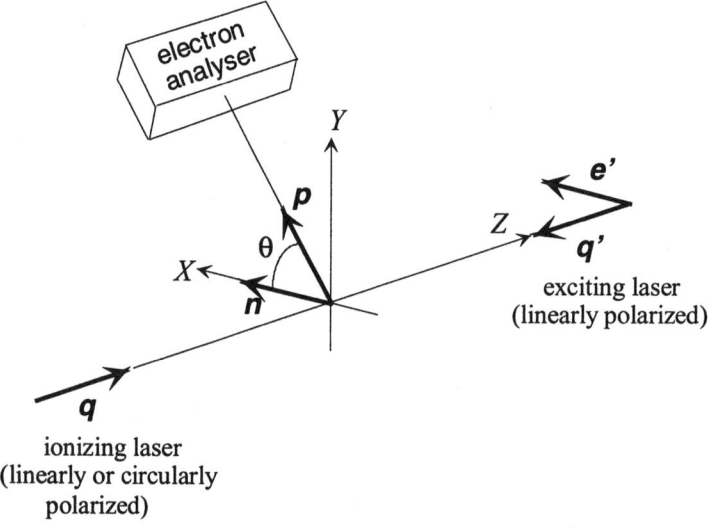

Figure 8.1 Schematic representation of the geometry of the (1+1) REMPI experiment performed in [17–19] for NO molecule.

molecules by two counterpropagating laser beams as is shown in Fig.8.1. The first linearly polarized laser beam excited via the $P_{21} + Q_1(22.5)$ transition the $A^2\Sigma^+(\nu_i = 1, N = 22)$ level of NO, which was ionized by the second (either linearly or circularly polarized) laser beam to the final state $NO^+ X^1\Sigma^+(v^+ = 1, N^+ = 20 - 24)$. In [19] the analogous processes have been investigated but with $v_i = v^+ = 0$. The second light beam with the wavelength $\lambda_2 = 313.84 nm$ was chosen so as to give slow photoelectrons

($132 - 187 meV$) which were energy analyzed by time–of–flight spectrometer. The energy resolution of $2.5 meV$ was sufficient to resolve single rotational levels. It was assumed that the dipole matrix elements and phase shifts do not depend on the molecular rotational level, and that they are constant over the range of photoelectron kinetic energies involved in the experiment.

For the particular geometry of experiment used in [18] (see Fig.8.1) and for linear polarization of the ionizing light beam the general equation 8.2 for the angular distribution of photoelectrons ejected from aligned molecules is simplified to

$$I^{NN^+}(\theta) = \beta_{00} Y_{00}(\theta, \phi) + \beta_{20} Y_{20}(\theta, \phi) + \beta_{40} Y_{40}(\theta, \phi) \quad (8.5)$$

where N and N^+ are the rotational quantum numbers of the intermediate excited state and the final ion state mentioned above, and β_{i0} are the parameters containing the dipole matrix elements and phase shifts. Since in practice the measurements of the angular distributions are usually relative, one can obtain from one measurement of that kind only two parameters which is not sufficient for a complete experiment. But since the parameters β_{i0} are different for different rotational transitions, one can perform the measurements of the angular distributions 8.5 for several rotational transitions which allows to obtain many more parameters provided the dipole matrix elements and phase shifts do not depend on the rotational quantum numbers. Also the measurements with different light polarization give an additional information. In particular, when the ionizing light beam is linearly polarized, one can perform measurements with two different relative orientations of the polarization vectors of two laser beams, either parallel or perpendicular to each other. For these two cases the angular distributions are described by the same eq. 8.5, but with different coefficients β_{i0}. When the ionizing light beam is circularly polarized, the angular distribution differs from 8.5 and is presented by the equation

$$I^{NN^+}(\theta) = C_0 + C_1 \cos^2 \theta + C_2 \sin \theta \cos \theta \quad (8.6)$$

with another coefficients C_i. From the measurements with circularly polarized light one can define also the CDAD, that is the difference between the intensities obtained with left and right circularly polarized light [30]

$$I^{NN^+}_{CDAD}(\theta) = I^{NN^+}_{left}(\theta) - I^{NN^+}_{right}(\theta) = 2C_2 \sin \theta \cos \theta. \quad (8.7)$$

Important is that the coefficient C_2 contains the sine function of the phase shift differences, while the coefficients in eq. 8.5 are proportional to the cosine function. Therefore the analysis of data obtained with linearly polarized light yields double-valued results for the relative phases, $\Delta \delta_i$ and $(2\pi - \Delta \delta_i)$, while the use of circularly polarized light allows to obtain a single-valued answer.

Parameter	Fit	Ab initio
$r_{s\sigma}$	0.204(2)	0.158
$r_{p\sigma}$	0.503(11)	0.278
$r_{p\pi}$	0.471(6)	0.537
$r_{d\sigma}$	0.166(30)	0.221
$r_{d\pi}$	0.073(15)	0.020
$r_{f\sigma}$	0.321(25)	0.358
$r_{f\pi}$	0.244(13)	0.268
$d_{p\pi-p\sigma}$	12.4(1.5)	9.9
$d_{d\pi-d\sigma}$	−68(13)	−92.8
$d_{f\pi-f\sigma}$	−1(18)	−1.6
$d_{d\sigma-s\sigma}$	−157(9)	+173.7
$d_{f\sigma-p\sigma}$	−59(14)	−76.3

Table 8.1 Dynamical parameters resulting from the fit of experimental data for photoionization of NO via the $A^2\Sigma^+(\nu = 0)$ state, compared to the results of *ab initio* calculation of Rudolph and McKoy [30]. The $r_{l\lambda}$ values are normalized so that their squares sum to unity. The phase shift differences $d_{l\lambda-l'\lambda'}$ are given in degrees. The values in parentheses represent 1σ uncertainties. (From Ref. [19]).

From the angular distributions measured in [19] for five rotational levels of NO molecule with circularly polarized light, together with the angular distributions reported in [17] for linearly polarized light, the authors were able to extract 12 dynamical parameters, 7 dipole matrix elements and 5 phase shift differences shown in Table 8.1. The results of *ab initio* calculations of Rudolph and McKoy [31] in the HF approximation are also given in the table for comparison. The partial wave expansion of the photoelectron wave function has been truncated at $l = 3$. The relative phase shifts between even and odd l waves were unobservable and therefore are not presented. From Table 8.1 it is evident that the photoionization cross section is dominated by the contribution of the p wave. The agreement between theory and experiment is reasonable. The most important difference between them is in the ratio $r_{p\pi}/r_{p\sigma}$ which is equal to approximately one in the experiment compared to two in the theory. Also the experimental value of the magnitude of the dipole matrix element $r_{p\sigma}$ is 1.8 times larger than the corresponding theoretical value. To exploit the advantages of the complete experiment, the authors of [19] have calculated the three-dimensional angular distributions of photoelectrons and the ion angular momentum polarization (alignment and orientation) for specific final rotational levels using the results of fitting presented in Table 8.1.

5. PHOTOIONIZATION OF ORIENTED (FIXED–IN–SPACE) MOLECULES

5.1 GENERAL CONSIDERATION

Consider now photoionization of oriented diatomic molecules with a fixed direction of molecular axis in space taken as Z-axis of a molecular coordinate system. We will imply in the following that K-shells are ionized so that the dissociation process is fast enough for the molecular rotation to be neglected in the coincidence experiment with gaseous molecular target. Suppose that

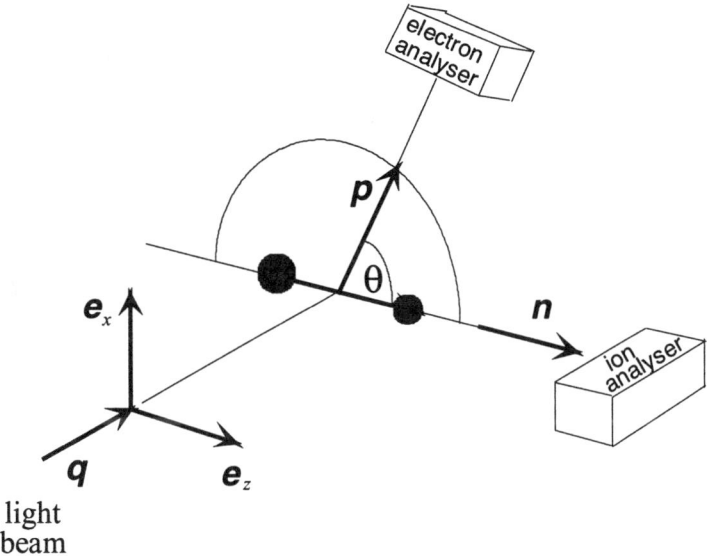

Figure 8.2 Schematic representation of the geometry of the AR-PEPICO experiment performed in [32] for CO molecule.

the incoming photon beam is perpendicular to the molecular axis, and light is linearly polarized (see Fig. 8.2). When the polarization vector $\mathbf{e}_z$ of the light beam is parallel to the molecular axis, the process acquires the axial symmetry about the molecular axis, and the spherical harmonics $Y_{J-M}(\hat{\mathbf{e}})$ in 8.4 are different from zero only when $M = 0$. Due to that the general equation 8.4 for the angular distribution of photoelectrons is greatly simplified and can be presented not as an expansion in spherical harmonics, but as an expansion in Legendre polynomials

$$I^0_{\Lambda i}(\mathbf{p}, \mathbf{e_z}) = \frac{\sigma_{\lambda i}(\omega)}{4\pi} \frac{3}{B} \sum_{L=0}^{L_{max}} A_L P_L(\theta) \qquad (8.8)$$

where θ is the polar angle of the electron ejection, and the parameters A_L, as usual, contain the products of dipole matrix elements and cosine or sine functions of the phase shift differences. From the symmetry arguments it follows that only the dipole matrix elements corresponding to the $\sigma \to \sigma$ transitions are contributing to these parameters.

Determination of the dipole matrix elements and phase shift differences from the measured angular distributions for the $\sigma \to \sigma$ transitions is performed in two steps. At first one should define the ratios of the parameters A_L/A_0 by fitting the measured angular distributions (which are usually not absolute but relative) by eq. 8.8. Then one can determine the ratios of dipole matrix elements $d_{l\sigma}/d_{0\sigma}$ and phase shift differences $(\delta_{l\sigma} - \delta_{0\sigma})$ from the analytical equations for A_L presented in [26].

In heteronuclear molecules both even and odd L contribute to 8.8, and for any even value of L_{max} the number of parameters A_L is just equal to the number of dipole matrix elements and phase shift differences entering the equations for A_L, provided the partial wave expansion of the photoelectron wave function is restricted by the maximal value $l_{max} = L_{max}/2$. It means that in this approximation one can perform the complete experiment and extract the dipole matrix elements and phase shift differences directly from the measured angular distributions. This opens the possibility of a direct probe of the orbital angular momentum composition of molecular photoelectrons, in particular in the σ^* shape resonance which reveal itself only in $\sigma \to \sigma$ transitions.

Consider now the case when the light beam is linearly polarized along the X axis of the molecular frame, that is $\theta_e = 90°$, $\phi_e = 0°$ (see Fig. 8.2). Then it follows from 8.4 that

$$I^0_{\Lambda i}(\mathbf{p}, \mathbf{e_x}) = \frac{\sigma_{\Lambda i}(\omega)}{4\pi} \sum_{L=0}^{L_{max}} C_L P_L(\cos\theta) \tag{8.9}$$

where the parameters C_L contain only the matrix elements and phase shift differences corresponding to the $\sigma \to \pi$ transitions.

Suppose that from fitting of the experimental data by eq. 8.9 one has defined 8 parameters C_L, which corresponds to taking into account the partial waves up to $l_{max} = L_{max}/2 = 4$. But since for the $\sigma \to \pi$ transitions the partial wave with $l = 0$ is not contributing, one needs to define only 6 values, 3 ratios of dipole matrix elements $d_{i\pi}/d_{1\pi}$, and 3 phase shift differences $(\delta_{ip} - \delta_{1p})$, $i = 2, 3, 4$. It means that not all the parameters C_L are mutually independent, and actually there are the following two relations between them

$$\sum_{L=0}^{L_{max}} C_L = 0, \quad \sum_{L=0}^{L_{max}} (-1)^L C_L = 0, \tag{8.10}$$

or equivalently

$$\sum_{L_{even}}^{L_{max}} C_L = 0, \quad \sum_{L_{odd}}^{L_{max}} C_L = 0. \tag{8.11}$$

Eqs. 8.10 and 8.11 are valid for any L_{max} provided L_{max} is even and $l_{max} = L_{max}/2$. Inserting them into 8.9 one obtains that the electron intensities along the molecular axis for $\mathbf{e} \perp \mathbf{n}$ are always equal to zero

$$I^0_{\Lambda i}(\theta = 0, \mathbf{e_x}) = I^0_{\Lambda i}(\theta = \pi, \mathbf{e_x}) = 0. \tag{8.12}$$

Therefore for any even L_{max} the number of independent coefficients C_L is equal to $L_{max} - 2$, and the same is the number of ratios of dipole matrix elements and phase shift differences to be determined from them in the complete experiment.

Measurements for $\mathbf{e} \parallel \mathbf{n}$ and $\mathbf{e} \perp \mathbf{n}$ where $\mathbf{n}$ is the direction of molecular axis enable one to define separately two independent sets of matrix elements and phase shifts for $l\sigma$ and $l\pi$ channels. The ratios of matrix elements belonging to different sets, for example, $d_{i\pi}/d_{0\sigma}$, and the corresponding phase shift differences $(\delta_{i\pi} - \delta_{0\sigma})$ remain undefined. Actually it is sufficient to define only one ratio, say $d_{1\pi}/d_{0\sigma}$, and one phase shift difference, say $(\delta_{1\pi} - \delta_{0\sigma})$. The ratio $d_{i\pi}/d_{0\sigma}$ can be obtained from the relative measurements of the cross sections corresponding to the σ and π ionization channels like those performed in [33], while for determining the phase shift difference $(\delta_{i\pi} - \delta_{0\sigma})$ one needs to perform, for example, one measurement at the angle between the molecular axis and light polarization vector $\mathbf{e}$ different from 0 and 90 degrees.

There is another modification of the experiment aimed at investigation of oriented molecules when the angular distribution of residual ions is measured with the direction of photoelectron ejection being fixed [27]. For a theoretical consideration of that process one must use the photon coordinate system as a laboratory one, and the final equation will be similar to eq. 8.2, only the coefficients C^J_{kLx} will be different. It is very convenient to present the angular distribution of residual ions as an expansion in bipolar spherical harmonics $Y^{LN}_{kq}(\mathbf{p}, \mathbf{n})$ [34] of two directions, the direction of photoelectron momentum $\mathbf{p}$, and the direction of molecular axis $\mathbf{n}$

$$I^{\Lambda^+}(\mathbf{p}, \mathbf{n}) = \sigma^{\Lambda^+}(\omega) \sum_{k,L,N} A^k_{LN} \sum_q \rho^\gamma_{kq} Y^{LN}_{kq}(\mathbf{p}, \mathbf{n}) \tag{8.13}$$

The coefficients A^k_{LN} depend on the dipole matrix elements and phase shift differences of both $\sigma \to \sigma$ and $\sigma \to \pi$ transitions, and one could not separate the determination of that dynamical parameters into two steps as it was possible in the cases of measurements of the angular distribution of photoelectrons for $\mathbf{e} \parallel \mathbf{n}$ and $\mathbf{e} \perp \mathbf{n}$ discussed above. That is a disadvantage of the experiment with determination of the angular distribution of ions.

5.2 PARTICULAR RESULTS FOR O K–SHELL OF CO

The first determination of the dipole matrix elements and phase shifts for the $\sigma \to \sigma$ transitions for C K-shell of CO molecule from the measured photoelectron angular distributions has been made in [26]. In that work the measurements have been performed for the light polarization parallel to the molecular axis, therefore the dynamical parameters for the $\sigma \to \pi$ transitions remained not defined. Only very recently a really complete experiment has been reported for O K-shell of CO molecule in [32]. Photoionization of fixed–in–space molecules in a gas phase has been realized there by using the Angle Resolved Photoelectron-Photoion Coincidence technique (AR-PEPICO) at the soft X-ray undulator beamline BL-2C of the Photon Factory. The experimental apparatus described in detail elsewhere [35] consists of two electron-ion coincidence circuits (instead of one in [26]). Two ion analyzers were used to detect the ions moving in two mutually perpendicular directions. Compared to earlier measurements [26] it was possible to obtain simultaneously the angular distributions of photoelectrons for light polarization parallel and perpendicular to the molecular axis. Fig. 8.3 shows two angular distributions of photoelectrons

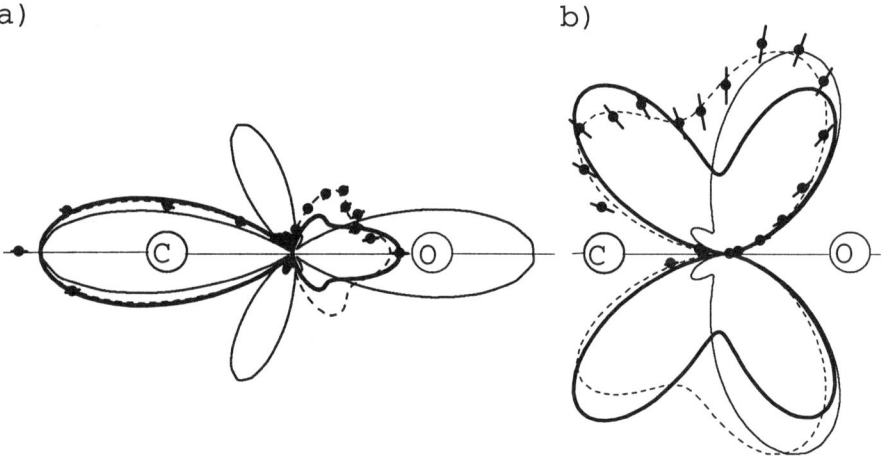

Figure 8.3 Angular distributions of photoelectrons from O K-shell of fixed–in–space CO molecules ejected by light linearly polarized parallel (a) and perpendicular (b) to the molecular axis. Photon energy is $552.3eV$, the angular distributions are normalized to unity at maximum. Dots: experiment; dashed line: the fit to the experimental data; thick solid line: RCHF calculation; thin solid line: calculation in the MS approximation [36]. (From Ref. [32]).

measured in [32] at photon energy $552.3eV$ (closest to the maximum of the σ^* shape resonance) compared to a previously available calculation [36] in the multiple scattering (MS) approximation, and the new calculation of the authors in the Relaxed Core Hartree-Fock (RCHF) approximation. The RCHF result is in a much better agreement with the experiment though discrepancies still remain.

Starting from the measured angular distributions, the authors proceed towards the complete experiment and defined the coefficients A_L of Eq. 8.8 with $L_{max} = 10$, and the coefficients C_L of Eq. 8.9 with $L_{max} = 8$. Then, using the analytical expressions for these coefficients given in [26] they determined 5 ratios of dipole matrix elements $d_{l\sigma}/d_{0\sigma}$ and 5 phase shift differences $(\delta_{l\sigma} - \delta_{0\sigma})$ for the $1s\sigma \to \varepsilon l\sigma$ transitions with $0 < l \leq 5$, and 3 ratios of dipole matrix elements $d_{l\pi}/d_{1\pi}$ and 3 phase shift differences $(\delta_{l\pi} - \delta_{1\pi})$ for the $1s\sigma \to \varepsilon l\pi$ transitions with $1 < l \leq 4$. Using the relative cross sections for the $1s\sigma \to \varepsilon l\sigma$ and $1s\sigma \to \varepsilon l\pi$ channels measured in [33], and normalizing the relative experimental cross section for the $1s\sigma \to \varepsilon l\pi$ channels to the theoretical cross section at $561eV$ photon energy, it was possible to obtain from the experiment also the absolute values of the dipole matrix elements. As a result, as many as 18 dynamical parameters have been determined from the experimental data. But since the equations for the coefficients A_L and C_L are quadratic in dipole matrix elements, and the phase shift differences are arguments of sine or cosine functions, the solution is not unique. There are 32 different solutions for 10 values (5 ratios of dipole matrix elements and 5 phase shift differences) of the $1s\sigma \to \varepsilon l\sigma$ transitions, and 8 different solutions for 6 values of the $1s\sigma \to \varepsilon l\pi$ transitions. One could not select unambiguously one of these solutions without additional information on photoionization process. The authors used the results of the RCHF calculations to select one set of solutions giving the closest agreement with the theory.

Fig. 8.4 shows the comparison between the dipole matrix elements for the first six $1s\sigma \to \varepsilon l\sigma$ transitions ($l \leq 5$) extracted from the experimental data and calculated in RCHF. At about $550eV$ there is a σ^* shape resonance where, according to the MS calculations [37], the f-wave contribution to the cross section is expected to be predominant. Both the RCHF calculations and experiment show that the situation in reality is quite different and more complicated. Though the matrix elements with $l = 3$ is the largest, all other matrix elements are of the same order of magnitude (only the theoretical matrix element with $l = 5$ is small). Therefore the appearance of the shape resonance could not be attributed to the contribution of the $l = 3$ partial wave alone as it was supposed in [37] on the basis of a relatively simple MS calculations (there is no data in Refs. [36] and [37] on the contributions of different partial waves for CO molecule). The phase shift differences for the $1s\sigma \to \varepsilon l\sigma$ channels are presented in Fig. 8.5. At some energies the discrepancies between theory and experiment are large. Since only the differences of phase shifts have been extracted from the experimental data, it was impossible to prove that the sum of phase shifts is increasing by about π in the σ^* shape resonance as it follows

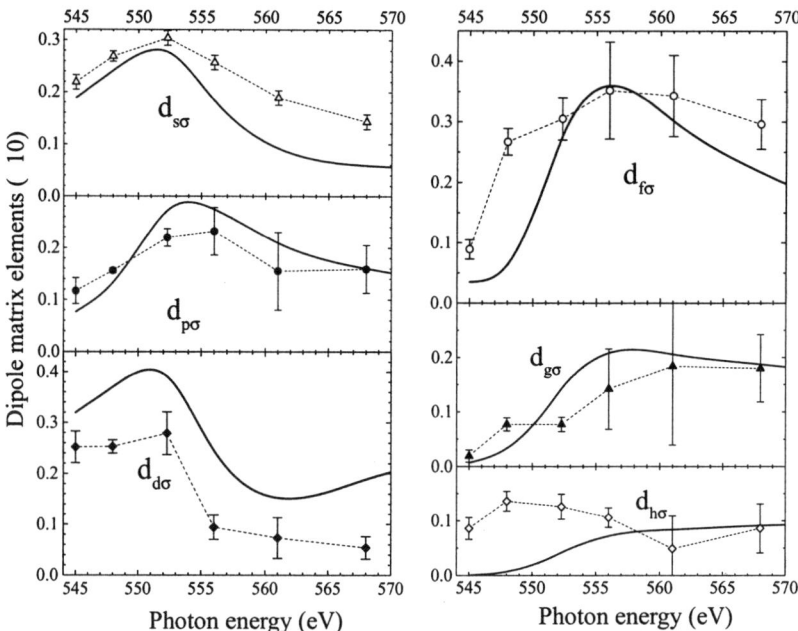

Figure 8.4 Dipole matrix elements $d_{l\sigma}$ (in atomic units, multiplied by 10) for the $1s\sigma \to \varepsilon l\sigma$ transitions with $0 \le l \le 5$ extracted from the experimental data (points with error bars connected by straight lines) and calculated in the RCHF approximation (curves). (From Ref. [32]).

from theoretical consideration in [37]. An important contribution of at least four partial waves in the region of the σ^* shape resonance in the valence shell photoionization of CO molecule has been mentioned in [38].

Fig. 8.6 demonstrates the matrix elements and phase shift differences for the first four $1s\sigma \to \varepsilon l\pi$ transitions with $l \le 4$. Theory and experiment are in a reasonable agreement showing that the largest is the dipole matrix element with $l = 2$. But the energy dependencies of the matrix elements with $l = 1$ and $l = 3$ in theory and experiment are different. On the average the agreement between the RCHF calculations and experiment could not be considered as satisfactory. One should keep it in mind that more advanced theory (including electron correlations) can differ from the RCHF results, and may be another set among $32 + 8$ sets of dynamical parameters determined from the experiment will better fit that theory.

The number of terms retained in the expansions 8.8-8.9 in the analysis performed in [32] was limited by the angular resolution of the experiment. Figs. 8.4-8.6 show that a better convergence can be reached if one includes at least one additional partial wave. Therefore further improvement of experimental technique is necessary for a more exact determination of the dipole matrix elements and phase shifts.

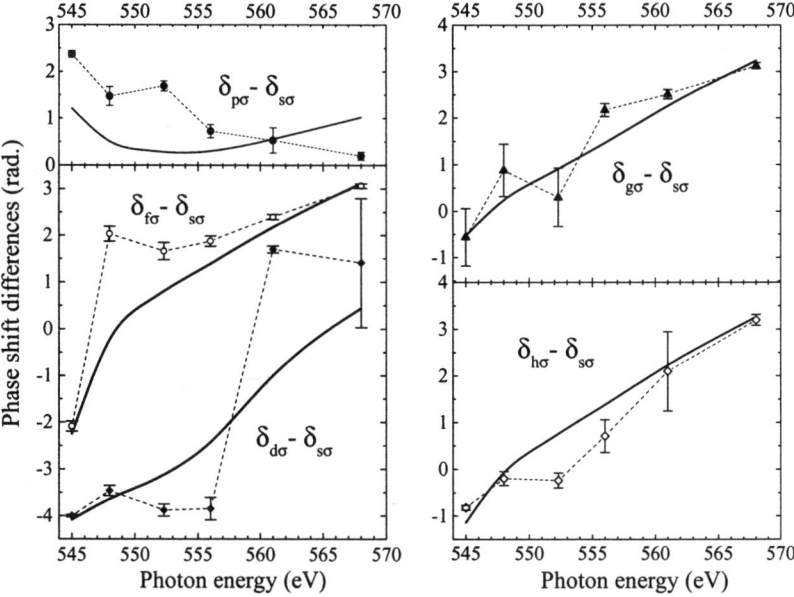

Figure 8.5 Phase shift differences $(\delta_{l\sigma} - \delta_{0\sigma})$ for the $1s\sigma \to \varepsilon l\sigma$ transitions with $l \leq 5$ extracted from the experimental data (points with error bars connected by straight lines) and calculated in the RCHF approximation (curves). (From Ref. [32].

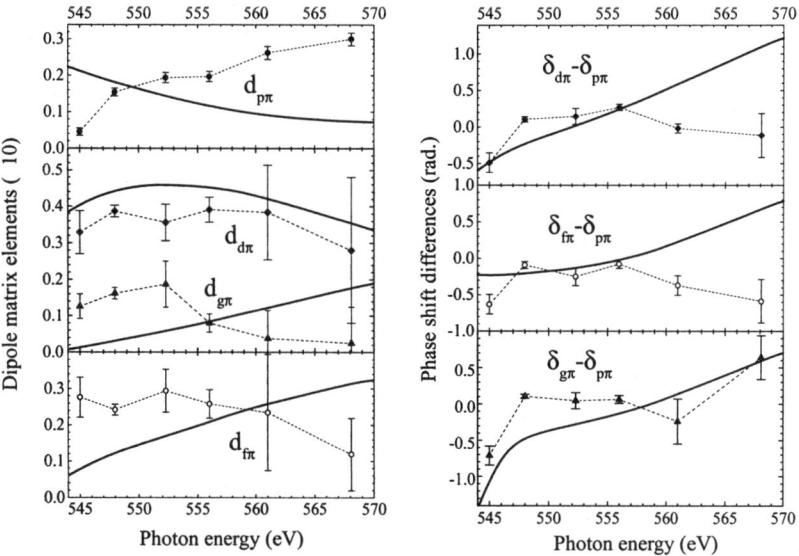

Figure 8.6 Dipole matrix elements $d_{l\pi}$ and phase shift differences $(\delta_{l\pi} - \delta_{1\pi})$ for the $1s\sigma \to \varepsilon l\pi$ transitions with $l \leq 4$ (the same notations as in Figs. 8.4 and 8.5). (From Ref. [32]).

6. CONCLUSIONS

The main goal of a complete experiment is the most detailed comparison between theory and experiment. The first complete experiment with CO molecule [32] clearly demonstrated that the existing theory is not satisfactory and more advanced theory is needed. It is implied that both the improvement of the RCHF method and the inclusion of many-electron correlations can be important. A recent calculations for N_2 molecule in the Random Phase Approximation (RPA) [39] have shown that due to the contribution of many-electron correlations the σ^* shape resonance in the K-shell photoionization reveals itself not solely in the $1\sigma_g \to \varepsilon\sigma_u$ channel as it was believed earlier on the basis of MS calculations [37] but in both $1\sigma_g \to \varepsilon\sigma_u$ and $1\sigma_u \to \varepsilon\sigma_g$ channels. This conclusion is supported by the comparison between the theoretical and experimental angular distributions of photoelectrons ejected from fixed-in-space N_2 molecules by light linearly polarized along the molecular axis. It can be correctly described theoretically only provided the contribution of the $1\sigma_u \to \varepsilon\sigma_g$ channel in the σ^* shape resonance is greatly enchanced in accord with the RPA calculations, while in the Hartree-Fock approximation the cross section for the $1\sigma_u \to \varepsilon\sigma_g$ channels in the shape resonance is constant like in the MS approximation [37]. The analogous behaviour in the σ^* shape resonances is expected also in other homonuclear diatomic molecules provided the two conditions are fulfilled, namely, there are two closely spaced shells, and the photoionization cross section of one of them is much larger than of the other one. Though in heteronuclear diatomic molecules like CO the two K-shells have strongly different ionization thresholds and due to that the role of many-electron correlations is not expected to be as strong as in N_2, this question needs to be examined.

Another advantage of a complete experiment is the possibility to predict the result of any other photoionization experiment without performing new measurements. The first steps along this line were made in [19, 40]. In particular, in [40] the angular distribution of ions for a given direction of photoelectron emission observed in the AR-PEPICO experiment was calculated based partly on the matrix elements and phase shifts extracted from the experiment [26] and partly on the results of RCHF calculations [32]. The calculated angular distribution of ions is in a reasonable agreement with the first experimental measurement also presented in [40]. One can use the results of the complete experiment to determine the direction of molecular axis relative to the surface normal for molecules adsorbed on surfaces. Earlier the results of the MS or Hartree-Fock calculations for isolated oriented molecules have been used for that purposes (see [23, 41–43] and references therein). Now from the comparison between the angular distributions of photoelectrons for free oriented molecules determined from the complete experiment, with that measured for adsorbed molecules, one can extract a rather detailed information on the character of the interaction between the molecule and the surface. So, the complete experiments with molecules will definitely lead in a future to new developments in both basic and applied research.

Acknowledgments

The author is greatly indebted to the Max-Planck-Gesellschaft for financial support and the Fritz-Haber-Institut for the hospitality extended to him during the work on this paper. He acknowledges also the financial support of the Ministry of Education, Science and Culture of Japan and the hospitality of the Photon Factory extended to him at the stage of the preparation of the manuscript.

References

[1] B. Bederson, Comments At. Mol. Phys. **1**, 65 (1969).

[2] B. Bederson, Comments At. Mol. Phys. **1**, 41 (1969).

[3] J. Kessler, Comments At. Mol. Phys. **10**, 47 (1981).

[4] N. A. Cherepkov, Adv. At. Mol. Phys. **19**, 395 (1983).

[5] U. Heinzmann, J. Phys. B: At. Mol. Opt. Phys. **13**, 4353 (1980).

[6] U. Heinzmann and N. A. Cherepkov, in *VUV- and soft X-ray photoionization*, edited by U. Becker and D. A. Shirley (Plenum Press, New York, 1996), p. 521.

[7] H. Klar and H. Kleinpoppen, J. Phys. B: At. Mol. Opt. Phys. **15**, 933 (1982).

[8] N. A. Cherepkov, V. V. Kuznetsov, and V. A. Verbitskii, J. Phys. B: At. Mol. Opt. Phys. **28**, 1221 (1995).

[9] J. A. Duncanson, Jr., M. P. Strand, A. Lindgard, and R. S. Berry, Phys. Rev. Lett. **37**, 987 (1976).

[10] C. Kerling, N. Bowering, and U. Heinzmann, J. Phys. B: At. Mol. Opt. Phys. **23**, L629 (1990).

[11] M. Pahler *et al.*, Phys. Rev. Lett. **68**, 2285 (1992).

[12] C. S. Feigerle *et al.*, Phys. Rev. A **53**, 4183 (1996).

[13] S. Schohl *et al.*, J. Phys. B: At. Mol. Opt. Phys. **31**, 3363 (1998).

[14] C. N. Yang, Phys. Rev. **74**, 764 (1948).

[15] B. Schmidtke, M. Drescher, N. A. Cherepkov, and U. Heinzmann, J. Phys. B: At. Mol. Opt. Phys. , to be published.

[16] J. Kessler, *Polarized Electrons*, 2 ed. (Springer, Berlin, 1985).

[17] S. W. Allendorf, D. J. Leahy, D. C. Jacobs, and R. N. Zare, J. Chem. Phys. **91**, 2216 (1989).

[18] D. J. Leahy, K. L. Reid, and R. N. Zare, J. Chem. Phys. **95**, 1757 (1991).

[19] D. J. Leahy, K. L. Reid, H. Park, and R. N. Zare, J. Chem. Phys. **97**, 4948 (1992).

[20] K. Blum, *Density Matrix Theory and Applications* (Plenum Press, New York and London, 1981).

[21] D. Dill, J. Chem. Phys. **65**, 1130 (1976).

[22] N. A. Cherepkov and V. V. Kuznetsov, Z. Phys. D **7**, 271 (1987).

[23] C. Westphal et al., Surf. Sci. **253**, 205 (1991).

[24] A. Golovin, N. Cherepkov, and V. Kuznetsov, Z. Phys. D **24**, 371 (1992).

[25] E. Shigemasa, J. Adachi, M. Oura, and A. Yagishita, Phys. Rev. Lett. **74**, 359 (1995).

[26] E. Shigemasa et al., Phys. Rev. Lett. **80**, 1622 (1998).

[27] F. Heiser et al., Phys. Rev. Lett. **79**, 2435 (1997).

[28] D. Dowek et al., Nucl. Instr. Meth. , to be published.

[29] P. Downie and I. Powis, Phys. Rev. Lett. **82**, 2864 (1999).

[30] R. L. Dubs, S. N. Dixit, and V. McKoy, J. Chem. Phys **85**, 656 (1986).

[31] H. Rudolph and V. McKoy, J. Chem. Phys **91**, 2235 (1989).

[32] K. Ito et al., Phys. Rev. Lett. , in press.

[33] E. Shigemasa, T. Hayaishi, T. Sasaki, and A. Yagishita, Phys. Rev. A **47**, 1824 (1993).

[34] D. A. Varshalovich, A. N. Moskalev, and V. K. Khersonskii, *Quantum Theory of Angular Momentum* (World Scientific, Singapore, 1988).

[35] A. Yagishita, in *Photonic, Electronic and Atomic Collisions (Invited Papers of the XX ICPEAC)*, edited by F. Aumayr and H. Winter (World Scientific, Singapore, 1998), p. 149.

[36] D. Dill, J. Siegel, and J. Dehmer, J. Chem. Phys. **65**, 3158 (1976).

[37] J. L. Dehmer and D. Dill, J. Chem. Phys. **65**, 5327 (1976).

[38] B. Leyh and G. Raseev, Phys. Rev. A **34**, 2920 (1986).

[39] N. A. Cherepkov et al., Phys. Rev. Lett. **84**, 250 (2000).

[40] O. Geßner et al., J. Electr. Spectrosc. Relat. Phenom. **101-103**, 113 (1999).

[41] J. W. Davenport, Phys. Rev. Lett. **36**, 945 (1976).

[42] A. M. Bradshaw and F. M. Hoffman, Surf. Sci. **72**, 513 (1978).

[43] P. Budau, M. Buchner, and G. Raseev, Surf. Sci. **292**, 67 (1993).

III
EINSTEIN–PODOLSKY–ROSEN EXPERIMENTS

Chapter 9

EXPERIMENTAL TESTS OF BELL'S INEQUALITIES WITH CORRELATED PHOTONS

Alain Aspect
Institut d'Optique Théorique et Appliquée Bâtiment 503–Centre universitaire d'Orsay–BP 147 91403 ORSAY Cedex – France

1. INTRODUCTION

The organizers of this meeting have suggested that professor Hans Kleinpoppen would appreciate to hear about experimental tests of Bell's inequalities using correlated photons. I am very happy to have the opportunity to present this work in his honour. We all know that he has done a lot in scattering experiments involving electrons and photons. Among his most recognized work howerver, was the first test of Bells inequalities [1] by the two–photon decay of metastable hydrogen. Nobody better than him could appreciate the fact that their violation allows us to point out situations where the predictions of Quantum Mechanics are so far from our usual images, that they cannot be mimicked by classical models.

This presentation is partly based on a paper that was published 15 years ago as a proceedings of a conference, not so easy to find nowadays [2]. The first part of the paper aims at explaining what are Bell's theorem and Bell's inequalities, and why I find it so important. It is followed by a rapid review of the *first generation of experimental tests of Bell's inequalities with correlated photons*{*Bell's inequalities, with correlated photons*, carried out between 1971 and 1976. I give then a more detailed description of the three experiments of second generation, that we performed in the Institut d'Optique d'Orsay between 1976 and 1982, with pairs of correlated photons emitted in a calcium radiative cascade excited by two photon absorption. The last part gives an overview of the experiments of third generation, developed since the late 80's, and carried out with pairs of correlated photons produced in parametric down conversion: these experiments have the potentiality to close most of the loopholes still left open in the second generation experiments.

In the first part of this presentation (sections 2. to 6.), we will see that Bell's Inequalities provide a quantitative criterion to test «reasonable» Supplementary Parameters Theories versus Quantum Mechanics. Following Bell, I will first explain the motivations for considering supplementary parameters theories the argument is based on an analysis of the famous Einstein–Podolsky–Rosen (EPR) Gedankenexperiment [3]. Introducing a reasonable Locality Condition, we will then derive Bell's theorem, which states:

i. that Local Supplementary Parameters Theories are constrained by Bell's Inequalities;

ii. that certain predictions of Quantum Mechanics violate Bell's Inequalities, and therefore that Quantum Mechanics is incompatible with Local Supplementary Parameters Theories.

We will then point out that a fundamental assumption for this conflict is the Locality assumption. And we will show that in a more sophisticated version of the E.P.R. thought experiment («timing experiment»), the Locality Condition may be considered a consequence of Einstein's Causality, preventing faster–than–light interactions.

The purpose of this first part is to convince the reader that the formalism leading to Bell's Inequalities is very general and reasonable. What is surprising is that such a reasonable formalism conflicts with Quantum Mechanics. In fact, situations exhibiting a conflict are very rare, and it is remarkable that Quantum Optics is the domain where the most significant tests of this conflict have been carried out (sections 7. to 11.).

2. WHY SUPPLEMENTARY PARAMETERS? THE EINSTEIN–PODOLSKY–ROSEN–BOHM GEDANKENEXPERIMENT

2.1 EXPERIMENTAL SCHEME

Let us consider the optical variant of the Bohm's version [4] of the E.P.R. Gedankenexperiment (Fig. 9.1). A source S emits a pair of photons with different frequencies ν_1, and ν_2, counterpropagating along Oz. Suppose that the polarization part of the state vector describing the pair is:

$$|\Psi(\nu_1, \nu_2)\rangle = \frac{1}{\sqrt{2}} \{|x, x\rangle + |y, y\rangle\} \quad (9.1)$$

where $|x\rangle$ and $|y\rangle$ are linear polarizations states. This state is remarkable: it cannot be factorized into a product of two states associated to each photon, so we cannot ascribe any well defined state to each photon. In particular, we cannot assign any polarization to each photon.

We perform linear polarization measurements on these photons, with analysers I and II. The analyser I, in orientation **a**, is followed by two detectors, giving results + or −, corresponding to a linear polarization found parallel or perpendicular to **a**. The analyser II, in orientation **b**, acts similarly[1]. It is

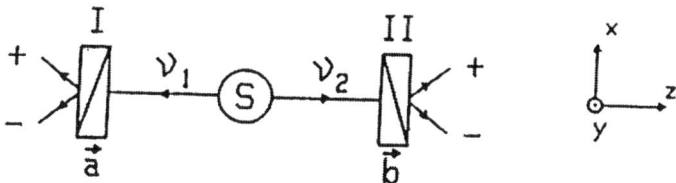

Figure 9.1 Einstein–Podolsky–Rosen–Bohm Gedankenexperiment with photons. The two photons ν_1, and ν_2, emitted in the state $|\Psi(1,2)\rangle$ of Equation 9.1, are analyzed by linear polarizers in orientations **a** and **b**. One can measure the probabilities of single or joint detections in the output channels of the polarizers.

easy to derive the Quantum Mechanical predictions for these measurements of polarization, single or in coincidence. Consider first the singles probabilities $P_\pm(\mathbf{a})$ of getting the results ± for the photon ν_1. Similarly, the singles probabilities $P_\pm(\mathbf{b})$ caracterize the results of measurements on photon ν_2. Quantum Mechanics predicts:

$$P_+(\mathbf{a}) = P_-(\mathbf{a}) = 1/2$$
$$P_+(\mathbf{b}) = P_-(\mathbf{b}) = 1/2 \quad \text{(Q.M.)} \qquad (9.2)$$

These results are in agreement with the remark that we cannot assign any polarization to each photon, so that each individual polarization measurement gives a random result.

Let us now consider the probabilities $P_{\pm\pm}(\mathbf{a},\mathbf{b})$ of joint detections of ν_1, and ν_2 in the channels + or − of polarizers I or II, in orientations **a** and **b**. Quantum mechanics predicts:

$$P_{++}(\mathbf{a},\mathbf{b}) = P_{--}(\mathbf{a},\mathbf{b}) = \frac{1}{2}\cos^2(\mathbf{a},\mathbf{b})$$
$$P_{+-}(\mathbf{a},\mathbf{b}) = P_{-+}(\mathbf{a},\mathbf{b}) = \frac{1}{2}\sin^2(\mathbf{a},\mathbf{b}) \quad \text{(Q.M.)} \qquad (9.3)$$

We are going to show that these quantum mechanical predictions have far reaching consequences.

---

[1] There is a one-to-one correspondence with the EPR Bohm Gedankenexperiment dealing with a pair of spin 1/2 particles, in a singlet state, analysed by two orientable Stern–Gerlach filters.

2.2 CORRELATIONS

Consider first the particular situation $(\mathbf{a}, \mathbf{b}) = 0$, where polarizers are parallel. The Quantum Mechanical predictions for the the joint detection probabilities (equations 9.3) are:

$$P_{++}(\mathbf{a}, \mathbf{a}) = P_{--}(\mathbf{a}, \mathbf{a}) = \frac{1}{2}$$
$$P_{+-}(\mathbf{a}, \mathbf{a}) = P_{-+}(\mathbf{a}, \mathbf{a}) = 0 \qquad (9.4)$$

According to this result, and taking into account (9.3), when the photon ν_1, is found in the + channel of polarizer I, ν_2 is found with certainty in the + channel of II (and similarly for the channels). For parallel polarizers, there is thus a *strong correlation* between the results of measurements of polarizations on the two photons ν_1, and ν_2.

A convenient way to measure the amount of correlations between random quantities, is to calculate the correlation coefficient. For the polarization measurements considered above, it is equal to

$$E(\mathbf{a}, \mathbf{b}) = P_{++}(\mathbf{a}, \mathbf{b}) + P_{--}(\mathbf{a}, \mathbf{b}) - P_{+-}(\mathbf{a}, \mathbf{b}) - P_{-+}(\mathbf{a}, \mathbf{b}) \qquad (9.5)$$

Using the prediction (9.3) of Quantum Mechanics, we find a correlation coefficient

$$E_{QM}(\mathbf{a}, \mathbf{b}) = \cos 2(\mathbf{a}, \mathbf{b}) \qquad (9.6)$$

In the particular case of parallel polarizers $((\mathbf{a}, \mathbf{b}) = 0)$, we find $E_{QM}(0) = 1$: this confirms that the correlation is total.

In conclusion, the quantum mechanical predictions suggest that although each individual measurement gives random results, these random results are correlated, as expressed by equation (9.6). For parallel (or perpendicular) orientations of the polarizers, the correlation is total ($|E_{QM}| = 1$).

2.3 DIFFICULTY OF AN IMAGE DERIVED FROM THE FORMALISM OF QUANTUM MECHANICS

We may ask the question of finding a simple image to understand these strong correlations. The most natural way to find an image may seem to follow the quantum mechanical calculations leading to (9.3). In fact, there are several ways to do this calculation. A very direct one is to project the state vector (9.1) onto the eigenvector corresponding to the relevant result. This give immediately the joint probabilities (9.3). However, since this calculation bears on state vectors describing globally the two photons, it is difficult to extract a picture in our ordinary space.

In order to identify separately the two measurements happening on both ends of the experiment, we can split the joint measurement in two steps. Suppose for instance that the measurement on photon ν_1 takes place first, and gives the result +, with the polarizer I in orientation **a**, that we take parallel to x (**a** = **e**$_x$) to simplify the reasoning. The + result has a probability of $1/2$. To proceed with the calculation, we must then use the postulate of reduction of the state vector, which states that after this measurement, the new state vector $|\Psi'(\nu_1, \nu_2)\rangle$ describing the pair is obtained by projection of the initial state vector $|\Psi(\nu_1, \nu_2)\rangle$ (equation 9.1) onto the eigenspace associated to the result +: this two dimensional eigenspace has a basis $\{|x, x\rangle, |x, y\rangle\}$. Using the corresponding projector, we find

$$|\Psi'(\nu_1, \nu_2)\rangle = |x, x\rangle \qquad (9.7)$$

This means that immediately after the first measurement, photon ν_1, takes the polarization $|x\rangle$: this is obvious because it has been measured with a polarizer oriented along x, and the result + has been found. More surprisingly, the distant photon ν_2, which has not yet interacted with any polarizer, has also been projected into the state $|x\rangle$ with a well defined polarization, parallel to the one found for photon ν_1. This surprising conclusion however leads to the correct final result (9.3), since a straightforward application of Malus law shows that a subsequent measurement performed on photon ν_2 will lead to

$$P_{++}(\mathbf{e}_x, \mathbf{b}) = \frac{1}{2}\cos^2(\mathbf{e}_x, \mathbf{b}) \qquad (9.8)$$

The calculation in two steps therefore gives the same result as the direct calculation. But in addition it suggests a picture for the two steps measurement:

i. Photon ν_1, which had not a well defined polarization before its measurement, takes the polarization associated to the obtained result, at the moment of its measurement: this is not surprising.

ii. When the measurement on ν_1 is done, photon ν_2, which had not a well defined polarization before this measurement, is projected into a state of polarization parallel to the result of the measurement on ν_1. This is very surprising, because this change in the description of ν_2 happens instantaneously, whatever the distance between ν_1, and ν_2 at the moment of the first measurement.

This picture seems in contradiction with relativity. According to Einstein, what happens in a given region of space–time cannot be influenced by an event happening in a region of space–time that is separated by a space like interval. It is therefore not surprising that one tries to find a more acceptable pictures for «understanding» the EPR correlations. It is such a picture that we consider now.

2.4 SUPPLEMENTARY PARAMETERS

Correlations between distant measurements on two separated systems that had previously interacted are common in the classical world. For instance, if a mechanical object with a null linear (or angular) momentum is split in two parts by some internal repulsion, the linear (or angular) momentum of the two separated parts remain equal and opposite in the case of a free evolution. In the general case where each fragment is submitted to some interaction, the two momenta remain correlated since they are at each moment determined by their initial values, which had a perfectly defined sum.

It is tempting to use such a classical picture to render an account of the EPR correlations, in term of common properties of the two systems. Let us consider again the perfect correlation of polarization measurements in the case of parallel polarizers $(\mathbf{a}, \mathbf{b}) = 0$. When we find $+$ for ν_1, we are sure to find also $+$ for ν_2. We are thus led to admit that there is some property (Einstein said «an element of physical reality») pertaining to this particular pair, and determining the result $++$. For another pair, when the results is $--$, we can similarly invoke a common property, determining the result $--$. It is then sufficient to admit that half the pairs are emitted with the property $++$, and half with the property $--$, to reproduce all the results of measurement in this configuration. Note however that such properties, differing from one pair to another one, are not taken into account by the Quantum Mechanical state vector $|\Psi(\nu_1, \nu_2)\rangle$ which is the same for all pairs. This is why *Einstein concluded that Quantum Mechanics is not complete*. And this is why such additional properties are referred to as «supplementary parameters», or «hidden variables»[2].

As a conclusion, it seems possible to «understand» the E.P.R. correlations by such a classical–looking picture, involving supplementary parameters differing from one pair to another one. It can be hoped to recover the statistical Quantum Mechanical predictions when averaging over the supplementary parameters. It seems that so was Einstein's position [5–7]. Note that at this stage of the reasoning, a commitment to this position does not contradict quantum mechanics: there is no logical problem to fully accept the predictions of quantum mechanics, *and* to invoke supplementary parameters giving an acceptable picture of the EPR correlations.

---

[2] Einstein actually did not speak of «hidden variables» or «supplementary parameters» but rather of «elements of the physical reality». This is why many authors refer to «realistic theories» rather than to a «hidden variable theories», or to «supplementary variable theories».

3. BELL'S INEQUALITIES

3.1 FORMALISM

Bell translated into mathematics the consequences of the preceding discussion, and he explicitly introduced supplementary parameters, denoted λ. Their distribution on an ensemble of emitted pairs is specified by a probability distribution $\rho(\lambda)$, such that

$$\rho(\lambda) \geq 0$$
$$\int \rho(\lambda) \, d\lambda = 1 \qquad (9.9)$$

For a given pair, characterised by a given supplementary parameter λ, the results of measurements are given by the bivalued functions

$$A(\lambda, \mathbf{a}) = \pm 1 \quad \text{at analyzer I (in orientation } \mathbf{a}\text{)}$$
$$B(\lambda, \mathbf{b}) = \pm 1 \quad \text{at analyzer II (in orientation } \mathbf{b}\text{)} \qquad (9.10)$$

A particular Supplementary Parameter Theory is completely defined by the explicit form of the function $\rho(\lambda)$, $A(\lambda, \mathbf{a})$ and $B(\lambda, \mathbf{b})$. It is then easy to express the probabilities of the various results of measurements. For instance, noting that the function $\frac{1}{2}[A(\lambda, \mathbf{a}) + 1]$ assumes the value $+1$ for the $+$ result, and 0 otherwise (and similarly $\frac{1}{2}[1 - B(\lambda, \mathbf{b})]$ assumes the value $+1$ for the $-$ result, and 0 otherwise), we can write

$$P_+(\mathbf{a}) = \int \rho(\lambda) \frac{[A(\lambda, \mathbf{a}) + 1]}{2} \, d\lambda$$
$$P_{+-}(\mathbf{a}, \mathbf{b}) = \int \rho(\lambda) \frac{[A(\lambda, \mathbf{a}) + 1]}{2} \frac{[1 - B(\lambda, \mathbf{b})]}{2} \, d\lambda. \qquad (9.11)$$

The correlation function assumes the simple form

$$E(\mathbf{a}, \mathbf{b}) = \int \rho(\lambda) A(\lambda, \mathbf{a}) B(\lambda, \mathbf{b}) \, d\lambda \qquad (9.12)$$

3.2 A (NAIVE) EXAMPLE OF SUPPLEMENTARY PARAMETERS THEORY

As an example of Supplementary Parameter Theory, we present a model where each photon travelling along Oz is supposed to have a well defined linear polarization, determined by its angle (λ_1 or λ_2) with the x axis. In order to account for the strong correlation, we assume that the two photons of a pair are emitted with the same linear polarization, defined by the common angle λ (figure 9.2). The polarization of the various pairs is randomly distributed,

196 COMPLETE SCATTERING EXPERIMENTS

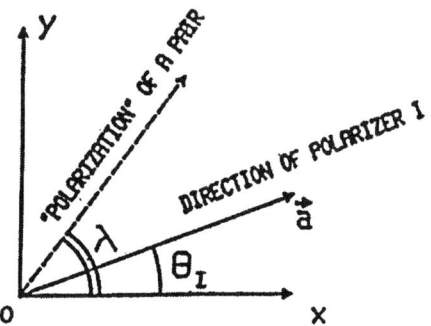

Figure 9.2 The example. Each pair of photons has a «direction of polarization», defined by λ which is the supplementary parameter of the model.

according to the probability distribution $\rho(\lambda)$ that we take isotropic:

$$\rho(\lambda) = \frac{1}{2\pi} \tag{9.13}$$

To complete our model, we must give an explicit form for the functions $A(\lambda, \mathbf{a})$ and $B(\lambda, \mathbf{b})$. We take the following form

$$\begin{aligned} A(\lambda, \mathbf{a}) &= \mathrm{sign}\,\{\cos 2(\Theta_I - \lambda)\} \\ B(\lambda, \mathbf{b}) &= \mathrm{sign}\,\{\cos 2(\Theta_{II} - \lambda)\} \end{aligned} \tag{9.14}$$

where the angles Θ_I, and Θ_{II}, indicate the orientations of the polarizers. Note that these forms are very reasonable: $A(\lambda, \mathbf{a})$ assumes the value $+1$ when the polarization of photon ν_1, makes an angle less than $\frac{\pi}{4}$ with the direction of analysis $\mathbf{a}$, and -1 for the complementary case (polarization closer to the perpendicular to $\mathbf{a}$).

With this explicit model, we can use equations (9.11) to calculate the probabilities of the various measurements. We find for instance joint probabilities

$$P_+(bfa) = P_-(\mathbf{a}) = P_+(\mathbf{b}) = P_-(\mathbf{b}) = \frac{1}{2} \tag{9.15}$$

identical to the Quantum Mechanical results. The model also allows us to calculate the correlation function, and we find, using (9.12):

$$E(\mathbf{a}, \mathbf{b}) = 1 - 4\frac{|\Theta_I - \Theta_{II}|}{\pi} = 1 - 4\frac{|(\mathbf{a}, \mathbf{b})|}{\pi} \tag{9.16}$$
$$\text{for} \quad -\frac{\pi}{2} \leq \Theta_I \leq \Theta_{II}$$

This is a remarkable result. Note first that $E(\mathbf{a}, \mathbf{b})$ depends only on the relative angle $(\mathbf{a}, \mathbf{b})$, as the Quantum Mechanical prediction (9.6). Moreover, as shown on figure 9.3, the difference between the predictions of the simple supplementary parameters model and the quantum mechanical predictions remains moderate. There is an exact agreement for the angles 0 and $\pi/2$, i.e. cases of total correlation. This result, obtained with an extremely simple supplementary parameters model, is very encouraging, and it might be hoped that a more sophisticated model could be able to reproduce exactly the Quantum Mechanical predictions. *Bell's discovery is the fact that the search for such models is hopeless*, as we are going to show now.

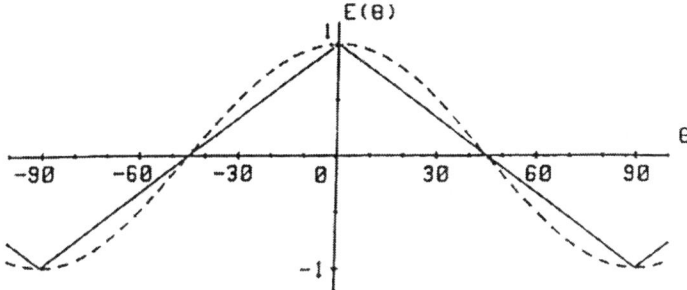

Figure 9.3 Polarization correlation coefficient, as a function of the relative orientation of the polarizers: (i) Dotted line: Quantum Mechanical prediction (ii) solid line: our simple model.

3.3 BELL'S INEQUALITIES

There are many different forms, and demonstrations of Bell's inequalities. We give here a very simple demonstration leading to a form directly applicable to the experiment[3].

Let us consider the quantity

$$\begin{aligned} s &= A(\lambda, \mathbf{a})B(\lambda, \mathbf{b}) - A(\lambda, \mathbf{a})B(\lambda, \mathbf{b}') \\ &\quad + A(\lambda, \mathbf{a}')B(\lambda, \mathbf{b}) + A(\lambda, \mathbf{a}')B(\lambda, \mathbf{b}') \\ &= A(\lambda, \mathbf{a})\left[B(\lambda, \mathbf{b}) - B(\lambda, \mathbf{b}')\right] + A(\lambda, \mathbf{a}')\left[B(\lambda, \mathbf{b}) + B(\lambda, \mathbf{b}')\right] \end{aligned} \qquad (9.17)$$

Remembering that the four numbers A and B take only the values ± 1, a simple inspection of the second line of (9.17) shows that

$$s(\lambda, \mathbf{a}, \mathbf{a}', \mathbf{b}, \mathbf{b}') = \pm 2 \qquad (9.18)$$

---

[3]It is important to distinguish between inequalities which show a mathematical contradiction with quantum mechanics, but without the possibility of an experimental test with (necessarily) imperfect apparatus, and inequalities allowing an experimental test provided that the experimental imperfections remain in certain limits.

The average of s over λ is therefore comprised between $+2$ and -2

$$-2 \leq \int \rho(\lambda) s(\lambda, \mathbf{a}, \mathbf{a'}, \mathbf{b}, \mathbf{b'}) \, d\lambda \leq 2 \qquad (9.19)$$

According to (9.12), we rewrite these inequalities

$$-2 \leq S(\mathbf{a}, \mathbf{a'}, \mathbf{b}, \mathbf{b'}) \leq 2 \qquad (9.20)$$

with

$$S(\mathbf{a}, \mathbf{a'}, \mathbf{b}, \mathbf{b'}) = E(\mathbf{a}, \mathbf{b}) - E(\mathbf{a}, \mathbf{b'}) + E(\mathbf{a'}, \mathbf{b}) + E(\mathbf{a'}, \mathbf{b'}) \qquad (9.21)$$

These are B.C.H.S.H. inequalities, ie. Bell's inequalities generalized by Clauser, Home, Shimony, Holt [8]. They bear upon the combination S of the four polarization correlation coefficients, associated to two directions of analysis for each polarizer (**a** and **a'** for polarizer I, **b** and **b'** for polarizer II.

4. CONFLICT WITH QUANTUM MECHANICS

4.1 EVIDENCE

We can use the predictions (9.6) of Quantum Mechanics for EPR pairs, to evaluate the quantity $S(\mathbf{a}, \mathbf{a'}, \mathbf{b}, \mathbf{b'})$ defined by equation (9.21). For the particular set of orientations shown on Figure 9.4, the result is

$$S_{QM} = 2\sqrt{2} \qquad (9.22)$$

This quantum mechanical prediction definitely conflicts with the Bell's inequality (9.20) which is valid for any supplementary Parameter Theory of the general form defined in § 3.1.

We have thus found a situation where the quantum mechanical predictions cannot be reproduced (mimicked) by Supplementary Parameters Theories. This is the essence of Bell's theorem: it is impossible to find a Supplementary Parameter Theory, of the general form defined in § 3.1, that reproduces **all** the predictions of quantum mechanics. This statement is the generalisation of what appears on Figure 9.3, for the particular supplementary parameter model considered in § 3.2: the model exactly reproduces the predictions of quantum mechanics for some particular angles $(0, \pi/4, \pi/2)$, but it somewhat deviates at other angles. The importance of Bell's theorem is that it is not restricted to a particular supplementary parameters model, but it is general.

4.2 MAXIMUM CONFLICT

It is interesting to look for the maximum violation of Bell's inequalities by the quantum mechanical predictions. Let us take the quantum mechanical value

Experimental Tests of Bell's Inequalities with Correlated Photons

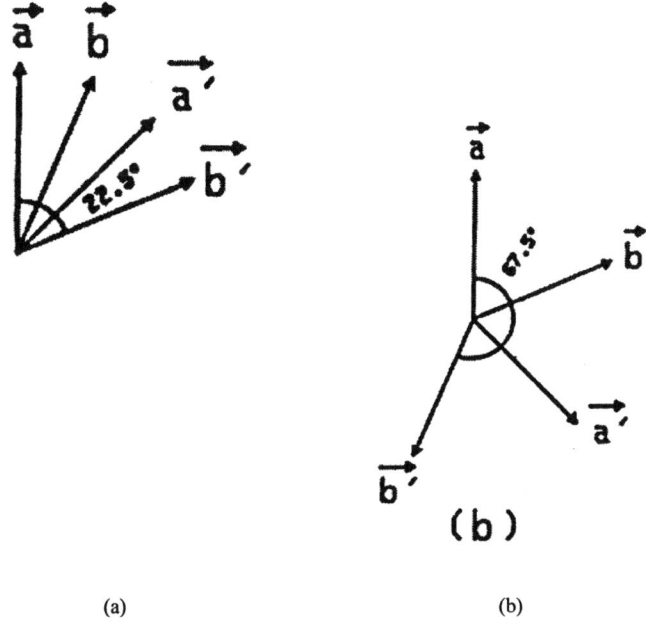

(a)   (b)

Figure 9.4 Orientations yielding the largest conflict between Bell's inequalities and Quantum Mechanics.

of S:

$$S_{QM}(\mathbf{a}, \mathbf{a}', \mathbf{b}, \mathbf{b}') = \cos(\mathbf{a}, \mathbf{b}) - \cos(\mathbf{a}, \mathbf{b}') \\ + \cos(\mathbf{a}', \mathbf{b}) + \cos(\mathbf{a}', \mathbf{b}') \tag{9.23}$$

It is a function of the three independent variables $(\mathbf{a}, \mathbf{b})$, $(\mathbf{a}', \mathbf{b})$, and $(\mathbf{a}', \mathbf{b}')$. Note that

$$(\mathbf{a}, \mathbf{b}') = (\mathbf{a}, \mathbf{b}) + (\mathbf{a}', \mathbf{b}) + (\mathbf{a}', \mathbf{b}')$$

In order to find the extremum values of S_{QM}, we write that the three partial derivatives are null, and we find

$$(\mathbf{a}, \mathbf{b}') = (\mathbf{a}', \mathbf{b}) = (\mathbf{a}', \mathbf{b}') = \Theta \tag{9.24}$$

and

$$\sin \Theta = \sin 3\Theta \tag{9.25}$$

We have plotted on Figure 9.5 the function $S_{QM}(\Theta)$ evaluated in the case of condition (9.24). It shows that the absolute maximum and minimum of S_{QM}

are

$$S_{QM} = 2\sqrt{2} \quad \text{for} \quad \Theta = \pm\frac{\pi}{8} \tag{9.26}$$

$$S_{QM} = -2\sqrt{2} \quad \text{for} \quad \Theta = \pm\frac{3\pi}{8}. \tag{9.27}$$

These values are solutions of (9.25). The corresponding sets of orientations are displayed on Figures 9.4. They give the maximum violations of Bell's inequalities.

More generally, Figure 9.5 shows that there is a broad range of orientations leading to a conflict with Bells inequalities. However, it is also clear that not all the sets of orientations entail such a conflict.

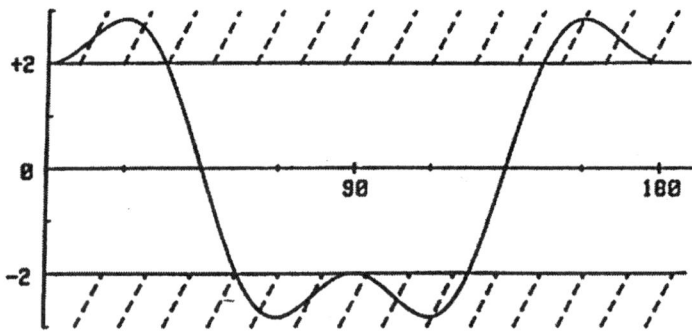

Figure 9.5 $S(\Theta)$ as predicted by Quantum Mechanics for EPR pairs. The conflict happens when $|S|$ is larger than 2, and it is maximum for the sets of orientations of Figure 9.4.

5. DISCUSSION: THE LOCALITY CONDITION

We have now established Bell's theorem: Quantum Mechanics conflicts with any Supplementary Parameter Theory as defined in § 3.1, since it violates a consequence of such a theory (Bell's inequalities). Clearly, it is interesting at this stage to look for the hypotheses underlying the formalism introduced in § 3.1. One may then hope to point out a specific hypothesis responsible for the conflict.

A first hypothesis is the existence of supplementary parameters. As we have seen, they have been introduced in order to render an account of the correlations at a distance. This hypothesis is strongly linked to a conception of the world, as expressed by Einstein, where the notion of separated physical realities for separated particles has a meaning. It is even possible to derive the existence of supplementary parameters from general statements about the Physical Reality, in the spirit of Einstein's ideas [9]. But an hypothesis in this spirit seems absolutely necessary to obtain inequalities conflicting with quantum mechanics.

The second considered hypothesis is based on the remark that the formalism of section 3.1 is deterministic: once λ is fixed, the results $A(\lambda, \mathbf{a})$ and $B(\lambda, \mathbf{b})$ of the polarization measurements are certain. One has speculated that it may be a good reason for a conflict with the non–deterministic formalism of quantum mechanics. In fact, as first shown by Bell [10], and subsequently developed [11], it is easy to generalize the formalism of section 3.1 to *Stochastic* Supplementary Parameter Theories where the deterministic measurement functions $A(\lambda, \mathbf{a})$ and $B(\lambda, \mathbf{b})$ are replaced by probabilistic functions. One then finds that the Bell's inequalities still hold, and that the conflict does not disappear. It is therefore generally accepted that the deterministic character of the formalism is not the reason for the conflict [4].

The most important hypothesis, stressed by Bell in all his papers, is the local character of the formalism of section 3.1. We have indeed implicitly assumed that the result $A(\lambda, \mathbf{a})$ of the measurement at polarizer I, does not depend on the orientation $\mathbf{b}$ of the remote polarizer II, and vice–versa. Similarly, it is assumed that the probability distribution $\rho(\lambda)$ (i.e. the way in which pairs are emitted) does not depend on the orientations $\mathbf{a}$ and $\mathbf{b}$. This *locality assumption* is crucial: Bell's Inequalities would no longer hold without it. It is indeed clear that the demonstration of § 3.3 fails with quantities such as $A(\lambda, \mathbf{a}, \mathbf{b})$ or $\rho(\lambda, \mathbf{a}, \mathbf{b})$.

To conclude, there are two hypothesis that seem to be necessary to obtain Bell's inequalities, and consequently a conflict with quantum mechanics:

- distant correlations can be understood by introduction of supplementary parameters carried along by the separated particles, in the spirit of Einstein's ideas that separated objects have separated physical realities.

- the quantities $A(\lambda)$, $B(\lambda)$, and $\rho(\lambda)$ obey the locality condition, i.e. they do not depend on the orientations of the distant polarizers.

6. GEDANKENEXPERIMENT WITH VARIABLE ANALYZERS: THE LOCALITY CONDITION AS A CONSEQUENCE OF EINSTEIN'S CAUSALITY

In static experiments, in which the polarizers are held fixed for the whole duration of a run, the Locality Condition must be stated as an assumption. Although highly reasonable, this condition is not prescribed by any fundamental physical law. To quote J. Bell, «the settings of the instruments are made sufficiently in advance to allow them to reach some mutual rapport by exchange of signals with velocity less than or equal to that of light». In that case, the result

---

[4]This conclusion is not shared by all authors. For instance, it has been argued that the stochastic theories of Bell [10] or of Clauser and Horne [11] achieve no further generality, since they can be mimicked by a deterministic theory [12]

$A(\lambda)$ of the measurement at polarizer I could depend on the orientation **b** of the remote polarizer II, and vice-versa. The Locality Condition would no longer hold, nor would Bell's Inequalities.

Bell thus insisted upon the importance of «experiments of the type proposed by Bohm and Aharonov [5], in which the settings are changed during the flight of the particles»[5]. In such a *timing–experiment, the locality condition would become a consequence of Einstein's Causality that prevents any faster–than–light influence.*

As shown in our 1975 proposal [13, 14], it is sufficient to switch each polarizer between two particular settings (**a** and **a'** for I, **b** and **b'** for II). It then becomes possible to test experimentally a large class of Supplementary Parameters Theories: those obeying Einstein's Causality. In such theories, the response of polarizer I at time t, is allowed to depend on the orientation **b** (or **b'**) of polarizer II at times anterior to $t - L/c$ (L is the distance between the polarizers). A similar retarded dependence is considered for the probability distribution $\rho(\lambda)$, i.e. the way in which pairs are emitted at the source. For random switching times, with both sides uncorrelated, the predictions of these more general «separable supplementary parameters theories» are constrained by generalized Bell's Inequalities [13, 14], based on Einstein's causality and not on Bell's locality condition.

On the other hand, it is easy to show [15] that the polarization correlations predicted by Quantum Mechanics depend only on the orientations a and b at the very time of the measurements, and do not involve any retardation terms L/c. For a suitable choice of the set of orientations (**a**, **a'**, **b**, **b'**) – for instance the sets displayed on Figure 9.4 – the Quantum mechanical predictions still conflict with generalized Bell's Inequalities.

In an experiment with time varying polarizers, Bell's theorem therefore states that Quantum Mechanics is incompatible with Supplementary Parameters theories obeying Einstein's causality. Note that Einstein's causality already played an important role in the discussions leading to the notion of supplementary parameters, or equivalently of an independent physical reality for each separated subsystem [6]. It therefore does not seem exaggerate to conclude that in a scheme with time varying polarizers, Bell's theorem establishes a contradiction between Quantum Mechanics and a description of the world in the spirit of Einstein's ideas. Note however that Einstein did not know Bell's theorem, and he could logically think that his world view was compatible with all the algebraic predictions of quantum mechanics. It is of course impossible to know what would have been his reaction in front of the contradiction revealed by Bell's inequalities.

---

[5]The idea was already expressed in Bohm's book [4].

7. FROM BELL'S THEOREM TO A REALISTIC EXPERIMENT

7.1 EXPERIMENTALLY TESTING BELL'S INEQUALITIES

With Bell's theorem, the debate on the possibility (or necessity) of completing quantum mechanics changed dramatically. It was no longer a matter of philosophical position (realism versus positivism ...), or of personal taste. It became possible to settle the question by an experiment. If one can produce pairs of photons (or of spin $1/2$ particles) in an EPR state, and measure the 4 coincidence rates $N_{\pm\pm}(\mathbf{a},\mathbf{b})$ with detectors in the output channels of the polarizers (or Stern–Gerlach filters), one obtains the polarization correlation coefficient, for polarizers in orientations $\mathbf{a}$ and $\mathbf{b}$:

$$(\mathbf{a},\mathbf{b}) = \frac{N_{++}(\mathbf{a},\mathbf{b}) - N_{+-}(\mathbf{a},\mathbf{b}) - N_{-+}(\mathbf{a},\mathbf{b}) + N_{--}(\mathbf{a},\mathbf{b})}{N_{++}(\mathbf{a},\mathbf{b}) + N_{+-}(\mathbf{a},\mathbf{b}) + N_{-+}(\mathbf{a},\mathbf{b}) + N_{--}(\mathbf{a},\mathbf{b})} \quad (9.28)$$

By performing four measurements of this type in orientations $(\mathbf{a},\mathbf{b})$, $(\mathbf{a},\mathbf{b}')$, $(\mathbf{a}',\mathbf{b})$ and $(\mathbf{a}',\mathbf{b}')$, one obtains a measured value $S_{exp}(\mathbf{a},\mathbf{a}',\mathbf{b},\mathbf{b}')$ for the quantity S defined in equation (9.21). Choosing a situation where quantum mechanics predicts that this quantity violates Bell's inequalities (9.20), we have a test allowing one to discriminate between quantum mechanics and any local supplementary parameter theory. If in addition we use a scheme with variable polarizers, we even test the more general class of «separable» (or causal in the relativistic sense) supplementary parameters theories.

7.2 SENSITIVE SITUATIONS ARE RARE

Quantum Mechanics has been upheld in such a great variety of experiments, that Bell's Theorem might just appear as a proof of the impossibility of supplementary parameters. However, situations in which the conflict revealed by Bell's inequalities arises (sensitive situations) are so rare that, in 1965, none had been realized.

To better appreciate this point, let us first note that Bell's inequalities are compatible with the whole classical physics, namely classical (relativistic) mechanics and classical electrodynamics, which can be embedded into the Supplementary Parameters formalism obeying Einstein's causality. For instance, in classical mechanics, the λ's would be the initial positions and velocities of the particles, from which the future evolution can be derived. Similarly, in classical electrodynamics, the λ's would be the currents and charges motions in the sources, from which one can deduce the electromapetic fields, and their action on the measuring apparatus.

Moreover, in situations usually described by quantum mechanics, it does not often happen that there is a conflict with Bell's inequalities. More precisely, for situations in which one looks for correlations between two separated subsystems (that may have interacted in the past), we can point out two conditions necessary to have the possibility of a conflict with Bell's inequalities:

- The two separated subsystems must be in a non–factorizing state, such as (9.1) (or the singlet state for two spin 1/2 particles).

- For each subsystem, it must be possible to choose the measured quantity among at least two non–commuting observables (such as polarization measurements along directions **a** and **a'**, neither parallel nor perpendicular).

Even in such cases, we have seen that the conflict exists only for well chosen measured quantities (sets of orientations). But, as shown on figure 9.5, there are many orientations sets for which the quantum mechanical predictions do not violate Bell's inequalities.

It was realized in 1965 that there was no experimental evidence of a violation of Bell's inequalities. Since these inequalities are derived from very reasonable hypothesis, one could consider the possibility that the violation of Bell's inequalities indicate a situation where quantum mechanics fails. It was thus tempting to design a sensitive experiment, i.e. an experiment where the predictions of quantum mechanics for the real situation definitely violate Bell's inequalities. The experiment would then give a clearcut result between Quantum Mechanics, and Supplementary Parameter Theories obeying Bell's locality condition.

7.3 PRODUCTION OF PAIRS OF PHOTONS IN AN EPR STATE

As pointed out by C.H.S.H. [8], pairs of photons emitted in suitable atomic radiative cascades are good candidates for a sensitive test. Consider for instance a $J = 0 \to J = 1 \to J = 0$ atomic cascade (Figure 9.6). Suppose that we select, with the use of wavelengths filters and collimators, two plane waves of frequencies ν_1, and ν_2 propagating in opposite directions along the z axis (Figure 9.7). It is easy to show, by invoking parity and angular momentum conservation, that the polarization part of the state vector describing the pair (ν_1, ν_2) can be written :

$$|\Psi(\nu_1, \nu_2)\rangle = \frac{1}{\sqrt{2}} \{|R, R\rangle + |L, L\rangle\} \tag{9.29}$$

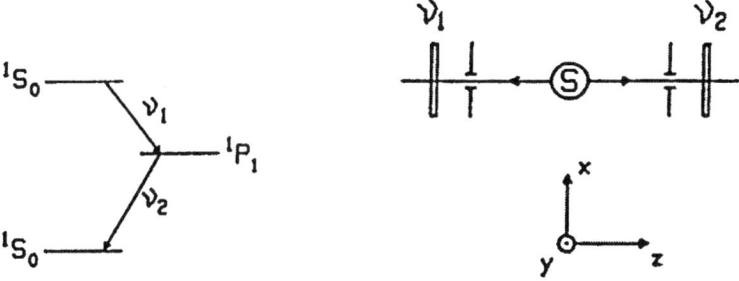

Figure 9.6 Radiative cascade emitting pairs of photons correlated in polarization.

Figure 9.7 Ideal configuration photons correlated in polarization. (infinitely small solid angles).

where $|R\rangle$ and $|L\rangle$ are circularly polarized states. By expressing $|R\rangle$ and $|L\rangle$ on a linear polarization basis, we obtain the state (9.1)

$$|\Psi(\nu_1,\nu_2)\rangle = \frac{1}{\sqrt{2}}\{|x,x\rangle + |y,y\rangle\}.$$

With this EPR state, one can envisage a sensitive experiment.

7.4 REALISTIC EXPERIMENT

A real experiment differs from the ideal one in several respects. For instance, the light should be collected in finite angles $2u$, as large as possible (Figure 9.8). In this situation, one can show [16] that the contrast of the correlation function decreases, since (9.6) is replaced by :

$$E_{QM}(\mathbf{a},\mathbf{b}) = F(u)\cos 2(\mathbf{a},\mathbf{b}) \tag{9.30}$$

where

$$F(u) \leq 1.$$

Figure 9.9 displays $F(u)$ for a $J = 0 \to J = 1 \to J = 0$ cascade (alkaline–earth atom, with no hyperfine structure). Fortunately, one can use large angles without great harm. For $u = 32°$ (our experiments), one has $F(u) = 0.984$.

All experimental inefficiencies (polarizers defects, accidental birefringences etc ...) will similarly lead to a decrease of the correlation function $E(\mathbf{a},\mathbf{b})$. The function $S_{QM}(\Theta)$ (Figure 9.5) is then multiplied by a factor less than 1, and the conflict with Bell's Inequalities decreases, and even may vanish. Therefore, an actual experiment must be carefully designed and every auxiliary effect must be evaluated. All relevant parameters must be perfectly controlled,

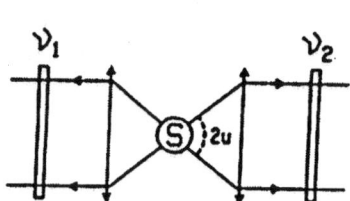
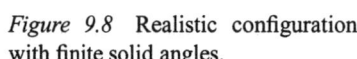

Figure 9.8 Realistic configuration, with finite solid angles.

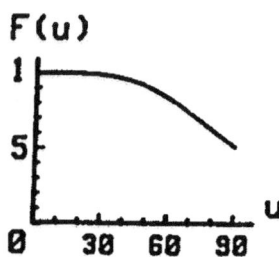

Figure 9.9 Reduction factor $F(u)$ for a $J = 0 \to J = 1 \to J = 0$ cascade.

since one may guess that a forgotten effect would similarly lead to a decrease of the conflict. For instance, one knows that an hyperfine structure dramatically decreases $F(u)$, so that only even isotopes can be used [16].

7.5 TIMING CONDITIONS

As we have seen in section 6., Bell's Locality Condition may be considered a consequence of Einstein's Causality, if the experiment fulfils requirements, that can be split in two conditions:

i. The distant measurements on the two subsystems must be space–like separated.

ii. The choices of the quantities measured on the two separated subsystems must be made at random, and must be space–like separated.

The second condition is obviously more difficult to fulfil.

8. FIRST GENERATION EXPERIMENTS

The C.H.S.H. paper [8], published in 1969, had shown the possibility of sensitive experiments with correlated photons produced in certain atomic cascades. Two groups started an experiment, one in Berkeley, one in Harvard. After their conflicting results, a third experiment was carried out in College station (Texas). All the three experiments used a simplified experimental scheme, somewhat different from the ideal one since it involved one–channel polarizers.

8.1 EXPERIMENTS WITH ONE CHANNEL POLARIZER

In this simplified experimental scheme, one uses polarizers that transmit light polarized parallel to **a** (or **b**), but blocks the orthogonal one. Compared to the scheme of Figure 9.1, one thus only detects the + results, and the coincidence measurements only yield the coincidence rates $N_{++}(\mathbf{a}, \mathbf{b})$ between the + channels. In order to recover the missing – data, auxiliary runs are performed with one or both polarizers removed (the «orientation», of a removed polarizers is conventionally denoted ∞). We can then write relations between the measured coincidence rates $N_{++}(\mathbf{a}, \mathbf{b})$, $N_{++}(\mathbf{a}, \infty)$, and $N_{++}(\infty, \mathbf{b})$ and coincidence rates which are not measured.

$$\begin{aligned} N_{++}(\infty, \infty) &= N_{++}(\mathbf{a}, \mathbf{b}) + N_{-+}(\mathbf{a}, \mathbf{b}) \\ &\quad + N_{+-}(\mathbf{a}, \mathbf{b}) + N_{--}(\mathbf{a}, \mathbf{b}) \\ N_{++}(\mathbf{a}, \infty) &= N_{++}(\mathbf{a}, \mathbf{b}) + N_{+-}(\mathbf{a}, \mathbf{b}) \\ N_{++}(\infty, \mathbf{b}) &= N_{++}(\mathbf{a}, \mathbf{b}) + N_{-+}(\mathbf{a}, \mathbf{b}) \end{aligned} \quad (9.31)$$

By substitution into the expression (9.28) of the polarization correlation coefficient, and into inequalities (9.21), one can eliminate all the quantities which are not measured, and obtain new B.C.H.S.H. inequalities

$$-1 \le S' \le 0 \quad (9.32)$$

where the quantity S'

$$S' = \frac{N(a, b) - N(a, b') + N(a', b) + N(a', b') - N(a', \infty) - N(\infty, b)}{N(\infty, \infty)} \quad (9.33)$$

is expressed as a function of *measured* coincidence rates only (we have omitted the implicit subscripts $_{++}$ in the expression above). For the orientation sets shown on Figure 9.4, the Quantum Mechanical predictions violate the Bell's inequalities 9.32:

$$\begin{aligned} S'^{Max}_{QM} &= \frac{\sqrt{2} - 1}{2} \quad \text{for} \quad \Theta = \pi/8 \\ S'^{Min}_{QM} &= \frac{-\sqrt{2} - 1}{2} \quad \text{for} \quad \Theta = 3\pi/8 \end{aligned} \quad (9.34)$$

It is therefore possible to make a sensitive test with one channel polarizers.

Note however that the derivation of the Bell's inequalities (9.32) requires a supplementary assumption. Since the detection efficiencies are low (due to small angular acceptance and low photomultipliers efficiencies), the probabilities involved in the expression of $E(\mathbf{a}, \mathbf{b})$ must be redefined on the ensemble of pairs that would be detected with polarizers removed. This procedure is valid only if one assumes a reasonable hypothesis about the detectors. The C.H.S.H. assumption [8] states that, «given that a pair of photons emerges from the polarizers, the probability of their joint detection is independent of the polarizer orientations» (or of their removal). Clauser and Horne [11] have exhibited another assumption, leading to the same inequalities. The status of these assumptions has been thoroughly discussed in reference [17].

8.2 RESULTS

In the Berkeley experiment [18], Clauser and Freedman built a source where calcium atoms were excited by ultraviolet radiation to highly lying states. The atom would then decay, and among the various desexcitation routes, it had some probability to emit a couple of green and violet correlated photons ($4p^2\,{}^1S_0 \to 4s4p^1\,{}^1P_1 \to 4s^2\,{}^1S_0$). Since the signal was weak, and spurious cascades occurred, it took more than 200 hours of measurement for a significant result. The results were found in agreement with Quantum Mechanics, and a violation of the relevant Bell's inequalities (9.32) was observed (by 5 standard deviations).

At the same time, in Harvard, Holt and Pipkin [19] found a result in disagreement with Quantum Mechanics, and in agreement with Bell's Inequalities. Their source was based on the $9\,{}^1P_1 \to 7\,{}^3P_1 \to 6\,{}^3P_0$ cascade of Mercury – (isotope 200), excited by electron bombardment. The data accumulation lasted 150 hours. Clauser subsequently repeated their experiment, but with Mercury 202. He found an agreement with Quantum Mechanics, and a significant violation of Bell's Inequalities [20].

In 1976, in Houston, Fry and Thompson [21] built a much improved source of correlated photons, emitted in the $7^3S_1 \to 6^3P_1 \to 6^3S_0$ cascade in Mercury 200. They could selectively excite the upper level of the cascade, by use of a C.W. single line tunable laser (a quite rare instrument at that time). The signal was several order of magnitude larger than in previous experiments, allowing them to collect the relevant data in a period of 80 minutes. Their result was in excellent agreement with Quantum Mechanics, and they found a violation, by 4 standard deviations, of the Bell's inequalities (9.32) for single channel polarizers experiments.

9. ORSAY EXPERIMENTS (1980–1982)

9.1 THE SOURCE

From the beginning of our programme, our goal was to implement more sophisticated experimental schemes [13, 14], so we devoted a lot of efforts to develop a high–efficiency, very stable, and well controlled source of correlated photons. This was achieved (Figure 9.10) by a two photon selective excitation [22] of the $4p^2\,{}^1S_0 \to 4s4p\,{}^1P_1 \to 4s^2\,{}^1S_0$ cascade of calcium already used by Clauser and Freedman. This cascade is very well suited to coincidence experiments, since the lifetime τ_r, of the intermediate level is rather short ($5ns$). If one can reach an excitation rate of about $1/\tau_r$, then an optimum signal–to–noise ratio for coincidence measurements with this cascade is reached. We have been able to obtain this optimum rate with the use of a Krypton laser $\lambda_K = 406nm$) and a tunable dye laser ($\lambda_D = 581nm$) tuned to resonance for the twophoton process. Both lasers are single–mode operated. They are

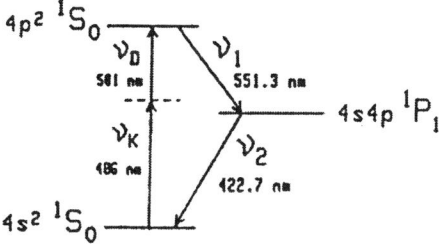

Figure 9.10 Two–photon selective excitation of the Calcium cascade delivering the pairs of correlated photons.

focused onto a Calcium atomic beam (laser beam waists about $50\mu m$). Two feedback loops provide the required stability of the source (better than 0.5% for several hours): the first loop controls the wavelength of the tunable laser to ensure the maximum fluorescence signal; a second loop controls the angle between the lasers polarizations, and compensates all the fluctuations of the cascade rate. With a few tens of milliwatts from each laser, the cascade rate is about $N = 4 \cdot 10^7 s^{-1}$. An increase beyond this rate would not significantly improve the signal–to–noise ratio for coincidence counting, since the accidental coincidence rate increases as N^2, while the true coincidence rate increases as N. At this cascade rate, the coincidence rate with parallel polarizers was about $10^2 s^{-1}$, several orders of magnitude larger than in the first experiments. A statistical accuracy of 1% could then be achieved in each individual run of duration $100s$.

9.2 DETECTION – COINCIDENCE COUNTING

The fluorescence light is collected by two large–aperture aspherical lenses ($u = 32°$ defined on figure 9.8), followed in each leg by an interference filter (respectively at $551.3 nm$ and $422.7 nm$), a transport optical system, a polarizer, and a photomultiplier tube. The photomultipliers feed the coincidence–counting electronics, that includes a time–to–amplitude converter and a multichannel analyzer, yielding the time–delay spectrum of the two–photon detections (Figure 9.11). This spectrum shows a flat background due to accidental coincidences (between photons emitted by different atoms). The true coincidences (between photons emitted by the same atom) are displayed in the peak rising at the null–delay, and exponentially decaying with a time constant $\tau = 5ns$ (lifetime of the intermediate state of the cascade). The measured–coincidence signal is thus the area of the peak. Additionally, a standard coincidence circuit with a $19ns$ coincidence window monitors the rate of coincidences around null delay, while a delayed–coincidence channel monitors the accidental rate. It is

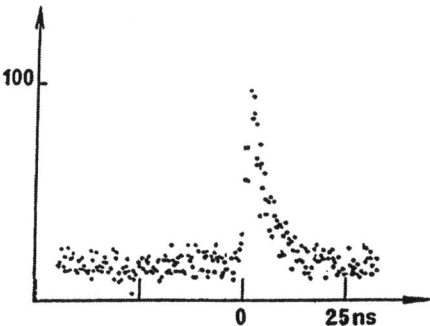

Figure 9.11 Time delay spectrum. Number of detected pairs as a function of the delay between the detections of two photons.

then possible to check that the true coincidence rate obtained by subtraction is equal to the signal in the peak of the time–delay spectrum.

In the second and third experiments, we used a fourfold coincidence system, involving a fourfold multichannel analyzer and four double–coincidence circuits. The data were automatically gathered and processed by a computer.

9.3 EXPERIMENT WITH ONE–CHANNEL POLARIZERS

Our first experiment was carried out using one channel pile of plates polarizers, made of ten optical grade glass plates at Brewster angle, ensuring an excellent rotational invariance [23]. For a fully polarized light, the maximum and minimum transmission were 0.975 ± 0.005 and 0.030 ± 0.005 respectively.

Thanks to our high–efficiency source, the statistical uncertainly was less than 2% in a $100s$ run (with polarizers removed). This allowed us to perform various statistical checks, as well as physical checks, for instance on the rotational invariance of the signals (for all these measurements, the long term stability of the source, at the level of 0.5%, was found crucial).

A direct test of the Bell's inequalities for single channel polarizers (9.32) has been performed. We have found for the quantity S' (equation 9.33)

$$S'_{exp} = 0.126 \pm 0.014 \qquad (9.35)$$

violating inequalities (9.32) by 9 standard deviations, and in good agreement with the Quantum Mechanical predictions for our polarizers efficiencies and lenses aperture angles:

$$S'_{QM} = 0.118 \pm 0.005 \qquad (9.36)$$

The error in S'_{QM} accounts for the uncertainty in the measurements of the polarizers efficiencies. The agreement between the experimental data and the

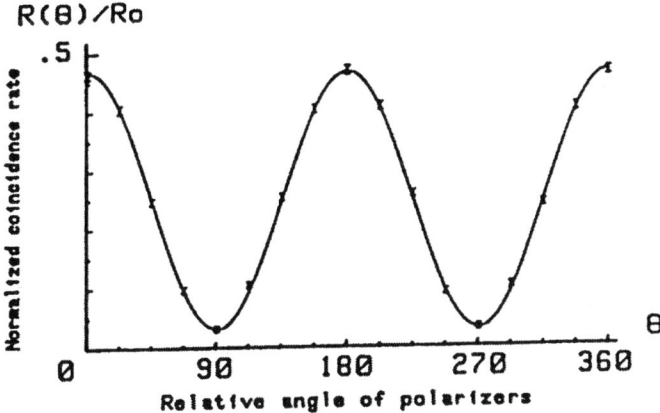

Figure 9.12 Experiment with one channel polarizers: Normalized coincidence rate as a function of the relative polarizers orientation. Indicated errors ±1 standard deviation. The solid curve is not a fit to the data but the prediction by Quantum Mechanics.

Quantum Mechanical predictions has been checked in a full 360° range of orientations (Figure 9.12).

We have repeated these measurements with the polarizers moved at $6.5 in$ from the source. At such a distance (four coherence–lengths of the wave packet associated with the lifetime τ_r) the detection events are space–like separated (we therefore fulfil the first time condition of section 7.5). No modification of the experimental results was observed, and the Bell's inequality was violated by the same amount.

9.4   EXPERIMENT WITH TWO–CHANNEL ANALYZERS

With single–channel polarizers, the measurements of polarization are inherently incomplete. When a pair has been emitted, if no count is obtained at one of the photomultipliers, there is no way to know if «it has been missed» by the detector or if it has been blocked by the polarizer (only the later case corresponds to a − result for the measurement) [24–26]. This is why one had to resort to auxiliary experiments, and indirect reasoning using supplementary assumptions, in order to test Bell's inequalities.

With the use of two–channel polarizers, it is possible to follow much more closely the ideal scheme of Figure 9.1 [24, 25, 27]. We have performed such an experiment, with polarizing cubes with dielectric layers transmitting one polarization and reflecting the orthogonal one. Such a polarization splitter, and the two corresponding photomultipliers, are fixed on a rotatable mount.

This device (polarimeter) yields the + and − results for a linear polarization measurement. It is an optical analog of a Stem–Gerlach filter for a spin 1/2 particle.

With polarimeters I and II in orientations **a** and **b**, and a fourfold coincidence counting system, we are able to measure in a single run the four coincidence rates $N_{++}(\mathbf{a},\mathbf{b})$, and to obtain directly the polarization correlation coefficient $E(\mathbf{a},\mathbf{b})$ (see equation 9.28). It is then sufficient to repeat the same measurement for three other orientations, and the Bell's inequality (9.20) can be directly tested.

This experimental scheme is much closer to the ideal scheme of figure 9.1 than previous experiments with one channel polarizers, and we do not need the strong supplementary assumption on the detectors. However, the detection efficiency in each channel is well below unity, first because of the detection solid angle, and second because of the efficiency of the photomultiplier. An advocate of hidden variable theories could then argue that we are not sure that the sample on which the measurement bears remains the same when the orientations of the polarimeters are changed. In order to be logically allowed to compare our measurements to Bell's inequalities, we therefore also need a supplementary assumption: we must assume that the ensemble of actually detected pairs is independent of the orientations of the polarimeters. This assumption is very reasonable with our symmetrical scheme, where the two orthogonal output channels of a polarizer are treated in the same way (the detection efficiencies in both channels of a polarimeter are equal). Moreover, we have checked that the sum of the four coincidence rate $N_{\pm\pm}(\mathbf{a},\mathbf{b})$ remains constant when the orientations change, although each coincidence rate is 100% modulated. This shows that the size of the selected sample of pairs is constant. Of course, it is not a proof of the validity of the assumption, but at least it is consistent with it. Note that it is possible to use a stronger assumption, the «fair sampling assumption», in which one assumes that the ensemble of detected pairs is a fair sample of the ensemble of all emitted pairs. Our assumption is clearly a logical consequence of the fair sampling assumption, which is therefore stronger.

The experiment has been done at the set of orientations of Figure 9.4, for which a maximum conflict is predicted. We have found

$$S_{exp} = 2.697 \pm 0.015 \tag{9.37}$$

violating the inequalities 9.21 ($|S| \leq 2$) by more than 40 standard deviations! Note that this number is in excellent agreement with the predictions of Quantum Mechanics for our polarizers efficiencies and lenses apertures:

$$S_{QM} = 2.70 \pm 0.05 \tag{9.38}$$

The uncertainty indicated for S_{QM} accounts for a slight lack of symmetry of the two channels of a polarizer ($\pm 1\%$). The effect of these asymmetries; has been computed and cannot create a variation of S_{QM} greater than 2%.

We have also performed measurements of the polarization correlation coefficient $E(\mathbf{a}, \mathbf{b})$ in various orientations, for a direct comparison with the predictions of Quantum Mechanics (Figure 9.13). The agreement is clearly excellent.

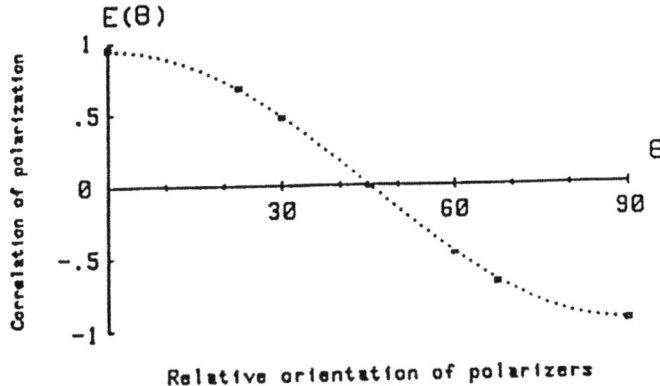

Figure 9.13 Experiment with two–channels polarizers. Polarization correlation of as a function of the relative angle of the polarimeters. The indicated errors are ± 2 standard deviations. The dashed curve is not a fit to the data, but Quantum Mechanical predictions for the actual experiment. For ideal polarimeters, the curve would exactly reach the values ± 1.

9.5 TIMING EXPERIMENT

As stressed in sections 6. and 7.5, an ideal test of Bell's inequalities would involve the possibility of switching at random times the orientation of each polarizer [13,14,28], since the locality condition would become a consequence of Einstein's causality. We have done a step towards such an ideal experiment by using the modified scheme shown on Figure 9.14. Each (single–channel) polarizer is replaced by a setup involving a switching device followed by two polarizers in two different orientations: **a** and **a'** on side I, **b** and **b'** on side II. The optical switch is able to rapidly redirect the incident light from one polarizer to the other one in a different orientation. Each setup is thus equivalent to a variable polarizer switched between two orientations. In our experiment, the distance L between the two switches is $13m$, and L/c has a value of $43ns$.

The switching of the light is effected by acousto–optical interaction of the light with an ultrasonic standing wave in water. The incidence angle (Bragg angle) and the acoustic power are adjusted for a complete switching between

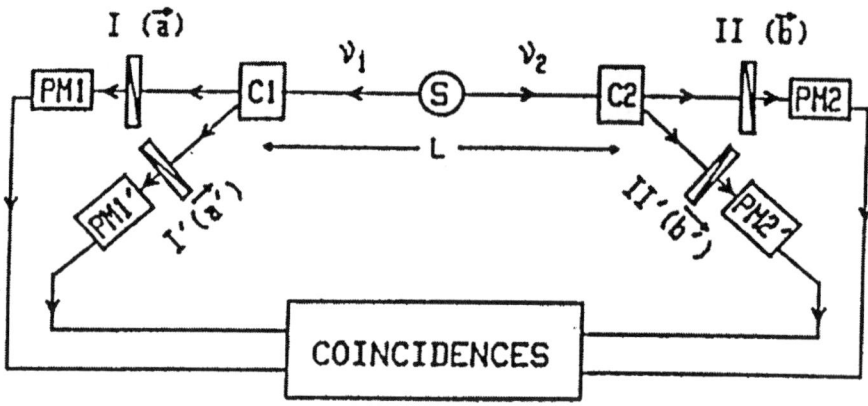

Figure 9.14 Timing–experiment with optical switches (C1 and C2). A switching occurs each 6.7 ns and 13.3 ns. The two switches are independently driven.

the 0th and 1st order of diffraction. The switching function is then of the form $\sin^2\left(\frac{\pi}{2}\cos\Omega_a t\right)$, with the acoustic frequency $\Omega_a/2\pi$ of the order of $25 MHz$. The change of orientation of the equivalent variable polarizer then occurs after inequal intervals of $6.7 ns$ and $13.3 ns$. Since these intervals as well as the delay between two photons of a pair (average value of $\tau_r = 5 ns$) are small compared to L/c ($43 ns$), a detection event on one side and the corresponding change of orientation on the other side are separated by a space–like interval. The second time condition is thus partially fulfilled[6], although we note that the switching was quasiperiodic and not truly random (we discuss this point below).

The experiment was far from ideal on several other points. First, in order to match the photon beams to the switches, we had to reduce their size by a factor of 3, entailing a reduction of the coincidence rates by one order of magnitude. Accordingly, to achieve a significant statistical accuracy, the duration of data accumulation was much longer than in previous experiments, and we had to face problems of drifts. It was then necessary to average out the various measured quantities.

Second, for not infinitely small beams, the commutation by the switches is incomplete, because the incidence angle is not exactly the Bragg angle for all rays. In our experiment, the minimum of the light intensity in each channel was 20%, so that not all photons were submitted to forced switching.

---

[6]The first time condition is clearly fulfilled.

Third, we used single channel polarizers, which allowed us to carry out this experiment with the same fourfold coincidence system as in the the static experiment of section 9.4.

Our test of Bell's Inequalities involved a total of $8000s$ of data accumulation with the 4 polarizers in the orientations of Figure 9.4. A total of $16000s$ was devoted to the measurements with half or all the polarizers removed. In order to compensate the effects of systematic drifts, the data accumulation was alternated between the various configurations each $400s$, and the data were averaged out. We finally obtained

$$S'_{exp} = 0.101 \pm 0.020 \qquad (9.39)$$

violating the upper limit of the Bell's inequality (9.32) by 5 standard deviations, and in good agreement with the Quantum Mechanics predictions

$$S'_{QM} = 0.113 \pm 0.005. \qquad (9.40)$$

Other measurements of the coincidence rate have been carried out for a comparison with Quantum Mechanics at different angles. As shown on Figure 9.15, the results are in excellent agreement with the predictions of Quantum Mechanics.

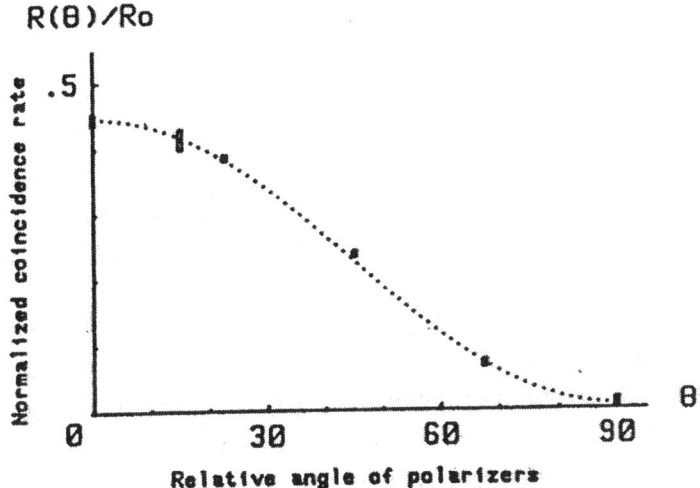

Figure 9.15 Timing experiment: average normalized coincidence rate as a function of the relative orientation of the polarizers. Indicated errors are ± 1 standard deviation. The dashed curve is not a fit to the data but the predictions by Quantum Mechanics for the actual experiment.

According to these results, Supplementary–Parameters Theories obeying Einstein's Causality seem to be untenable. However, as indicated earlier, our experiment is far from ideal, from several points of view, and several loopholes are left open for an advocate of hidden variable theories. First, because we used single channel polarizers, the experiment is significant only if one accepts a strong version of the «fair sampling» assumption. Addressing more specifically the timing aspect of this experiment, an advocate of hidden variable theories might argue that the switching was not complete. However, a large fraction of the pairs undergo forced switching. If Bell's Inequalities were obeyed by these pairs, while the other ones would follow Quantum Mechanics, it is hard to believe that we would not have observed a significant discrepancy between our results and the Quantum Mechanical predictions.

The most important point to address is the fact that the switches were not truly random, since the acousto–optical devices were driven by periodic generators. Note however that the two generators on the two sides were functioning in a completely uncorrelated way, since they were operated at different frequencies ($23.1 MHz$ and $24.2 MHz$), and they had uncorrelated frequency drifts. Moreover, another random feature can be found in the delays between the two photons of a pair (exponentially decaying distribution of constant $\tau_r = 5ns$, as shown on figure 9.11) which are distributed on an interval larger than the time between two successive switchings.

In conclusion, this experiment, which is still[7] the only one involving fast changes of the settings of the analysers, has certainly enough imperfections to allow for ad hoc supplementary parameter models not violating Einstein's causality. However, several models that we have tried are eliminated by our experimental results, which are constituted not only by the measured value (9.39) of S', but also by the time delays spectra without any accident observable on the exponential decay, and with an integrated value in good agreement with quantum mechanics as shown on figure 9.15.

10. THIRD GENERATION: EXPERIMENTS WITH PAIRS OF PHOTONS PRODUCED IN PARAMETRIC DOWN CONVERSION

As we have already noted, the calcium radiative cascade that we have used in our experiments was excited to an optimum rate beyond which there is not much possibility of gain in signal to noise ratio. Since the life time of the intermediate stage is quite short ($5ns$) the situation is very favourable for coincidence counting, and it does not seem that there is much room for improvement with sources based on atomic radiative cascades [29].

In the late 80's, new sources of pairs of correlated photons have been developed simultaneously by two groups [30, 31]. In these sources, a pair of red

---

[7]In 1997.

photons is produced by parametric down conversion of a U.V. photon. Because of the phase matching condition in the non linear crystal used for this process, there is a strong correlation between the directions of emission of the two photons of a pair, so that, by spatial selection with two diaphragms positioned in conjugate positions, one can in principle be sure to get the two photons of a pair. This is in contrast with the atomic radiative cascades which produce photons almost not correlated in direction [16]: since each photon is collected in solid angle Ω small compared to 4π, the probability to get the second photon of a pair, once a first one is detected, is of the order of $\Omega/4\pi$, so that the sample of detected pairs is smaller by this factor, than the sample of selected pairs. The new scheme allows one to get rid of this reduction factor, and this has far reaching consequences, both practical and fundamental. On the practical side, it allows larger coincidence rates to be obtained, for similar cascade rates: in the most favourable case [32] the coincidence rate may be more than one order of magnitude larger than in our best experiments. Moreover, such large coincidence rates can be obtained with narrow photon beams (with a small Fresnel number). Such beams can easily be matched into small optical components, or even optical fibers, which opens many new possibilities.

These new sources can produce pairs or photons correlated in polarization in states analogous to (9.1). But they can also produce entangled states exhibiting EPR type correlations between observables other than polarization. An interesting case [33] considers pairs of photons where each photon is emitted «at two different times». Here, the relevant observable is the time of emission of the two photons of the pair. Corresponding experiments have been carried out [34–36]. Note that this scheme, where polarization is not the relevant observable, is specially interesting for experiments with optical fibers, in which polarization control may be a crucial issue. Another interesting scheme considers the directions of emission as observables [37]: each photon of a pair involves the superposition of two different directions of emissions, strongly correlated to two directions of emission associated to the second photon. An experiment of this type has also been carried out [38].

As emphasised in reference [37] these new schemes can be embedded in the general framework of «two particles interferences»: indeed, the joint measurements probabilities are the square of a sum of two amplitudes (each involving the two photons), with a relative phase that can be controlled experimentally. Although it was not pointed out in reference [37], the original EPRB scheme is a very clear example of this situation. For instance, for the version presented in section 2., the state (9.1) can also be rewritten (see equation 9.29) as the superposition of a state $|L, L\rangle$, where both photons have a left handed helicity, and a state $|R, R\rangle$ with two right handed helicities. For each of these two states, the amplitude for being detected in any couple of output channels behind the linear polarizers (see figure 9.1) has a value of $1/\sqrt{2}$ times a phase factor which depends on the orientation of the polarizers (Figure 9.1). The addition of the amplitudes associated to $|L, L\rangle$ and $|R, R\rangle$ thus leads to an interference term responsible for the sinusoidal variations of the joint probabilities (9.3) when the orientations change.

These new sources and schemes have lead to a series of tests of Bell's inequalities, *which have all confirmed quantum mechanics*. Clear violations of Bell's inequalities have been found, under the assumption that the «fair sampling hypothesis» holds. Among these results, it is worth pointing out a violation of Bell's inequalities by 100 standard deviations in a few minutes only [32]. Note also an experiment [35] where a clear violation of Bell's inequalities has been observed with one leg of the apparatus made of 4 kilometers of optical fiber. More recently, EPR correlations have been observed with photons propagating in several tens of kilometers of telecommunication fibers [39]. Third generation experiments should ultimately allow one to close most of the various loopholes still left open, since qualitative progress are in principle possible, by comparison to second generation experiments. First, the perfect correlation between the directions of emission offers the possibility to close the loophole related to the low detection efficiency [11], when detectors with quantum efficiency close to unity are available.

The second class of fundamental improvements is related to «timing experiments» (section 7.5, and 9.5). Ideally [13, 14], one needs polarizers that can be independently reoriented at random times, with a reorientation autocorrelation time shorter than the space separation L/c between the polarizers. Our third experiment (section 9.5), which was the first attempt in this direction, was severely limited by the large size of the beams carrying the correlated photons: this prevented us to use small size electrooptic devices suitable for random switching of polarization. With the new schemes using optical fibers, it becomes possible to work with small integrated electro optical devices, at distances of kilometers. At such separations (several microseconds), the time conditions become less difficult to fulfil, and truly random operation of the polarizers become possible at this time scale. Experiments of this type are under preparation [39, 40].

In conclusion, the experiments of third generation, using pairs of photons produced by parametric down conversion, definitively offer several possibilities of qualitatively improved tests of Bell's inequalities.

11. CONCLUSION

We have nowadays an impressive amount of sensitive experiments where Bell's inequalities have been clearly violated. Moreover, the results are in excellent agreement with the quantum mechanical predictions including all the known features of the real experiment. Although there are still some loopholes open, so that improved or qualitatively different experiments are highly desirable, it is legitimate to discuss the consequences of the rejection of supplementary parameter theories obeying Einstein's causality.

It can be concluded that *quantum mechanics has some non–locality in it*, and that this non–local character is supported by experiments [41]. It is very important however to note that such a non–locality has a very subtle nature, and in particular that it cannot be used for faster than light telegraphy. It is indeed easy to show [42] that in a scheme where one tries to use EPR correlations to send a message, it is necessary to send a complementary classical information (about the orientation of a polarizer) via a normal channel, which of course does not violate causality. This is reminiscent of the recent teleportation schemes [43] where a quantum state can be teleported via a non–local process, provided that one also transmits classical information via a classical channel. In fact, there is certainly a lot to understand into the exact nature of non–locality, by a careful analysis of these teleportations schemes [44].

Each time I think again on this subject, I have to consider a question: *do we have a real problem?* I must admit that my answer to this question may vary (at random !) with time. But I am in good company. In 1963, R. Feynman gave a first answer to the question, in his famous course [45]: «This point was never accepted by Einstein ... it became known as the Einstein–Podolsky–Rosen paradox. But when the situation is described as we have done it here, there doesn't seem to be any paradox at all ... ». Two decades later he expressed a remarkably different opinion [46], still about the EPR situation: « We always have had a great deal of difficulty in understanding the world view that quantum mechanics represents .... It has not yet become obvious to me that there is no real problem ... I have entertained myself always by squeezing the difficulty of quantum mechanics into a smaller and smaller place, so as to get more and more worried about this particular item. It seems almost ridiculous that you can squeeze it to a numerical question that one thing is bigger than another. But there you are – it is bigger ... ».

References

[1] J. F. Bell, Physics **1**, 195 (1964).

[2] A. Aspect, in *Atomic Physics 8, Proceedings of the Eighth International Conference on Atomic Physics* (1982).

[3] A. Einstein, B. Podolsky, and N. Rosen, Phys. Rev. **47**, 777 (1935), see also Bohr's answer: [47].

[4] D. Bohm, *Quantum Theory* (Prentice-Hall, Englewoods Cliffs, 1951), republished by Dover (1989).

[5] D. Bohm and Y. Aharonov, Phys. Rev. **108**, 1070 (1957).

[6] *A. Einstein philosopher scientist*, edited by P. A. Schilp (Open court and Cambridge university press, Cambridge, 1949).

[7] A. Einstein and M. Born, *Correspondance between A. Einstein and M. Born* (Seuil, Paris, 1972), french translation.

[8] J. F. Clauser, M. A. Horne, A. Shimony, and R. A. Holt, Phys. Rev. Lett. **23**, 880 (1969).

[9] B. d'Espagnat, Phys. Rev. D **II**, 1424 (1975), see also: [48].

[10] J. S. Bell, in *Foundations of Quantums Mechanics*, edited by B. d'Espagnat (Academic, New York, 1972).

[11] J. F. Clauser and M. A. Horne, Phys. Rev. D **10**, 525 (1974).

[12] L. Hardy, Phys. Rev. Lett. **48**, 291 (1982).

[13] S. Aspect, Phys. Lett. A **54**, 117 (1975).

[14] S. Aspect, Phys. Rev. D **14**, 1944 (1976).

[15] A. Aspect, *Trois tests expérimentaux des inégalités de Bell par mesure de corrélation de polarisation de photons* (thèse d'Etat, Orsay, 1983).

[16] E. Fry, Phys. Rev. A **8**, 1219 (1973).

[17] J. F. Clauser and A. Shimony, Rep. Progr. Phys. **41**, 1881 (1978).

[18] S. J. Freedman and J. F. Clauser, Phys. Rev. Lett. **28**, 938 (1972).

[19] F. M. Pipkin, in *Atomic Physics Tests of the Basics Concepts in Quantum Mechanics*, edited by D. R. Bates and B. Bederson (Academic Press, New York, 1978).

[20] J. F. Clauser, Phys. Rev. Lett. **36**, 1223 (1976).

[21] E. S. Fry and R. C. Thompson, Phys. Rev. Lett. **37**, 465 (1976).

[22] A. Aspect, C. Imbert, and G. Roger, Opt. Comm. **34**, 46 (1980).

[23] A. Aspect, P. Grangier, and G. Roger, Phys. Rev. Lett. **47**, 460 (1981).

[24] A. Aspect, P. Grangier, and G. Roger, Phys. Rev. Lett. **49**, 91 (1982).

[25] P. Grangier, *Corréllation de polarisation de photons émis dans la cascade $4p^2\ ^1S_0 \to 4s4p\ ^1P_1 \to 4s^2\ ^1S_0$ du calcium: test des inégalités de Bell* (Thèse de troisième cycle, Orsay, 1982).

[26] A. Aspect and P. Grangier, Lett. Nuovo Cimento **43**, 345 (1985).

[27] A. Garruccio and V. A. Rapisarda, Nuovo Cimento **A18**, 269 (1982).

[28] A. Aspect, J. Dalibard, and G. Roger, Phys. Rev. Lett. **49**, 1804 (1982).

[29] W. Perrie, A. J. Duncan, H. J. Beyer, and H. Kleinpoppen, Phys. Rev. Lett. **54**, 1790 (1985), although the statistical significance of the results was limited as a test of Bell's inequalities, this experiment was a remarkable « tour de force » in atomic physics.

[30] Y. H. Shih and C. . Alley, Phys. Rev. Lett. **61**, 2921 (1988).

[31] Z. Y. Ou and L. Mandel, Phys. Rev. Lett. **61**, 50 (1988).

[32] P. G. Kwiat, K. Mattle, H. Weinfurter, and A. Zeilinger, Phys. Rev. Lett. **75**, 4337 (1995), note that the reported violation of Bell's inequalities by 100 standard deviations relies on a stronger version of the « fair sampling hypothesis » than our second experiment (Section 9.4), since this experiment uses one channel polarizers and not two channel polarizers.

[33] J. D. Franson, Phys. Rev. Lett. **62**, 2205 (1989).

[34] J. Brendel, E. Mohler, and W. Martienssen, Europhys. Lett. **20**, 575 .

[35] P. R. Tapster, J. G. Rarity, and P. C. M. Owens, Phys. Rev. Lett. **73**, 1923 (1994).

[36] W. Tittel *et al.*, Europhys., Lett. **40**, 595 (1997).

[37] M. A. Horne, A. Shimony, and A. Zeilinger, Phys. Rev. Lett. **62**, 2209 (1989).

[38] J. G. Rarity and P. R. Tapster, Phys. Rev. Lett. **64**, 2495 (1990).

[39] N. Gisin, private communication.

[40] G. Weis and A. Zeilinger, private communication.

[41] J. S. Bell, Comments on Atom. Mol. Phys. **9**, 121 (1981).

[42] A. Aspect, J. Physique Colloque **C2**, 940 (1981).

[43] C. H. Bennet *et al.*, Phys. Rev. Lett. **70**, 1895 (1993), see also [49, 50].

[44] S. Popescu, Phys. Rev. Lett. **72**, 797 (1994).

[45] *The Feynman Lectures on Physics* (Addison–Wesley Publishing Company, Massachusetts, 1966), Vol. III, Chap. 18.

[46] R. P. Feynman, Intern. Journ. of Theoret. Phys. **21**, 467 (1982).

[47] N. Bohr, Phys. Rev. **48**, 696 (1935).

[48] L. Hardy, Phys. Rev. Lett. **68**, 2981 (1992).

[49] D. Bouwmeester *et al.*, Nature **390**, 575 (1997).

[50] D. Boschi *et al.*, Phys. Rev. Lett. (1997), submitted.

Chapter 10

POLARIZATION AND COHERENCE ANALYSIS OF THE OPTICAL TWO–PHOTON RADIATION FROM THE METASTABLE $2^2S_{1/2}$ STATE OF ATOMIC HYDROGEN

A. J. Duncan
Unit of Atomic and Molecular Physics, University of Stirling, Stirling, Scotland.

H. Kleinpoppen
Unit of Atomic and Molecular Physics, University of Stirling, Stirling, Scotland.

Z. A. Sheikh
Department of Physics, B.Z. University, Multan, Pakistan.

Abstract The Paper first summarizes fundamental aspects and results of the quantum electrodynamical theory of the two–photon radiation from the decay of the metastable $2^2S_{1/2}$ atomic hydrogen state. After a brief description of the second improved Stirling two–photon coincidence experiment polarization correlations of the two–photon decay are described in which both two or three linear polarizers are applied in order to test predictions of such correlations based upon quantum mechanics and local realistic theories (i.e., Einstein–Podolsky–Rosen type experiments). It is particularly noticeable that the three–polarizer coincidence measurement provided the largest difference (about 40%) between the Bell limits of local realistic theories and quantum mechanics so far. Apart from confirming in addition the correlations of right–right and left–left circularly polarized two–photon correlations a new type of coherence analysis of the two–photon radiation has been carried out experimentally and theoretically. A result of it is the measured coherence time of $\tau_{coh} = 1.2 \cdot 10^{-15} s$ and coherence length of $l_{coh} = c \cdot \tau_{coh} = 350nm$ of the two–photon emission. By applying a theoretical model of the two–photon radiation linked to cascade transitions the coherence length can be estimated to $l_{coh} \approx 100nm$ in agreement by order of magnitude with the experimental data.

Complete Scattering Experiments, Edited by Becker and Crowe
Kluwer Academic/Plenum Publishers, New York 2001

1. INTRODUCTION

It is generally recognised that the spectroscopy of atomic hydrogen has provided crucial tests of the foundation of basic quantum physics, quantum electrodynamics and even areas of elementary particle physics (Series, [1] 1988; Selleri, [2] 1988; Greenberger and Zeilinger, [3] 1995 and Scully and Zubairy, [4]). As early as 1887 the American physicists Michelson and Morley observed that the first spectral line H_α of the Balmer series of atomic hydrogen was split into two components which Sommerfeld subsequently interpreted as a relativistic effect, afterwards called spin–orbit coupling and described by introducing a further quantum number referred to as electron spin. The experimental detection of the fine structure of the hydrogen Balmer line was the beginning of the precision spectroscopy of atoms. The $2^2S_{1/2}$ and $2^2P_{1/2}$ states were predicted by Dirac's quantum mechanics to be degenerate but this was proved incorrect by the sensational detection of an energy difference between these states by Lamb and Retherford [5] (Lamb shift $\delta E\left(^2P_{1/2} -^2 S_{1/2}\right) \cong 1050 MHz$). The question as to whether the $2^2S_{1/2}$–state of atomic hydrogen would be metastable in practice was a source of controversy during the first part of this century as discussed in detail by Novick [6]. However, the successful radio–frequency experiment of Lamb and Retherford [5] which detected transition between the $2^2S_{1/2}$ and $2^2P_{1/2}$ states, depended on the metastability of the $2^2S_{1/2}$ state. The first experiments for the direct detection of two–photon radiation from metastable states was reported for the decay of $H^+(2^2S_{1/2})$ by Lipeles et al. [7] and of $H(2^2S_{1/2})$ by O'Connell et al. [8] and also by Krüger and Oed [9]. The possibility of a spontaneous two–photon transition in general had been predicted by Göppert–Mayer [10] based upon her pioneering theory of multiphoton processes of atomic systems. Following this theory Breit and Teller [11] estimated in 1940 that the dominant decay mode of the atomic hydrogen $2^2S_{1/2}$ state should be the two–photon emission and much subsequent theoretical work has been carried out on the subject (Series, [1] ed., 1988 and Drake, [12], in Series, [1]). In this paper the data from the Stirling two–photon experiment is summarized and evaluated. This summary will include reports of measurements of the geometric angular correlations of the two photons and their polarization correlation observed in coincidence (including confirmation of the Breit–Teller hypothesis). As will be discussed, the two–photon metastable atomic hydrogen source also provides a means of carrying out fundamental experiments of the Einstein–Podolsky–Rosen (EPR) type (Einstein et al. [13]).

In addition to experiments involving two linear polarizers, following proposals by Garuccio and Selleri [14], experiments with three linear polarizers are described which provide particularly sensitive tests to distinguish between the predictions of quantum mechanics and local realism. A Stokes parameter analysis of the coincident two photons which proves the coherent nature of the two–photon transition of metastable hydrogen is also discussed. An experiment based on a delay of one of the orthogonal polarization components of one photon of the two–photon pair by a multiwave plate leads to the measurement of the coherence length of a single photon of the two–photon pair which is shown to be extremely short.

A novel Fourier–transform spectroscopic method using a Stokes parameter analysis of the two–photon polarization to determine the spectral distribution of the two photons emitted in the spontaneous decay of metastable atomic hydrogen is described.

The theoretical analysis (Biermann et al., [15] of the two–photon correlation spectroscopy of metastable atomic hydrogen in comparison to two–photon cascade emission from a three–level atom will be discussed.

Attention is also drawn to more general summaries on the physics of atomic hydrogen (including collision processes) by Series [1], Bassani et al [16] Friedrich [17] and McCarthy and Weigold [18].

2. ON THE THEORY OF THE TWO–PHOTON DECAY OF THE METASTABLE STATE OF ATOMIC HYDROGEN

As mentioned above, the theory of the two–photon emission of the metastable $2^2S_{1/2}$ state of atomic hydrogen was initiated by a paper of Breit and Teller [11] which was based upon Maria Goeppert's [19] and Goeppert–Mayer's [10] quantum theory of multiple–photon processes in atomic spectroscopy. Since then substantial progress has been made both theoretically and experimentally with regard to the physics of the two–photon emission of metastable atomic hydrogen. The exact quantum mechanical description of the two–photon process is based upon the four–component Dirac equation (see e.g. Drake, [12]).

This theoretical approach is well discussed in terms of the matrix formalism of quantum electrodynamics (Akhiezer and Berestetskii, [20]). Before we describe some results of the theoretical formalism for the two–photon decay of the metastable $2^2S_{1/2}$ state we consider the competing electromagnetic transitions in the absence of any perturbation such as an electric field or atomic collisions shall be considered. Due to the Lamb shift the $2^2P_{1/2}$ state is the only one of the $n = 2$ states lower than the $2^2S_{1/2}$ state. However, since the energy difference is so small the possible spontaneous electric dipole transition from the $2^2S_{1/2}$ to the $2^2P_{1/2}$ state has a negligibly small transition probability (corresponding lifetime of about 20 years). Magnetic dipole and electric quadrupole transitions of the $2^2S_{1/2}$ state are forbidden in the Pauli approximation but magnetic dipole transitions are allowed if exact Dirac theory with Dirac wave functions are allowed with a decay rate of $2.496 \cdot 10^{-6} s^{-1}$ corresponding to a medium lifetime of about 2 days. The two photon transition probability is much greater and has been estimated to be about $14s^{-1}$ for the two–photon transition of metastable atomic hydrogen (corresponding to a mean life time of $1/7s$). Fig. 10.1 illustrates spontaneous and field–induced radiative transitions

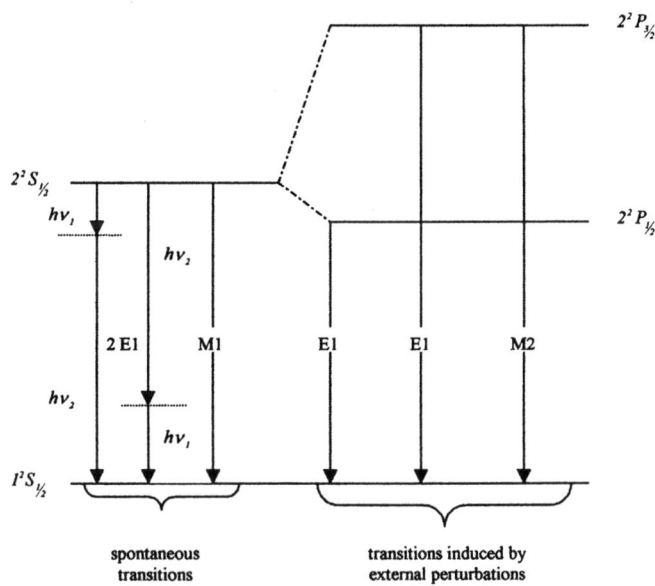

Figure 10.1 Spontaneous and field–induced radiative transitions of the metastable $2^2 S_{1/2}$ state of atomic hydrogen. 2E1 denotes the two–photon decay mode, M1 and M2 the magnetic dipole, magnetic quadrupole and E1 the electric dipole decay modes. The states between the metastable and the ground state represent "virtual" states (dotted lines) associated with the emission of the two photons of energies $h\nu_1$ and $h\nu_2$. Single photon decay modes may lead to cross terms to produce quantum beats and interference effects and also contributions to the two–photon decay rates (Drake, [12]). The dashed–dotted lines indicate the mixing of the metastable state with the $2^2 P_{1/2}$ and $2^2 P_{3/2}$ states by an external perturbation such as electric fields or atomic collisions. Note that the energy differences are not scaled.

modes of the metastable state of atomic hydrogen. While interest here is exclusively related to the two photon decay the study of the field induced quenching radiation reveals measurable interference effects and quantum beat phenomena applied in atomic spectroscopy (Andrä, [21, 22], van Wijngaarden et al., [23] and Drake et al. [24]).

The quantum electrodynamical theory of the simultaneous emission of the two photons with vector potentials $A_1(x)$ and $A_2(x)$ can be illustrated by second order Feynman diagrams shown in Fig. 10.2. There are several important consequences of the results of the theory:

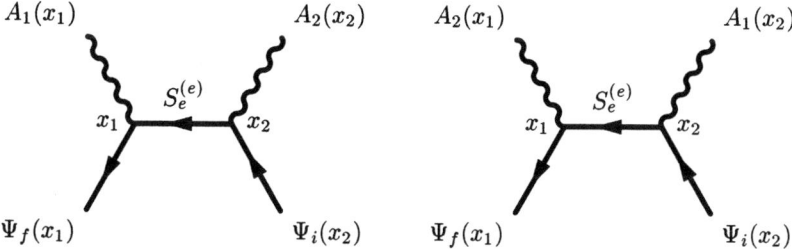

(a) $\psi_f(x_1)$ and $\psi_i(x_1)$ are the wave functions of the final ground state and the metastable state; $A_1(x_1)$ and $A_2(x_2)$ are the vector potentials of the two photons and $S_e^{(e)}(x_1, x_2)$ is the electron propagator in the external field of the nucleus (eq. 10.1); x are the relativistic four component co-ordinates (after Akheizer and Berestetskii, [20]).

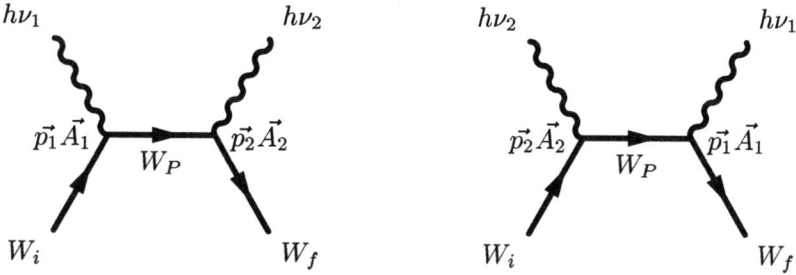

(b) In non–relativistic electric dipole approximation the energies of the ground, final, intermediate P state and metastable (initial) states, W_f, W_p and W_i are connected by the products $p_1 A_1$ and $p_2 A_2$ of the canonical momenta p_1 and p_2 and the vector potential A_1 and A_2 to result in the photon energies $h\nu_1$ and $h\nu_2$. It demonstrates that the emission of the two photons can be considered in either order as shown in the Feynman diagram.

Figure 10.2 Feynman diagrams for the two–photon decay of metastable atomic hydrogen.

1) Based upon the above replacement of the Dirac matrix α, the Feynman diagram of Fig. 10.2(a) can be drawn as shown in Fig. 10.2(b) with the energies of the states involved.

2) The energies of a complementary pair of two photons add up to the energy difference between the $2^2S_{1/2}$ and $1^2S_{1/2}$ states. Fig. 10.3 shows shapes of such energy distributions for $Z = 1$ and $Z = 92$.

3) According to Breit and Teller [11] the electric dipole operators in the interaction Hamiltonian are diagonal in nuclear and electron spins. As a result the presence of hyperfine and fine structure effects can be neglected in the two–photon emission process (see section 5.).

228   COMPLETE SCATTERING EXPERIMENTS

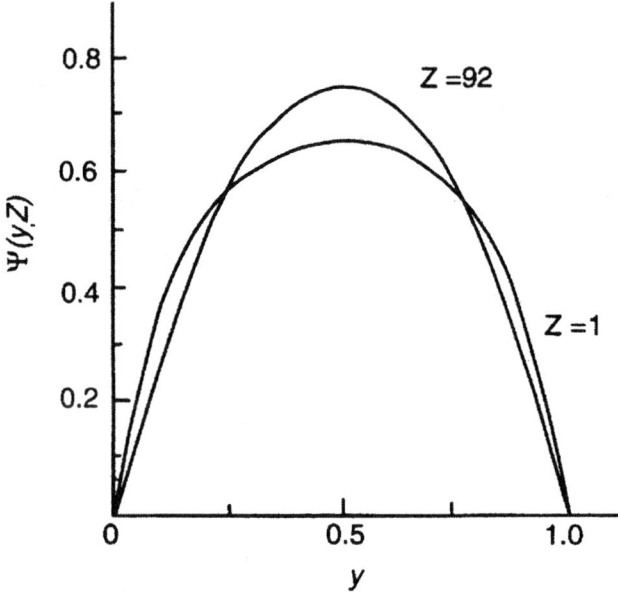

Figure 10.3 Energy distribution of the $2^2S_{1/2} \to 1^2S_{1/2}$ two–photon continuum of hydrogen–like atomic systems with nuclear charges $Z = 1$ and $Z = 92$ (uranium). $\psi(y, Z)$ is the spectral distribution function, and y is the fraction of the total transition energy transported by one of the two photons; correspondingly $l - y$ is the fraction of the energy of the other photon. The areas under the curves are normalized to unity (from Goldman and Drake, [25]).

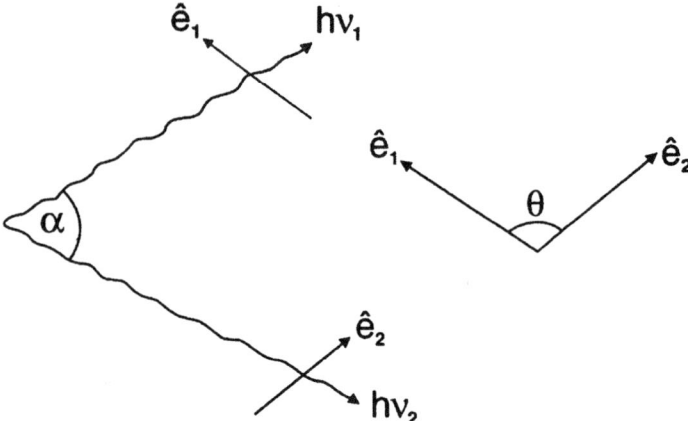

Figure 10.4 Diagram illustrating the angular correlation and the correlation angle between the polarization unit vectors $\hat{e}_1$ and $\hat{e}_2$ involved in the detection of the two–photon emission.

4) If $\hat{e}_1$ and $\hat{e}_2$ are the unit vectors of the linear polarisations for the two photons with energies $h\nu_1$ and $h\nu_2$ the transition probability has a cosine square dependence $I(\nu_1) \propto |\hat{e}_1\hat{e}_2| = cos^2\Theta$ where Θ is the angle between $\hat{e}_1$ and $\hat{e}_2$ (Fig. 10.4).

5) Averaging over these polarisations results in an angular correlation for the coincident detection of the two photons:

$$I(\nu_1) \propto \langle|\hat{e}_1 \cdot \hat{e}_2|\rangle_{av} \propto \left(1 + \cos^2 \alpha\right), \quad (10.1)$$

where α is the angle between the directions of coincident detection of the two photons. In chapter 4. such polarisation and angular correlations have been confirmed experimentally. In most of the experiments, however, the two photons are normally observed in diametrically opposite directions, i.e. with $\alpha = \pi$. A relevant and interesting form of describing the polarisation for such two photons can be based upon the formulation of a *two–photon state vector* for $\alpha = \pi$. Conservation of angular momentum and parity implies the following arguments: In the transition $H(2s) \to H(1s)$ no orbital angular momentum change occurs. Therefore the two photons with energies $h\nu_1$ and $h\nu_2$ must have equal *helicities* the photon pairs are either right–right $|R_1\rangle|R_2\rangle$ or left–left $|L_1\rangle|L_2\rangle$ handed circularly polarised or in a superposition of both kinds, i.e.,

$$|\psi\rangle_\pm = \frac{1}{\sqrt{2}} \{|R_1\rangle|R_2\rangle \pm |L_1\rangle|L_2\rangle\}$$

The parity operator P transforms the handedness of right and left circularly polarised light and the propagation of the two photons in opposite directions: $P|R_1\rangle = |L_2\rangle$ and $P|L_1\rangle = |R_2\rangle$. The two states of $2s$ and $1s$ have definite even parity ($l = 0$) so the two photons should also have even parity (otherwise the resulting parity could be a superposition of even and odd parity), i.e. the plus signs should be valid in the above equation:

$$|\psi\rangle_+ = \frac{1}{\sqrt{2}} \{|R_1\rangle|R_2\rangle + |L_1\rangle|L_2\rangle\}. \quad (10.2)$$

Because of the usual relations between circular and linear optical polarizations

$$|R_1\rangle = \tfrac{1}{\sqrt{2}}\{|x_1\rangle + i|y_1\rangle\}, \quad |R_2\rangle = \tfrac{1}{\sqrt{2}}\{|x_2\rangle - i|y_2\rangle\},$$
$$|L_1\rangle = \tfrac{1}{\sqrt{2}}\{|x_1\rangle - i|y_1\rangle\}, \quad |L_2\rangle = \tfrac{1}{\sqrt{2}}\{|x_2\rangle + i|y_2\rangle\} \quad (10.3)$$

the two photon state vector can be written alternatively as

$$|\psi\rangle = \frac{1}{\sqrt{2}} \{|x_1\rangle|x_2\rangle + |y_1\rangle|y_2\rangle\}. \quad (10.4)$$

The following implications of these *state vectors* should be noted:

a) They are invariant with respect to rotation about the detection axis (z–axis).

b) The state vector represents a pure quantum mechanical state, not a mixture of the $|R_1\rangle|R_2\rangle$ and $|L_1\rangle|L_2\rangle$ states (or alternatively not a mixture of the $|x_1\rangle|x_2\rangle$ and $|y_1\rangle|y_2\rangle$ states).

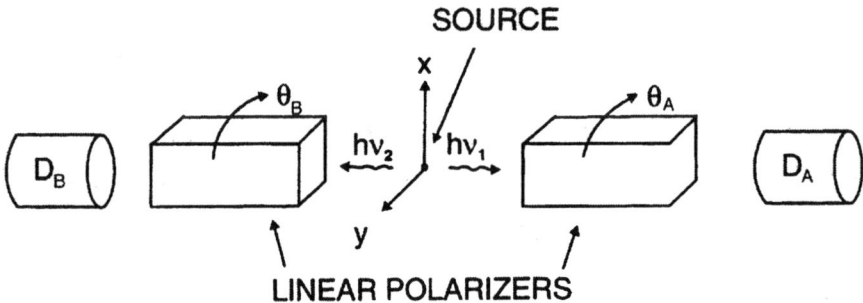

Figure 10.5 Co–ordinate system with reference to the emission and polarization correlations of the two–photon emission ($h\nu_1$ and $h\nu_2$) of metastable atomic hydrogen $H(2S)$ detected in the z and $-z$ directions by the two detectors D_1 and D_2. The transmission axes of the two polarizers are at angles Θ_A and Θ_B with respect to the directions (after Perrie et al., [26]).

c) Consider a typical ideal coincidence experiment for measuring linear or circular polarisation correlations in the opposite directions of z and $-z$ (Fig. 10.5). Making such measurements, the state vector $|\psi\rangle_+$ "collapses" into $|R_1\rangle|R_2\rangle$ or $|L_1\rangle|L_2\rangle$ or alternatively into $|x_1\rangle|x_2\rangle$ or $|y_1\rangle|y_2\rangle$, each possibility occuring with probability of one–half. This collapse of the state vector implies that detection af a photon, say in the z direction, with special choices for linear or circular polarisers determines the result of the measurement of the polarisation in the other detector say in the $-z$ direction, irrespective of the distance between them. The result obtained on one side of the source depends on the choice made for the setting of the polariser on the other side of the source. This situation clearly violates the "principle of locality" in classical and relativistic physics according to which the value obtained for a physical quantity at point A cannot be dependent on the choice of measurement made at point B as long as the physical quantities at point A are not correlated with the ones at point B. This discussion already in essence leads to the Bohm–Aharonov [27] version of the Einstein–Podolsky–Rosen–Paradox concerning the incompleteness of quantum mechanics (see section 4.).

3. THE STIRLING TWO–PHOTON APPARATUS

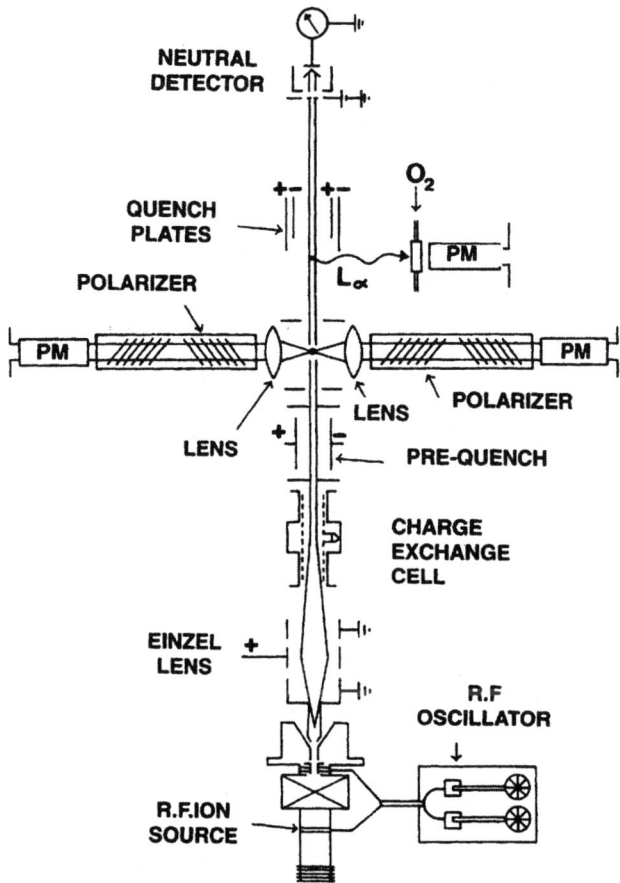

Figure 10.6 Schematic diagram of the second Stirling two–photon apparatus.

Figure 10.6 illustrates schematically the second improved two–photon apparatus built at Stirling University (Perrie [26]). Metastable $D(2s)$ atoms were produced as a result of the capture reaction $D^+ + Cs \to D(2s) + Cs^+$, which is favoured by a high resonance cross section $\approx 10^{-14} cm^2$. The deuterium ions D^+ were extracted from a radio frequency ion source and passed through a cesium vapour cell constructed after a design by Bacal et al [28] and Bacal and Reichelt [29]. Deuterium was used rather than hydrogen since the radiation noise generated by interaction of the deuterium beam with the background gas was less than with hydrogen. Best statistical data of the two–photon coincidences could be achieved at an energy of $1 keV$ for deuterium. The $D(2s)$ beam leaving the charge exchange cell passed through a set of electric field prequench plates allowing the metastables of the beam to be switched on and off by the effect of Stark mixing of the $2^2S_{1/2}$ and $2^2P_{1/2}$ states.

At the end of the beam apparatus the metastables were completely quenched by the electric field of another set of quench plates; the resulting Lyman–$\alpha(L_\alpha)$ radiation was used to normalise the two–photon coincidence signal from the metastables. The L_α signal was registered by a solar blind UV photomultiplier together with an oxygen filter cell with LiF windows through which dry oxygen flowed continuously. Tests were made regularly that the two–photon coincidence signal was proportional to the L_α–signal, which depends linearly on the density of the metastables.

The two–photon radiation was collimated and detected at right angles to the $D(2s)$–beam. Two lenses of $50mm$ diameter were used, each with a focal length of $43mm$ at a wavelength of $243nm$. For the linear polarization correlation measurements two high–transmission UV–polarizers were used each consisting of 12 amorphous silicon plates polished flat to 2λ at $243nm$ and set nearly at Brewster's angle as shown in Fig. 10.6. Additional optical elements such as quarter–wave, half–wave and multiwave plates could be inserted as required for other polarization correlation measurements. The material of the various lenses and plates was high–quality fused amorphous silica with a short–wavelength cut–off at $160nm$. These in turn correspond to a complementary long wavelength cut–off at $355nm$. Accordingly all photons in the wavelength range from 185 to $355nm$ can contribute to possible two–photon coincidence signals. The quantum efficiency of the photomultipliers was about 20% over this range. The transmission efficiencies ε_M and ε_m of the polarizers, for light respectively polarized parallel and perpendicular to the transmission axes of the polarizer, were measured by making use of the $253.7nm$ optical line from a mercury lamp. Two of the polarizers used had transmission efficiencies of $\varepsilon_M = 0.908 \pm 0.013$ and $\varepsilon_m = 0.0299 \pm 0.0020$; a third polarizer with plates from a different manufacturer had values of $\varepsilon_M = 0.938 \pm 0.010$ and $\varepsilon_m = 0.040 \pm 0.002$. The pulses detected by fast–rise–time photomultipliers were fed into a coincidence circuit described by O'Connell, Kollath, Duncan and Kleinpoppen [8]. It consisted of the common combination of constant–fraction discriminators, a time–to–amplitude converter and multichannel analyser operating in the pulse–height analysis mode. A typical run for acquiring coincidence signals lasted at least 20 hours. Spurious coincidence signals due to cosmic rays and residual radioactivity in the apparatus occurred at a rate of one every $100s$, and decreased as the distance between the photomultipliers increased.

The density of the metastable atomic deuterium $D(2s)$ for realistic coincidence measurements was about $10^4 cm^{-3}$ (equivalent to a partial gas pressure of about $0.3 \cdot 10^{-12} Torr$). A typical coincidence signal of the two–photon radiation is shown in Fig. 10.7. As can be seen the shape of the coincidence

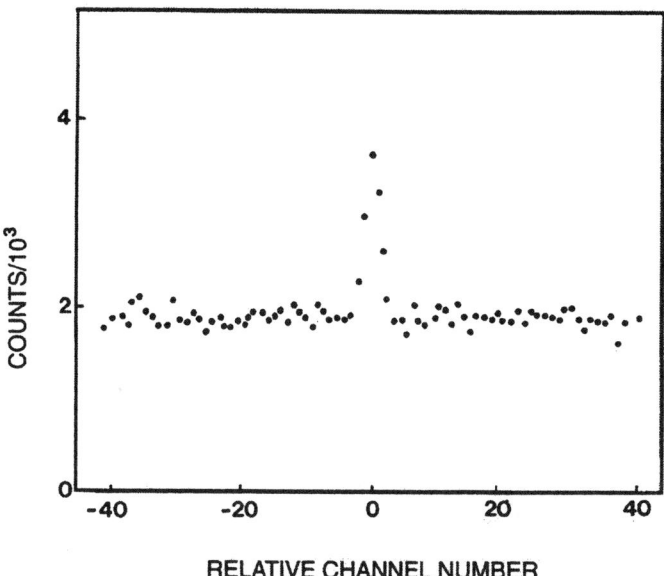

Figure 10.7 Typical coincidence spectrum for the two–photon emission from metastable atomic deuterium. The time–correlation spectrum is that which was obtained after subtraction of the spectrum produced when the metastable component of the atomic beam was quenched. Polarizers were removed for this example. Time delay differences between relative channel numbers are $0.8 ns$, total collection time $21.5 h$. Singles count rate with metastables present (quenched) is about $1.15 \cdot 10^4 s^{-1} (0.85 \cdot 10^4 s^{-1})$. The true two–photon coincidence rate from the decay of the metstable is $490 h^{-1}$ for this example.

peak is symmetrical which is expected for the effectively simultaneous emission process of the two photons from the decay of the metastables (see chap. 6.). The background signal mainly results from the single count rates of the order of $10^4 s^{-1}$ which were mainly due to radiation produced by the metastable atomic beam interacting with the background gas of the vacuum system at a pressure of $2 \cdot 10^{-7} Torr$; uncorrelated photons from the two–photon decay only contribute about 0.01% to the background coincidence signal.

4. ANGULAR AND POLARIZATION CORRELATION EXPERIMENTS

4.1 TWO–POLARIZERS EXPERIMENTS: POLARIZATION CORRELATION AND EINSTEIN–PODOLSKY–ROSEN–TESTS

The first coincidence measurement of the two–photon decay of atomic hydrogen was made by O'Connell et al [8] for three different angles between

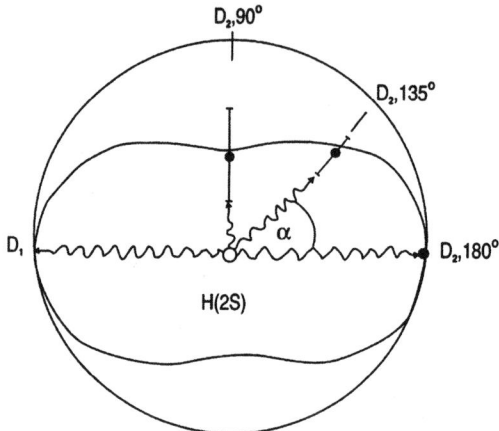

Figure 10.8 Angular correlation data for two–photon coincidences from $H(2s)$ with detectors D_1 and D_2 at the three correlation angles α equals $90°$, $135°$ and $180°$. The circle represents a symmetrical angular correlation with the other curve representing the theoretical prediction which has the form $1 + cos^2\alpha$.

directions of the detected photons (Fig. 10.8). While the accuracy of these early experiments was limited the data approached the shape of the theoretical prediction given in eq. 10.5 and clearly demonstrated a disagreement with a circularly symmetric angular correlation in the detection plane defined by the directions of the two coincident photons given by the equation proportional to $(1 + cos^2\alpha)$ which is expected by the theoretical prediction. In a subsequent improved experiment involving two linear polarizers (Fig. 10.5) for the detection of photons in diametrically opposite directions $\alpha = \pi$ the coincidence rate ratio $R(\Theta)/R_0$ was measured as a function of the angle Θ between the transmission axes of the polarizers; $R(\Theta)$ is the coincidence count rate with the two polarizers inserted while R_0 is the coincidence count rate with the two polarizers removed. For this case quantum mechanics predicts (Clauser et al., [30]) in the ideal case, a $(1 + cos2\Theta)$ dependence of the coincidence signal. In practice quantum mechanics predicts

$$\frac{R(\Theta)}{R_0} = \frac{1}{4}(\varepsilon_M + \varepsilon_m)^2 + \frac{1}{4}(\varepsilon_M - \varepsilon_m)^2 F(\delta) \cos 2\Theta, \qquad (10.5)$$

where in the present case the transmission efficiency $\varepsilon_M = 0.908 \pm 0.013$, $\varepsilon_m = 0.0299 \pm 0.0020$, the half–angle subtended by the lenses near the source of the two–photon radiation is $\delta = 23$ and $F(\delta) = 0.996$. As can be seen in Fig. 10.9 the quantum mechanical curve fits the data of our coincidence measurements of the two–photon radiation very well while the predictions of the two local realistic theories discussed below fail to do so.

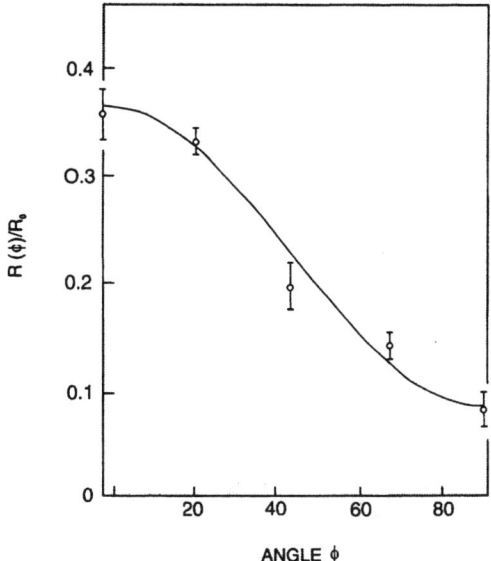

Figure 10.9 Coincidence signal $R(\phi)$ divided by the coincidence signal R_0 with the plates of both linear polarizers removed for the circular polarization experiment as a function of the angle ϕ made by the fast axis of the quarter–wave plate in one detection arm of the two–photon coincidence apparatus with respect to the fast axis of the quarter–wave plate in the other detection arm. The solid line represents the theoretical quantum mechanical curve for comparison.

The horizontal straight line (curve I) for $R(\Theta)/R_0$ in Fig. 10.10 is based upon the following local realistic model. In this model it is assumed that the source emits R_0 pairs of photons per second in the $+z$ and $-z$ directions with the photons emitted on either side independently of each other possessing an isotropic distribution of polarization vector directions. Since, in the ideal case, each polarizer only transmits one–half of the photons the coincidence rate R_0 observed is reduced to $R_0/4$ independently of angle Θ and hence, with corrections for the transmission efficiencies ε_M and ε_m of the polarizers, $R(\Theta)/R_0 = 0.22$.

Another specific example of a local realistic model (curve II in Fig 10.10) originally due to Holt [31] can be described as follows:
As above we assume an isotropic source emitting pairs of photons in the $+z$ and $-z$ directions but, in this case, each with the same polarization vector at an angle φ to the x–axis. These angles φ have values from 0 to π with equal probabilities. Taking R_0 decays per second and detectors D_A and D_B with 100% efficiency the coincidence signal would be

$$dS = R_0 \cos^2(\Theta_A - \varphi)\cos^2(\Theta_B - \varphi)d\varphi/\pi$$

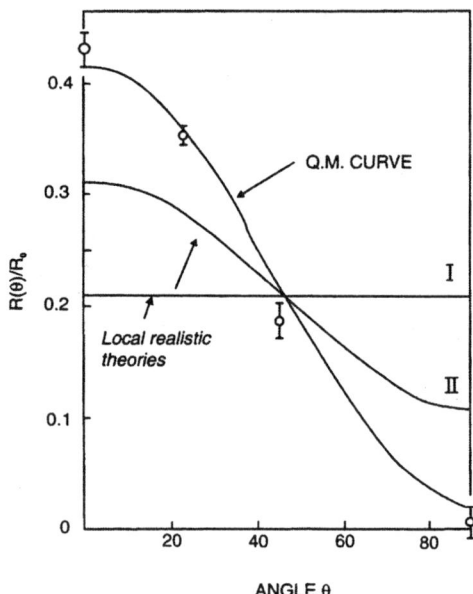

Figure 10.10 Linear polarization correlation of the two photons emitted in decay of metastable $D(2s)$ as measured and compared to quantum mechanics (Q.M.) and two local realistic models (I) and (II). The theoretical curves take account of the finite transmission efficiencies and the angles of acceptances of the lenses. The transmission axes of the polarizers are rotated by the angles Θ_A and Θ_B with $\Theta = \Theta_A - \Theta_B$.

for photons with polarization angles between φ and $\varphi + d\varphi$. Integrating over all angles φ from $0°$ to π gives

$$\frac{R(\Theta)}{R_0} = \frac{1}{4}\left(1 + \frac{1}{2}cos2\Theta\right)$$

with $\Theta = \Theta_A - \Theta_B$. This result compares to the quantum mechanical prediction of

$$\frac{R(\Theta)}{R_0} = \frac{1}{4}(1 + cos2\Theta)$$

in the ideal case, which does not have the factor $\frac{1}{2}$ in front of the cosine term found in the local realistic model of Holt [31].

This discussion leads already to the Einstein–Podolsky–Rosen (EPR) debate about the completeness of quantum mechanics. The literature on this topic has swollen dramatically over the past 15 years so that we only refer to some relevant reviews (Selleri, [2], Duncan and Kleinpoppen, [32] and Greenberger and Zeilinger [3]). The controversy between local realistic theories and quantum mechanics may be characterized as follow. In 1964 Bell [33] and later Clauser,

Horne, Shimony and Holt [30] showed that quantum mechanics predicts strong correlations in ideal two–photon experiments of which local realistic theories are incapable. In *local* theory a measurement of a physical quantity at some point A in a space–time representation is not influenced by a measurement made at another point B spatially separated from A in a relativistic sense. A *realistic* theory assumes that the world is made up of objects with physical properties, which exist independently of any observation made on them. Contrary to both local and realistic theory *quantum mechanics is neither local nor realistic*. Bell [33] showed in the form of his famous *inequality* that all local deterministic hidden variable theories (which are a sub–class of local realistic theories) predict a weaker correlation between photons than that given by quantum mechanics. Freedman's [34] form of Bell's inequality showed that, for local realistic theories, the quantity η must satisfy the following inequality:

$$\eta = \frac{R(22.5°) - R(67.5°)}{R_0} \leq 0.250$$

where $R(22.5°)$ is the coincidence rate for $\Theta_A - \Theta_B = 22.5°)$, $R(67.5°)$ the coincidence rate for $\Theta_A - \Theta_B = 67.5°$ and R_0 the coincidence rate with the two polarizers removed. Contrary to these limits quantum mechanics predicts, in the ideal case, a variation as $cos^2(\Theta_A - \Theta_B)$ which results in the value $\eta = 0.354$. Taking account of the solid angle of detection, the transmission efficiencies of the polarizers and applying eq. 5., quantum mechanically $\eta = 0.273 \pm 0.011$ for this experiment. Experimentally the two–photon data from the results in Fig. 10.10 provide a value of $\eta = 0.275 \pm 0.016(1\sigma)$ in agreement with quantum mechanics and beyond the limits of Bell's inequality.

A large number of experiments using two–photon radiation from atomic cascades (e.g., Fry and Thomson, [35] and Aspect et al. [36]) from positronium annihilation (Paramanande and Butt [37]) as well as interference experiments (Brendel et al. [38]) with laser photon pairs have also clearly confirmed the agreement with quantum mechanics. Positronium annihilation experiments can be criticized on the grounds that quantum mechanics itself must be assumed in order to analyze the experimental data. Doubts about atomic cascade experiments have been expressed (Garuccio and Selleri, [14], and Kleinpoppen et al. [39]), concerning the correctness of the results where rescattering effects may not be completely negligible. The metastable $D(2s)$ experiment is free of these objections but the low efficiency of the photon detectors here and in cascade experiments leaves the possibility that the results could be interpreted in terms of local realistic theories if the assumptions of Clauser, Horne, Shimony and Holt [30] are questioned.

In a further experiment the circular polarization correlation of the two photons from $D(2s)$ was measured by inserting a quarter–wave plate between the collimating lens and the linear polarizer in each detection arm of the apparatus.

238   COMPLETE SCATTERING EXPERIMENTS

These quarter–wave plates were achromatic with a retardation that varied by only about 10% in the wavelength range from 180 to 300nm. The presence of the plates reduced the solid angle and the overall sensitivity of the system so that considerably longer recording times were necessary to achieve a satisfactory statistical accuracy. In this experiment the transmission axes of the two linear polarizers were fixed perpendicular to each other and the fast axis of the quarter–wave plate in one detection arm was oriented at 45° to the transmission axis of the linear polarizer to form an analyzer for light of right–handed helicity. The fast axis of the other quarter–wave plate was then rotated from a position where it acted as an analyzer of light of right–handed helicity (i.e. fast axes parallel, $\varphi = 0°$) to a position for analyzing light of left–handed helicity (i.e. fast axes perpendicular, $\varphi = 90°$). In this way, and on the basis of the collapse of the state vector to $|R_1\rangle|R_2\rangle$ the measured coincidence ratio $R(\Theta)/R_0$ was expected to vary in the ideal case as $(1 + cos2\varphi)/4$. However, due to the various imperfections of the linear polarizers and in particular of the quarter–wave plates a satisfactory test of Bell's inequality using circular polarization was not possible, but the required left–left and right–right circular polarization correlation as required from the state vector of the two–photon radiation process could certainly be verified (Fig. 10.9). However, in a recent experiment using a down conversion source (Torgerson et al., [40]) it has been shown that Bell's inequality has been violated by about 40 standard deviations using circularly polarized light. This confirmation of the linear and circular polarization cor-

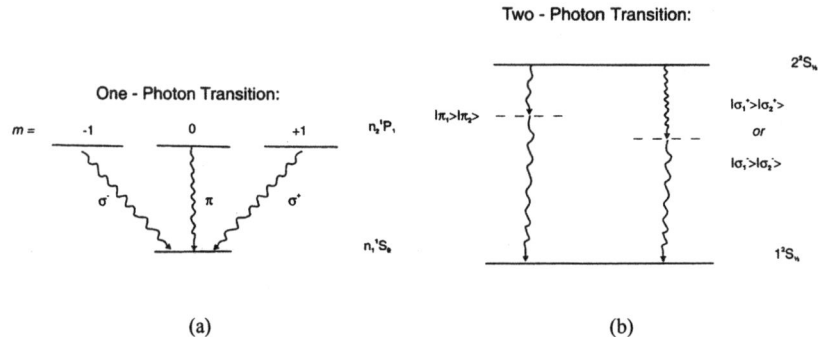

Figure 10.11   Comparison of linearly and circularly polarized π^- and σ^- components for one (a) and two–photon (b) transitions. Note the higher nP states which "contribute" to the two–photon transition as states represented in the dipole matrix elements of the second–order quantum mechanical approximation theory.

relations of the two–photon transition of metastable atomic hydrogen suggests comparison to traditional one–photon transitions with their linear π and circular (σ^+ and σ^-) polarization components as illustrated in Fig. 10.11 for a normal transition from a $n_2^1 P_1$ state to a $n_1^1 S_0$ state. In analogy to this picture the two–photon transitions can be associated with two linearly polarized and correlated π–transitions $|\pi_1\rangle|\pi_2\rangle$ or $|\pi_2\rangle|\pi_1\rangle$ or two right–right $|\sigma_1^+\rangle|\sigma_2^+\rangle$ and left–left $|\sigma_1^-\rangle|\sigma_2^-\rangle$ circularly polarized components. In comparing these one– and two–photon dipole transitions one has, however, to keep in mind that the two emitted photons are detected in coincidence.

4.2 TESTS OF GARUCCIO–SELLERI ENHANCEMENT EFFECTS

While the above linear and circular polarization experiments of the two–photon radiation appear to agree well with the predictions of quantum mechanics further critical considerations and suggestions have been made for testing all possible assumptions with regard to quantum mechanics versus local realistic theories. Important proposals were made by Garuccio and Selleri [14] involving single–photon physics and two– polarizer type experiments to be interpreted in local realistic terms. They assumed that in addition to a polarization vector l, a photon possessed a detection vector λ and, on this basis, were able to explain experimental results in local realistic terms. However, Haji–Hassan et al., [41] tested this concept in an extension of the previously described linear polarization correlation experiment. They inserted a half–wave plate in one detection arm between the linear polarizer P_1 and the photo–multiplier D_1. By adding this half–wave plate to the system the plane of polarization of the photons incident on photo–multiplier D_1 could be varied independently of the polarization axis of P_1. At first the transmission axes of the polarizers P_1 and P_2 in both arms of the apparatus and the fast axis of the half–wave plate were set parallel to each other. When polarizer P_1 was rotated by an angle Θ the fast axis of the half–wave plate was rotated by an angle $\varphi = \Theta/2$. By this procedure it was arranged that the orientation of the plane of polarization of the photon incident on the photo–multiplier D_1 did not change as the rotation of the polarizer and half–wave plate took place. Since the relative angle between the planes of polarization of the photons impinging on the detectors did not change significantly the results could not be distorted by *enhancement effects* due to the detection vector λ. Within the limits of experimental error the results were once again in agreement with quantum mechanics as shown in Fig. 10.12.

In a second experiment with the half–wave plate the polarization axes of the two linear polarizers were fixed parallel to each other and the fast axis of the half–wave plate was rotated by an angle ε relative to the axes of the polarizers. In this way the relative angle between the planes of polarization of the photons impinging on the photomultipliers could be varied continuously. It was also verified that the singles count rates did not vary as the half–wave plate was rotated.

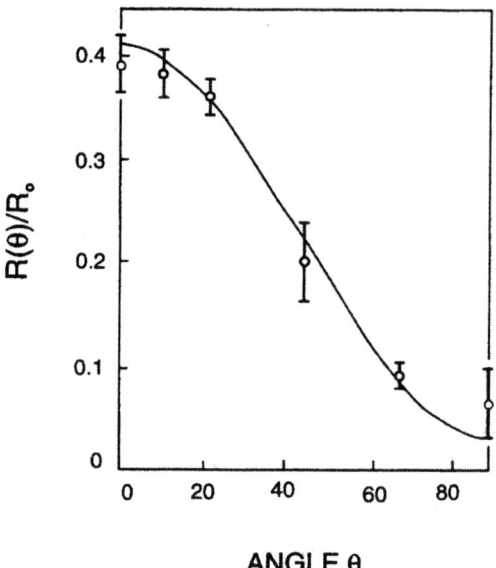

Figure 10.12 Ratios $R(\Theta)/R_0$ as a function of the orientation angle Θ of the fast axis of the half–wave plate inserted between the linear polarizer P_1 and the photomultiplier D_1. The coincidence signals $R(\Theta)$ results from the observations with the transmission axes of the two linear polarizers parallel to each other ($\Theta = 0°$) and the coincidence signal R_0 is obtained with the linear polarizers removed.

The two–photon coincidence measurements with the half–wave plate clearly establish the assumption that the relative angle between the planes of linear polarization of the two photons prior to detection plays no role in establishing experimental observation of polarization correlations. This statement confirms the assumption of Clauser, Horne, Shimony and Holt [30] that the probability of the joint detection of a pair of photons which emerge from two polarizers is independent of the relative angle between the polarization planes of the two photons just prior to their detection at the photomultipliers. There is no experimental evidence to support the above idea of Garuccio and Selleri [14] to introduce a detection vector and to consider an enhanced or modified photon detection depending on the combined action of the detection vector λ and polarization vector l.

4.3 THREE–POLARIZER EXPERIMENTS

The further proposal by Garuccio and Selleri [14] to introduce a second linear polarizer in one of the detection arms of the two–photon coincidence apparatus provided an opportunity to test quantum mechanics versus local realistic theories in a hitherto unexplored and novel procedure. The experiment and

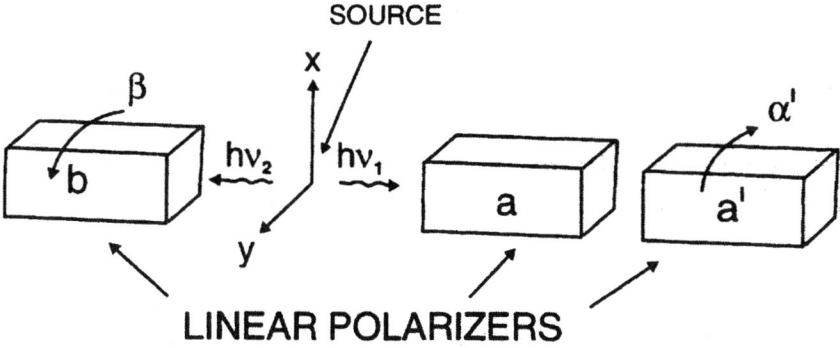

Figure 10.13 Schematic arrangement of the three-polarizer experiment (Haji–Hassan et al., [41]). The orientation of polarizer **a** is fixed with its polarizer transmission axis parallel to the x axis. The transmission axes of polarizers **b** and **a'** are rotated, respectively through angles β and α relative to the x-axis.

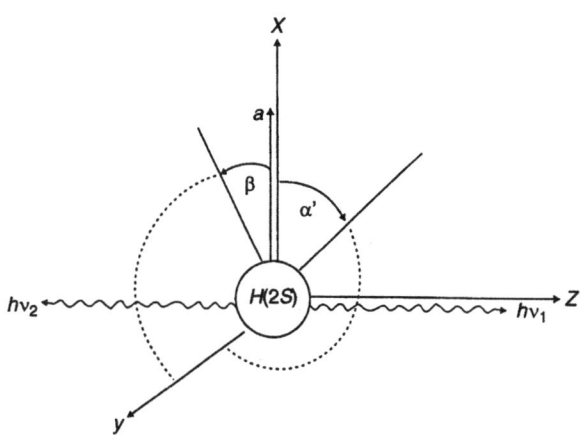

Figure 10.14 Geometry and angles of the transmission axes of the three polarizers used in the Stirling three-polarizer experiment.

the relevant geometries for the polarization directions of the three polarizers (Haji–Hassan et al., [41]) are shown in Figs. 10.13 and 10.14.

The orientation of the polarization plane of polarizer **a** was held fixed while that for polarizer **b** was rotated through an angle β in a clockwise sense and polarizer **a'** through an angle α' in an anticlockwise sense. The ratio $R(\beta, \alpha')/R(\beta, \infty)$ was measured where $R(\beta, \alpha')$ is the coincidence rate with all three polarizers in place and $R(\beta, \infty)$ the rate with polarizer **a'** removed.

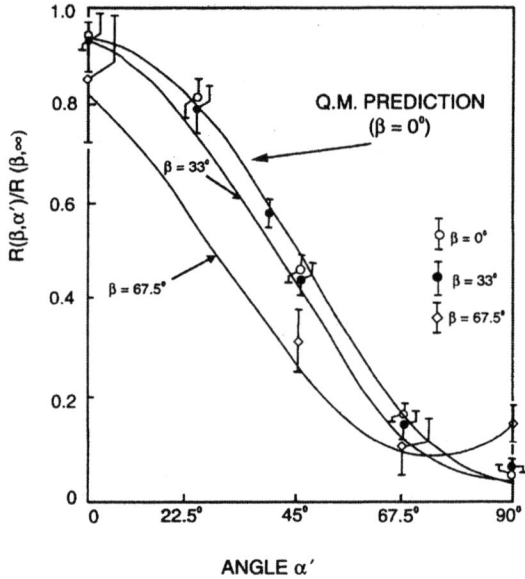

Figure 10.15 Ratios $R(\beta, \alpha')/R(\beta, \infty)$ as a function of β and α' in the experiment of Haji–Hassan et al. [41]. The solid lines represent quantum mechanical predictions for $\beta = 0°, 33°$ and $67.5°$.

Results of these ratios are shown in Figs. 10.15 and 10.16 along with quantum mechanical predictions and the limits of the Garuccio–Selleri local realistic model (1975). The quantum mechanical prediction for $\beta = 0°$ is close to the form $\varepsilon_M \cos^2\alpha' + \varepsilon_m \sin^2\alpha'$ with the transmission efficiencies of polarizer **a'**. These data confirm the validity of Malus' cosine–squared law for the transmission of polarized light from a very weak source through polarizer **a'**.

On the other hand the model of Garuccio and Selleri [14] showed that, for any angle $\Theta = 90°$ (note that Θ is not the angle between the polarizers in this case), arranging the angles of the transmission axes of the three polarizers to satisfy the relations $\beta = 30°$ and $\alpha' + \beta = \Theta$ the ratio of the quantum mechanical prediction to that of their local realistic model must always be greater than some minimum value γ_L as shown in Fig. 10.16. The range $58° < \Theta < 80°$ where $\gamma_L = l$ can be used in particular as a test between quantum mechanics and the local realistic theories of Garuccio and Selleri [14]. For the maximum of $\gamma_L = 1.447, \Theta = 71°$ and the corresponding values for $\beta = 33°$ and for α' is $38°$. For these values the approach of Garuccio and Selleri [14] sets an upper limit on the predictions of the local realistic theories of 0.413 for the ratio $R(\beta, \alpha')/R(\beta, \infty)$ while the experimental value, according to Fig 10.16 is 0.585 ± 0.029 violating the Garuccio–Selleri model by about six standard

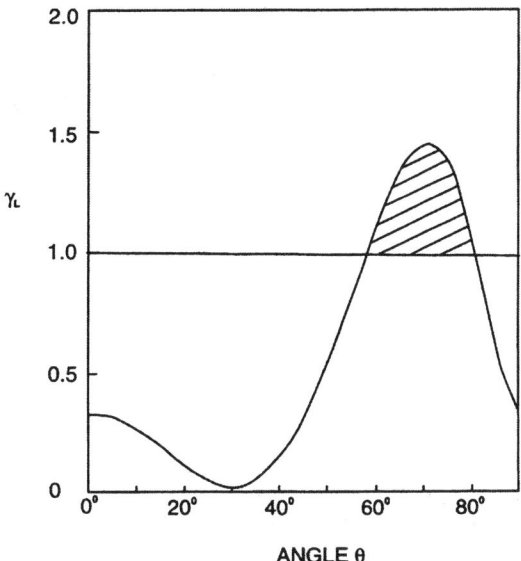

Figure 10.16 Lower limit γ_L of the ratio of the quantum mechanical prediction to that of local realistic theories with enhancement of the type proposed by Garuccio and Selleri [14].

deviations or a difference of more than approximately 40%. Even with some modification of the maximum value of γ_L to 0.162 suggested by Selleri [2], the above ratio will be only 0.514 which is still violated by the experimental data by almost three standard deviations or about 15%. Accordingly the three–polarizer experiment appears to rule out the class of local realistic theories of Garuccio and Selleri in a more convincing way than presently published two–polarizer coincidence experiments. The quantum–mechanical prediction for $\beta = 0°$ is close to the form $\varepsilon_M cos^2\alpha' + \varepsilon_m sin^2\alpha'$ with $\varepsilon_M, \varepsilon_m$ the transmission efficiencies of polarizer **a'**. These results thus confirm the validity of Malus' cosine–squared law for the transmission of polarized photons from a very weak light source passing through polarizer **a'**.

4.4 BREIT–TELLER HYPOTHESIS

It follows from considerations of parity and angular momentum conservation (section 2.) that the two–photon state vector can be written as described by eqs. 10.6 and 10.8 for circularly or linearly polarized components. In a coincidence experiment the "collapsed" components of the state vector are the right–right–hand $|R_1\rangle|R_2\rangle$ or left–left–hand $|L_1\rangle|L_2\rangle$ circularly polarized components or, alternatively, the linearly polarized components $|x_1\rangle|x_2\rangle$ and $|x_1\rangle|x_2\rangle$, respectively. These collapsed components of the state vector are com-

244   COMPLETE SCATTERING EXPERIMENTS

patible with the hypothesis of Breit and Teller [11] that the fine and hyperfine interaction due to electron and nuclear spin should not affect the two–photon polarization correlation since the Hamiltonian describing the two–photon emission has no off–diagonal components for fine and hyperfine interactions in second order. Figs. 10.17 demonstrates the various cases of the polarized com-

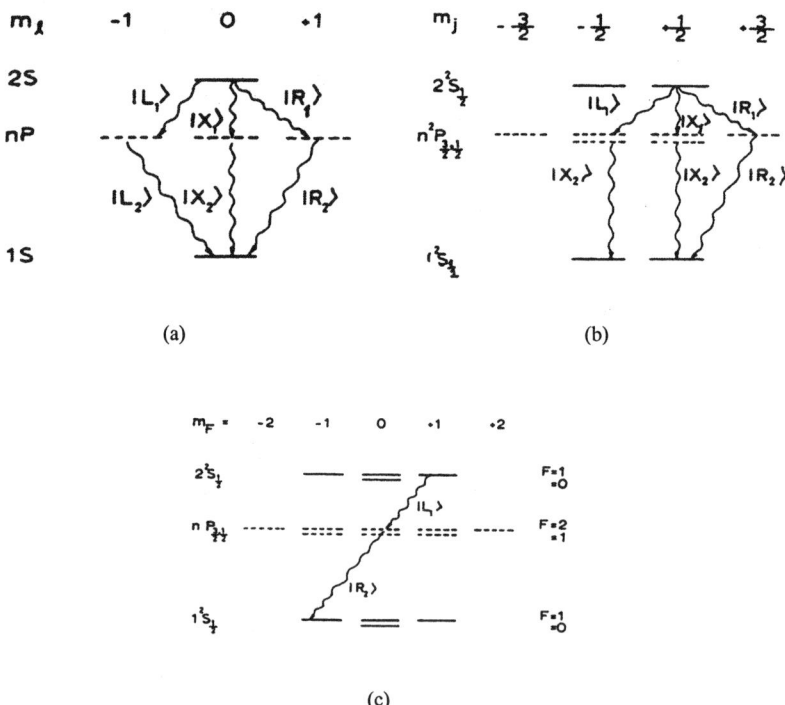

Figure 10.17 The "collapsed" state vectors for the two–photon transitions of metastable atomic hydrogen with 10.17(a) fine and hyperfine structure interactions neglected ($\Delta m_1 = 0$), 10.17(b) fine structure interaction included but hyperfine structure interaction neglected ($\Delta m_1 = 0, \pm 1$) and 10.17(c) fine and hyperfine structure interactions included ($\Delta m_1 = 0, \pm 1, \pm 2$).

ponents of the collapsed state vectors: with electron and nuclear spin neglected (Fig. 10.17(a)); including the electron spin (Fig. 10.17(b)): and including both nuclear and electron spin (Fig. 10.17(c)). For the case of zero nuclear and electron spin ($I = 0, S = 0$) only the collapsed state vectors $|R_1\rangle|R_2\rangle$, $|L_1\rangle|L_2\rangle$, $|x_1\rangle|x_2\rangle$ and $|y_1\rangle|y_2\rangle$ are compatible with the two–photon transitions through the intermediate virtual P states (Fig. 10.17(a)).

For the case $S = 1/2, I = 0$ (Fig. 10.17(b)) possible "collapsed" state vectors would be $|R_1\rangle|x_2\rangle$, $|R_2\rangle|x_1\rangle$, $|L_1\rangle|x_2\rangle$ or $|L_2\rangle|x_1\rangle$ with $\Delta m_j = \pm 1$ as a selection rule which is compatible with one photon carrying one unit of angular momentum away. A state vector of the corresponding polarization correlation for $\Delta m_j = +1$ would be

$$|\psi\rangle = \frac{1}{\sqrt{2}} \left(|R_1\rangle |x_2\rangle + |L_2\rangle |x_1\rangle \right), \tag{10.6}$$

which predicts, for example, that with a linear polarizer on one side of the two–photon source no variation in the coincidence signal would be observed when the transmission axis of a linear polarizer on the other side was rotated. However, the measured variation with rotation angle is described well by the state vectors in eqs. 10.2 and 10.3 ruling out to a high degree of accuracy state vectors of the type given in eq. 10.6. In other words the electron spin plays no role in the polarization correlation of the two–photon decay.

Including both the nuclear and electron spins of atomic hydrogen will again result in the above collapsed state vectors $|R_1\rangle|x_2\rangle$ with $\Delta m_F = \pm 1$ which are excluded by the experimental observations. However, as can be seen from Fig. 10.17(c) possible $\Delta m_F = \pm 2$ transitions are to be associated with the collapsed state vectors $|R_1\rangle|L_2\rangle$ and $|R_2\rangle|L_1\rangle$ which would require a state vector of the type

$$|\psi\rangle = \frac{1}{\sqrt{2}} \left(|R_1\rangle |L_2\rangle + |R_2\rangle |L_1\rangle \right). \tag{10.7}$$

The experiments of Perrie et al. [26], however, give effectively zero coincidence count rates for the two photons with opposite helicity.

In conclusion it can be stated that the polarization correlation experiments confirm the statement of Breit and Teller [11] that the influence of the nuclear and electron spins can be neglected in the two–photon emission of metastable hydrogen. The correct state vector model of the two–photon emission in diametrically opposite directions is well represented by the state vectors of equations 10.2 and 10.3 and not by those of eqs. 10.6 and 10.7. Based upon the "simultaneous" process of the two–photon emission of the metastable hydrogen atom (i.e. within a time interval of $\approx 10^{-15}$ s see next chapter) the available time during which the two–photon emission takes place is too short for setting up the coupling mechanisms of the fine $\approx 10^{-11} - 10^{-12}$ s and hyperfine interactions $\approx 10^{-8} - 10^{-9}$ s.

As can be seen in Fig. 10.18 for $\beta = 33°$ and $\alpha_1 \cong 40°$ there is a relative difference of more than about 40% between the prediction of quantum mechanics and the local realistic theory of Garrucio and Selleri which is so much larger than from experiments with photons from cascades (see Fry and Thomson, [35]; Aspect et al. [36]), positronium annihilation (Paramannada and Butt [37]) and in interference experiments with laser photon pairs (Brendel et al. [38], Tittel et al., [42] and Weihs et al., [43]).

We just note that the three–polarizer method which gives a much more sensitive test result as a large difference of at least 40% between quantum mechanics

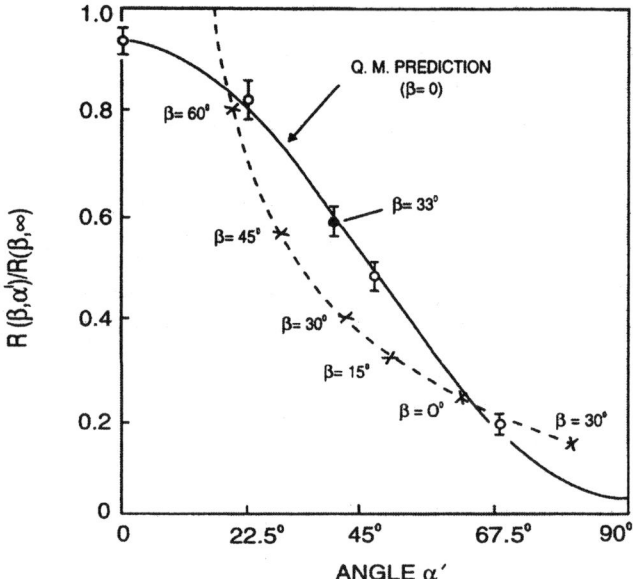

Figure 10.18 Variation of the ratio $R(\beta, \alpha')/R(\beta, \infty)$ as a function of the angle α' in the experiment of Haji–Hassan et al. [41] for $\beta = 0°, 15°, 30°, 45°$ and $60°$. The points marked (∘) correspond to the results for $\beta = 0°$, the point marked (•) to $\beta = 33°$. The solid curve represents the quantum mechanical prediction for $\beta = 0°$, while the broken curve shows the upper limit for the ratio set by the local realistic model of Garuccio and Selleri [14] for various angles β.

and Bell's limit for local realistic theories (Fig. 10.18) is not *simply based on a kind of grey filtering effect* by the additional polarizer, but is based on an interference effect of the observed polarization correlation. This can be seen as follows (see Fig. 10.18). With directions of polarization of polarizer **a** parallel to the x–direction, that of polarizer **b** at angle β and of polarizer **a'** at angle α' the polarization correlation of the coincident two–photon radiation becomes

$$P(\alpha', \beta) \cong \cos^2\alpha' \cos^2\beta$$

with $\alpha_1 = \beta + \alpha'$ and $\alpha' = \alpha_1 - \beta$

$$\begin{aligned}P(\alpha', \beta) &= \cos^2(\alpha' - \beta)\cos^2\beta \\ &= [\cos\alpha' + \sin(\alpha' - \beta)\sin\beta]^2 \\ &= \cos^2\alpha' + \sin^2(\alpha' - \beta)\sin^2\beta \\ &\quad + 2\cos\alpha'\sin(\alpha' - \beta)\sin\beta.\end{aligned} \qquad (10.8)$$

The last part of this equation represents kinds of interference term. Accordingly this type of the three polarizer system for the detection of the two–photon radiation in opposite directions leads to a different polarization correlation and, in a way, surprisingly larger difference between quantum mechanical and local realistic predictions for certain combinations of the directions of polarization of the polarizer as demonstrated in Fig. 10.18.

5. COHERENCE AND FOURIER SPECTRAL ANALYSIS

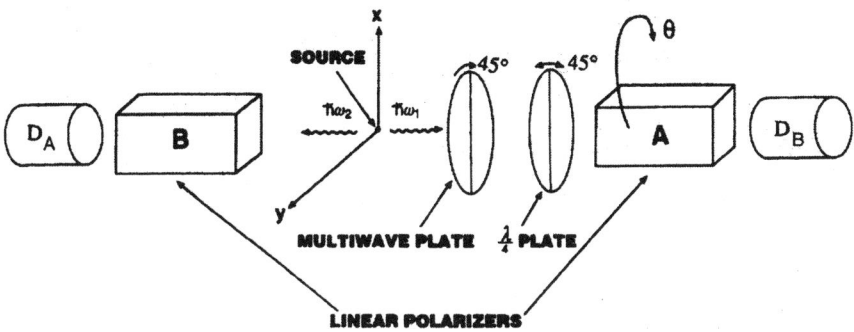

Figure 10.19 Experimental arrangement for the coherence and Fourier analysis of the two–photon radiation of metastable atomic hydrogen. While the detection arm B only contains the linear polarizer **b** the detection arm A includes a retardation plate of given thickness, a $\lambda/4$ wave plate and the linear polarizer **a** in order to measure Stokes parameters.

The coherence properties and spectral distribution of one–photon of the two–photon pair emitted by metastable atomic hydrogen were investigated in an experiment in which a series of birefingent retardation plates were placed in one arm of the detection system as shown in Fig. 10.19. The two photons emitted by the source are detected in coincidence by the detectors D_1 and D_2. A linear polarizer with its transmission axis orientated in the x direction is placed on the left hand side of the source (polarizer **b**). Thus, by virtue of the entangled nature of the two–photon state vector, equation 10.8 upon detection of a photon in the left, the complementary photon on the right, before being detected itself, can to all intents and purposes be regarded as also polarized in the x direction. In the absence of any wavelength filter on the left the frequency of the photon on the right is undeterminate, and the radiation on the right can be considered to consist of a sequence of single frequency photons with spectral distribution determined by the spectrum of the source or, alternatively, as a sequence of minimum uncertainty wavepackets with the spectrum and coherence length of each wavepacket determined by the spectral characteristics of the two–photon source.

If this radiation on the right is now passed through a uniaxial birefringent retardation plate with its axis at an angle of 45o to the x axis as shown in Fig. 10.19 its state of polarization will be changed and it will also be depolarized to an extent which depends on the thickness of the retardation plate. The state of polarization of the radiation emerging from the retardation plate may then be monitored by measuring the Stokes parameters P_1, P_2 and P_3 of the radiation on the right detected in coincidence with the radiation on the left. The Stokes parameters are defined (Born and Wolf, [44]) as:

$$P_1 = \frac{I(0°) - I(90°)}{I(0°) + I(90°)}, \quad P_2 = \frac{I(45°) - I(-45°)}{I(45°) + I(-45°)},$$

$$P_3 = \frac{I(RHC) - I(LHC)}{I(RHC) + I(LHC)},$$

where $I(0°)$ is the strength of the coincidence signal when the transmission axis of polarizer **a** is set at angle $\Theta = 0°$ to the axis with corresponding definitions for $I(90°), I(45°), I(-45°)$. Similarly $I(RHC)$ and $I(LHC)$ refer to the strength of the coincidence signal with the achromatic quarter–wave plate shown in Fig. 10.19 in place and orientated with its x axis at to the x axis so as to detect, respectively, right–hand–circularly (RHC) polarized light on left–hand–circularly (LHC) polarized light.

The degree of the total polarization P is then given by

$$P = \sqrt{P_1^2 + P_2^2 + P_3^2}, \tag{10.9}$$

which is also equal to the degree of coherence of the orthogonally polarized components if these components have equal intensity (Born and Wolf, [44]).

If the retardation of the retardation plate of thickness d is ϕ then

$$\phi = \frac{(n_e - n_o)d}{c}$$

, where n_e and n_o are, respectively, the extraordinary and ordinary refractive indices of the material of the retardation plate. It is then easy to show (Sheikh, [45]) that, for monochromatic radiation incident upon the retardation plate, the Stokes parameters of the emerging radiation are given by

$$P_1 = cos\phi, \quad P_2 = 0, \quad P_3 = -sin\phi.$$

However, since the two–photon radiation has a continuous spectral distribution $A(\omega)$, the expected values of P_1, P_2 and P_3 of the emerging radiation are

$$P_1 = \frac{\int_0^\infty \cos\phi(\Theta) A(\omega) d\omega}{\int_0^\infty A(\omega) d\omega}$$

$$P_2 = 0$$

$$P_3 = \frac{\int_0^\infty \sin\phi(\Theta) A(\omega) d\omega}{\int_0^\infty A(\omega) d\omega}$$

Examination of the above expression reveals that, if the birefringence $(n_e - n_o)$ were frequency independent, P_1 and P_3 would be precisely the Fourier cosine and Fourier sine transforms of the spectral distribution $A(\omega)$ with the quantity $(n_e - n_o)d/c$ acting as the "time" variable.

It has been shown (Duncan et al., [46]) that the frequency dependence of the birefringence can be allowed for by defining an effective spectral distribution $F(\omega)$ in terms of which

$$P_1 = \int_0^\infty F(\omega) \cos(\omega t) d\omega$$

$$P_2 = 0$$

$$P_3 = -\int_0^\infty F(\omega) \sin(\omega t) d\omega$$

which allows the easy calculation of P_1 and P_3, and also allows an inverse transformation to the performed which permits the easy calculation of the spectral distribution $F(\omega)$ and hence $A(\omega)$ from knowledge of P_1 and / or P_3.

In the experiment, a series of retardation plates were placed one at a time on the right–hand side of the source. These plates were in the form of zero–order half–wave plates at wavelengths of $200nm$, $300nm$ and $486nm$. They consisted of two flat pieces of crystal quartz of slightly different thicknesses cut parallel to the optical axis and placed in contact with their axes perpendicular to give "effective" thicknesses $d = 7.69\mu m$, $18.84\mu m$, $14.56\mu m$ and $26.27\mu m$ deduced from the known birefringence properties of quartz. Additional effective thickness of $d = 3.15\mu m$ and $18.53\mu m$ were obtained by placing the $200nm$ and $243nm$ plates in series with optical axes parallel. An effective thickness of $d = 37.11\mu m$ was obtained by placing the $243nm$ and $486nm$ plates in series with their axes parallel.

The experimental results, with allowance made for the imperfections of the polarizers are shown in Fig. 10.20 and show good agreement with the theoretical calculations based on the above Fourier analysis taking into account that the spectrum is cut off above an angular frequency of $1.02 \cdot 10^{16} rad \cdot s^{-1}$ ($\lambda = 185nm$) and below at $5.31 \cdot 10^{15} rad \cdot s^{-1}$ ($\lambda = 355nm$). The problem of

250   COMPLETE SCATTERING EXPERIMENTS

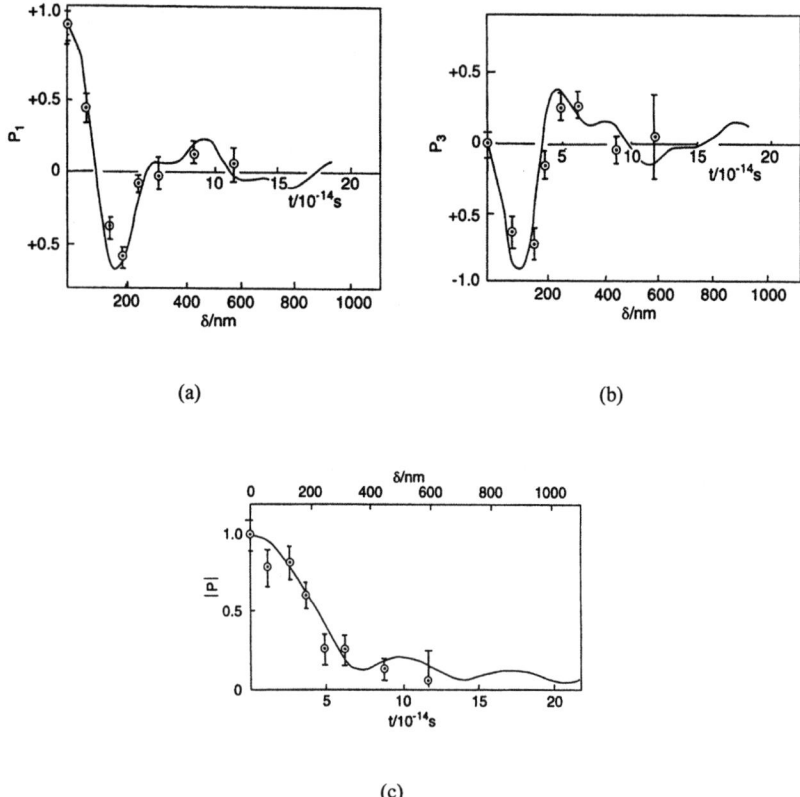

Figure 10.20   The Stokes parameters P_1, P_3 and the degree of total polarization $|P|$ as a function of the $t = d/c$ or the relative displacement of the photon wavepacket envelopes
$$\delta = d(n_e - n_o)\omega_e - \omega_c \left[\frac{\partial(n_e - n_o)}{\partial \omega}\right]_{\omega_e} d,$$
where d is the effective thickness of the retardation plate. The solid lines are calculated from Fourier transforms of the theoretical spectral distribution of the two–photon radiation.

determining the spectral distribution from the Stokes parameter measurements is straightforward in principle but, in the present case, difficult in practice, because of the small number of measured points.

However, by making a reasonable interpolation between the points and performing an inverse Fourier transform it has been possible to show (Duncan et al., [46]) that the spectral distribution has the expected broad–band form as show in Fig. 10.20.

A bandwidth extending between angular frequencies of $i1.02 \cdot 10^{16} rad \cdot s^{-1}$ and $5.31 \cdot 10^{15} rad \cdot s^{-1}$, on the basis of the usual Fourier bandwidth–time relationship, implies a coherence time of about $1.3 fs$ and coherence length of about $385 nm$ for a single photon of the two–photon pair. The coherence length is, therefore, less than two wavelengths of the predicted $243 nm$ centre wavelength of the spectral distribution. The coherence length and the coherence time of single photons of the two–photon pair are thus very short and, in the ideal case, in the absence of any filtering, determined by the lifetime of the virtual intermediate state of the decay rather than the long lifetime of the metastable state determines the coherence time and length of the two–photon pair. In fact an analogous situation arises in the case of certain experiments involving the two photons produced by parametric down conversion where single–photon second–order interference effects occur over distances determined by the short coherence lengths of the photons after conversion, whereas two–photon fourth–order interference effects take place over distances related to the usually much longer coherence length of the pump radiation (Franson [47], Ou et al. [48], Kwiat et al [49], Brendel et al [38]).

It should be pointed out, of course, that although the lifetime of an excited atomic state is related to the coherence time of the resulting radiation produced when it decays, it is not equal to it. If the excited state has, say an energy width ΔE then by Heisenberg's uncertainty principle, the lifetime $\tau_l \cong \hbar/2\Delta E$, whereas on the basis of the Fourier bandwidth–time relationship $\Delta \nu \tau_c \cong 1$ the coherence time $\tau_c \cong 1/\delta \nu$, where $\delta \nu$ is the frequency bandwidth. Hence

$$\tau_c = \frac{1}{\delta \nu} = \frac{h}{\Delta E} = \frac{2\pi \hbar}{\Delta E} = 4\pi \tau_l \qquad (10.10)$$

and the coherence time τ_c is greater than the lifetime τ_l by a factor 4π. An alternative interpretation of the results can be made in terms of photon wavepackets, which can be considered to exist on the right following detection of complementary photons on the left. As a result of their different group velocities, the two orthogonally polarized wavepackets passing through the retardation plate are separated by an amount δ where

$$\delta = (n_e - n_o)w_e \cdot d - w_c \left[\frac{d(n_e - n_o)}{dw} \right]_{w_e} d, \qquad (10.11)$$

w_e is the centre angular frequency of the wavepacket. For quartz from knowledge of its birefringence $\delta 1.667 \cdot 10^{-7} d$. When δ is large enough the polarization of the radiation emerging from the retardation plate is essentially zero, and the distance δ over which the degree of polarization or degree of coherence P reduces to near zero gives a measure of coherence length of the wavepackets. The degree of polarization P is plotted against δ in Fig. 10.20 from which it can be seen that the coherence length is approximately $350 nm$ with a corresponding coherence time of $1.2 fs$, in agreement with the previous observations.

Theoretical predictions of P_1, P_3 and P are calculated based upon the preceding Fourier analysis and the theoretical spectral distribution for hydrogen cut off above an angular frequency of $1.02 \cdot 10^{16} rad \cdot s^{-1}$ (corresponding to $\lambda = 185nm$) and below at the complementary frequency of $5.31 \cdot 10^{15} rad \cdot s^{-1}$ (corresponding to $\lambda = 355nm$). Allowance has been made for imperfections of the polarizers in the two arms for coincident photon detection. The results of Fig. 10.20 show good agreement between the calculated values and experimental data. The modulus $|P|$ of the total polarization vector falls rapidly to a low value at $\delta \cong 350nm$. The minimum length of the wave packets of the photons is thus approximately $350nm$, which can be considered as the experimentally determined coherence length $l_{coh} = ct_{coh}$ of a single photon of the coincident two–photon pair for which the corresponding coherence time t_{coh} is thus approximately $1.2 \cdot 10^{-15} s$ and the bandwidth $0.8 \cdot 10^{-15} s^{-1}$. The experimentally determined coherence length of $350nm$ is approximately one and a half times the wave length of the theoretically predicted centre and maximum of the spectral distribution of the two–photon radiation at $243nm$.

Some caution, of course, must be applied in using the wavepacket concept in the present case. Such a short, wide bandwidth classical wavepacket would undergo a significant amount of spreading and chirping in propagating through a dispersive medium. However, the maximum of the packet envelope would travel at the group velocity and the degree of overlap and hence the degree of polarization or coherence would be essentially unaffected. (See the discussion by Steinberg et al., [50]).

6. TIME CORRELATION

The existing theory of the two–photon decay implies that the spontaneous emission of the two photons takes place essentially simultaneously. However, the angular frequency bandwidth $\Delta\omega$ sets a lower limit on the correlation time T_c (as opposed to coherence time) between the detection of the two photons given by (Huang and Eberly, [51])

$$T_c = \frac{1}{\Delta\omega}. \qquad (10.12)$$

In the present case for an unfiltered signal the full bandwidth $\Delta\omega = 1.546 \cdot 10^{16} rad \cdot s^{-1}$ and hence $T_c = 6.47 \cdot 10{-16} s$. these values for the correlation time are very much less than the electronic resolving time of a few nanoseconds in the present experiment, (Fig. 10.19), and much less than could be achieved with the most modern photon detectors and counting electronics.

Biermann et al. [15] have tackled the problem of the time correlation between detection of the photons in the decay of metastable atomic hydrogen from the more sophisticated viewpoint fo quantum electrodynamics and have found a value $T_c \cong 3.33 \cdot 10^{-16} s$ which is not too different from the value found above. Their theory also makes a comparison of the nature of the two photon decay process with a cascade through a three–level atom. It successfully predicts the

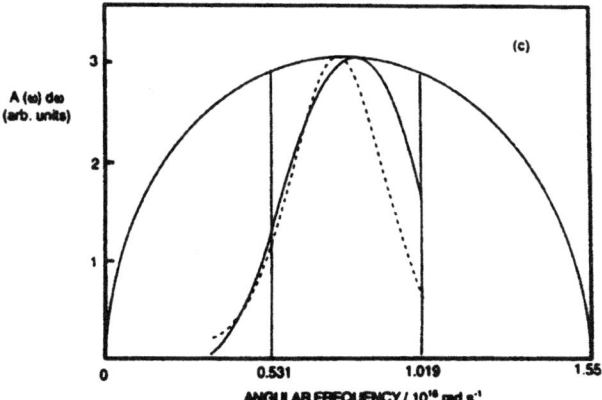

Figure 10.21 Spectral distribution of the two–photon radiation deduced from the P_1 data (narrow full curve) of Fig. 10.20 and P_3 data of Fig. 10.20 (broken curve) by a Fourier transformation. The upper and lower frequency cut–offs of the experimentally observed two–photon radiation are $0.531 \cdot 10^{16} rad \cdot s^{-1}$ and $1.019 \cdot 10^{16} rad \cdot s^{-1}$. The broad full curve is the theoretical prediction of the two–photon spectral distribution with all photons of energies from 0 to $10.2 eV$.

asymmetrical nature of the time correlation spectrum for a cascade source and the symmetrical nature of the time correlation spectrum (Fig. 10.21) for the two–photon source.

7. CORRELATED EMISSION SPECTROSCOPY OF METASTABLE HYDROGEN

Over the past few years it has been shown that atomic spectroscopic (Rathe and Scully, [52]; Rathe et al., [53]; Scully and Zubairy, [4]) properties such as the energies and width of atomic states involved in the production of correlated cascade radiation can be probed via intensity correlation interferometry (Hanbury–Brown and Twiss, [54]; Scully and Drühl, Scully:1982; Ou and Mandel, [55]; Shih and Alley, [56], Franson, [47], Rarity and Tapster, [57]; Kwait et al., [58]; Herzog et al., [59]). This correlated emission spectroscopy has, here–to–fore, been studied in the context of real states. However, recent intriguing experiments by Kleinpoppen et al., [39] on the $2s \rightleftarrows 1s$ transition in atomic hydrogen extend these considerations to the realm of virtual states. These experiments raise many questions, to wit: Can we interpret these studies as yielding a measure of a delay between emissions? And is there a virtual state between the $2s$ and $1s$ states, which is "like" a rapidly decaying real state?

Motivated by the preceding considerations, Biermann et al., [15] have calculated the two–photon correlation function for the radiation spontaneously emitted by metastable hydrogen. These studies taken together with the cascade

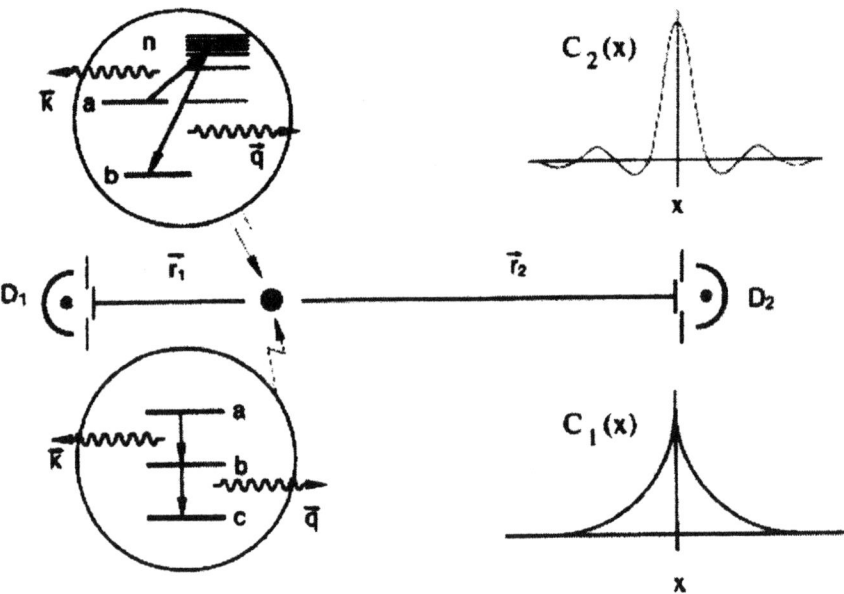

Figure 10.22 Figure illustrating Gedanken coincidence measurement of radiation from atoms, which are either of the hydrogen metastable (upper part) or of the cascade type (lower part). In both cases the two photons emitted travel to detectors located at distances $\vec{r_1}$ and $\vec{r_2}$ from the emitting atom. Both detectors are equipped with shutters such that the detection time is defined within narrow time windows about the times t_1 and t_2. The near detector at about $\vec{r_1}$ establishes the "start fate" and a coincidence is observed in detector 2 at a time which is determined by the position of detector 2 indicated by x. When one considers the cascade problem (see text) one finds that when the first photon goes to detector 1 coincidences are observed for $x < 0$ and when the second photon goes to detector 1 coincidences occur for $x > 0$. If the photons are indistinguishable then these two possibilities are added and this yields the "cusp" correlation function. In the case of hydrogen decay a similar logic prevails in which the functional form now changes from a cusp function to the smoother "Bessel function" result.

emission problem (Fig. 10.22) yield insight into what we mean by the "delay time" between emissions and the extent to which the virtual state concept is a useful one in the present context.

The derivation of the two photon correlation function, which describes the joint count probability for registering a count in both a detector located at $\vec{r}_1$ at time t_1 and one at $\vec{r}_2$ at time t_2 is given the Glauber second order correlation function. The comparison of the relevant two–photon radiation for the cascade emission with the two–photon radiation of metastable atomic hydrogen shows (Fig. 10.22) that the two share some important features with regard to a "sharp turn–on wave front" and the same propagation factor for the special case of the cascade emission with angular frequencies $\omega_{ab} = \omega_{bc} = \omega/2$ (frequencies between states a and b and b and c). Thus one could think that the cascade process and two–photon decay playing the role of intermediate state in the cascade problem. However, there is an important difference caused by a second–order correlation function in the two–photon emission of metastable hydrogen. While in the cascade case there is a sharp turn–on for the second photon (i.e. it can only be emitted after the first one) there is no such sharp turn for the second photon in the two–photon emission of metastable atomic hydrogen since it has a certain probability to be detected before the first one. Of course, in a measurement we cannot distinguish between the "first" and "second" photon so we would then label the photons according to their detection time, but still – independently of what we call them – only one of the two photons has a sharp wave front traveling with the light velocity and the other one can be detected earlier or later.

It follows from the theory of Biermann et al. [15] that the coherence length of the two–photon radiation emitted by metastable atomic hydrogen should be of the order of $100 nm (= 10^{-7} m)$, which is to be compared to the experimentally determined one of $350 nm$. While there appears to be a discrepancy it should be noted that a more detailed analysis of both theoretical and experimental facts of the two radiation may have to be taken into account, i.e. a detailed representation of the correlation functions related to the two–photon amplitudes coupled to the higher np states in the second order approximation, and a more exact measurement of the energy distribution of the two–photon spectrum observed. Apart from such remaining aspects it should be emphasized, however, that based upon the above polarizations and coherence analysis of the spontaneous two–photon emission of metastable atomic hydrogen, tests on some of the high–order radiation theory are now possible which were previously not yet available.

We would also like to refer the measured data for the coherence length and time to Heisenberg's uncertainty $\Delta E \Delta t = \hbar/2$ principle by considering an energy width ΔE corresponding to the wavelength limits of $\lambda_1 = 185 nm$ and $\lambda_2 = 355 nm$ for the spectral range of detection of the two–photon radiation from metastable atomic hydrogen. Assuming a square wave dependence of the two–photon energy distribution in the above spectral range Δt results in a value, which would be the relevant coherence time $\Delta t = t_{coh} = 1.03 \cdot 10^{-16} s$, with a corresponding coherence length of $l = c\Delta t = 30.8 nm$. These values differ from the foregoing experimental data being about an order of magnitude too short and too small, respeively. Confirmation of the linear and circular polarization correlation a of the two–photon transition of metastable atomic hydrogen suggests comparison to traditional one–photon transitions with their linear (π) and circular (σ^+ and σ^-) polarization components as illustrated in Fig. 10.23 for a normal transition from an $n_2^1 P_1$ state to an $n_1^1 S_0$ state. In analogy to this

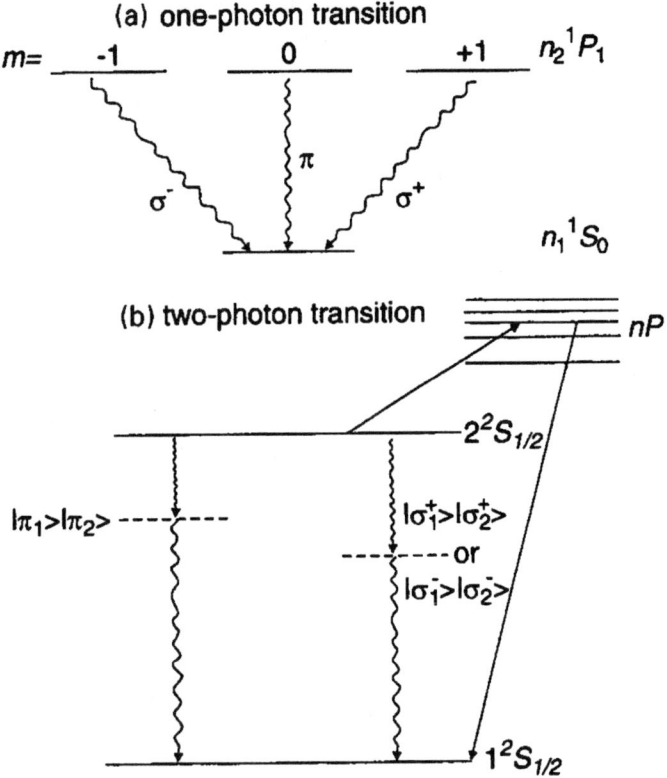

Figure 10.23 Comparison of linearly and circularly polarized π^- and σ^- components for one– (a) and two–photon (b) transitions. Note the higher nP states which "contribute" to the two–photon transition as states represented in the dipole matrix elements of the second–order quantum mechanical approximation theory.

picture the two–photon transitions can be associated with two linearly polarized and correlated π–transitions $|\pi_1\rangle|\pi_2\rangle$ or two right $|\sigma_1^+\rangle|\sigma_2^+\rangle$ or left $|\sigma_1^-\rangle|\sigma_2^-\rangle$ circularly polarized components assuming emission in opposite directions. In comparing these one– and two–photon dipole transitions, however, one has to keep in mind that the two emitted photons are detected in coincidence.

Whilst we have referred already to theoretical approximations for the inclusion of second–order transition amplitudes via the various higher np states, a proper energy scheme for the two–photon radiation of metastable atomic hydrogen has to take into account in these states in order to predict correctly the experimental observable coherence length and time. The influence of the spectral widths $\Delta\nu_1 = \Delta E_1/h$ and $\Delta\nu_2 = \Delta E_2/h$ of the two incident photons (with $E_1 \pm \Delta E_1 + E_2 \pm \Delta E_2 = E_{2s} - E_{1s}$) might also affect the test quantities for coherence length and time of the two–photon radiation. Anyway, while normal traditional fine and hyperfine spectroscopy of the $1^2S_{1/2}$ and $2^2S_{1/2}$ states can just be represented by these two states of the hydrogen energy scheme, only spectroscopy of the two–photon radiation from metastable atomic hydrogen requires the inclusion of the higher np states ($n \geq 3$ and possibly nl states ($n \geq 3, l \geq 2$) and continuum states. Accordingly a more proper atomic hydrogen energy scheme for the spectroscopy of the coincident two–photon radiation as indicated in Fig. 10.23.

8. CONCLUSIONS

The paper summarizes and highlights the activities of research on the two–photon coincidences of metastable atomic hydrogen with a view on both theoretical and experimental aspects. In comparison to similar two–photon investigations with one–electron hydrogen–like ions (see e.g. Gould and Marrus, [60]) the atomic hydrogen case has the disadvantage of low intensity of the two–photon radiation; however, there is the advantage of the possibilities to apply coincident optical polarization techniques and Fourier analysis in the near ultraviolet spectral region. The experiment with three polarizers, two polarizers in one and one polarizer in opposite direction for the detection of the two–photon emission appears to be the one of the many EPR–type measurements which has, for certain observational conditions, the largest difference ($\approx 40\%$) between predictions of non–local quantum mechanics and local classical theories. This effect results from an interference effect of the optical polarization analysis. One would have to judge future three or more particle EPR experiments based on the work by Greenberger, Horne and Zeilinger [61] and other possible techniques in comparison to the results of the three–polarizer EPR experiments for the two–photon radiation of metastable atomic hydrogen. The various physical characteristics of the two–photon radiation are linked to the properties of the energy scheme and the spectroscopic transitions of atomic and correlation hydrogen. While the static properties, such as the energies of the states and their various one–photon transitions of atomic hydrogen are well understood theoretically to a highest degree of accuracy, the newly measured coherence effects of the two–photon radiation require further more detailed developments of theory in order to get a better agreement with the results of the recent polarization correlation and coherence experiments. It is obvious that a modern representation of the hydrogen energy states and spectrum should take into account the influence of the "virtual states" between its real states (e.g. the $1s$ and $2s$ states but also any combination of higher states). The recent first approach to this representation by Biermann et al. [15] for considering coherence effects from cascade transitions of real higher states leads to a programme to extend and to consider different distributions of intensities and amplitudes for these transitions. This appears to be another important fundamental problem of quantum mechanical applications for the structure of the physical description of static and dynamical properties of atomic hydrogen.

Acknowledgments

H.K.Kleinpoppen would like to acknowledge his appreciation to The Leverhulme Trust (London) for their award of an Emeritus Fellowship. A. J. Duncan tragically died during the joint preparation of this manuscript and, in recognition of his outstanding contribution, we wish to dedicate this paper to his memory.

References

[1] G. W. Series, *The Spectrum of Atomic Hydrogen. Advances* (World Scientific, Singapore, 1988).

[2] F. Selleri, *Quantum mechanics Versus Local Realism - the Einstein-Podolsky-Rosen Paradox* (Plenum Press, New York, 1988).

[3] D. M. Greenberger and A. Zeilinger, *Fundamental Problems in Quantum theory* (New York Academy of Sciences, New York, 1995).

[4] M. O. Scully and M. S. Zubairy, *Quantum Optics* (Cambridge University Press, Cambridge, 1997).

[5] W. E. Lamb Jr. and R. C. Retherford, Phys. Rev. **72**, 241 (1947).

[6] R. Novick, in *Physics of the One - and Two - Electron Atoms*, edited by F. Bopp and H. Kleinpoppen (North-Holland, Amsterdam, 1969), pp. 296 – 325.

[7] M. Lipeles, R. Novick, and N. Tolk, Phys. Rev. Lett. **15**, 690 (1996).

[8] D. O'Connell, K. J. Kollath, A. J. Duncan, and H. Kleinpoppen, J. Phys. B: At. Mol. Opt. Phys. **8**, 214 (1975).

[9] H. Krüger and A. Oed, Phys. Lett. **54A**, 251 (1975).

[10] M. Goeppert-Mayer, Ann. Phys. **9**, 173 (1931).

[11] G. Breit and E. Teller, Astrophys. J. **91**, 215 (1940).

[12] G. W. F. Drake, in *The Spectrum of Atomic Hydrogen. Advances*, edited by G. W. Series (World Scientific, Singapore, 1988), p. 137.

[13] A. Einstein, B. Podolsky, and N. Rosen, Phys. Rev. **47**, 779 (1935).

[14] A. Garuccio and F. Selleri, Phys. Lett. **103A**, 99 (1984).

[15] S. Biermann, M. O. Scully, and A. H. Toor, Phys. Scr. T **72**, 45 (1997).

[16] G. F. Bassani, M. Inguscio, and T. W. Hänsch, *The Hydrogen Atom* (Springer Verlag, 1989).

[17] H. Friedrich, *Theoretical Atomic Physics*, 2 ed. (Springer Verlag, 1998).

[18] I. E. McCarthy and E. Weigold, *Electron-Atom Collisions* (Cambridge Monographs, 1995).

[19] M. Goeppert, Naturwiss. **17**, 932 (1929).

[20] A. I. Akheiser and V. B. Berestetskii, *Quantum Electrodynamics* (Wiley, New York, 1965).

[21] H. J. Andrä, Phys. Scr. **9**, 257 (1974).

[22] H. J. Andrä, in *Progress in Atomic Spectroscopy*, edited by W. Hanle and H. Kleinpoppen (Plenum Press, New York, 1979), pp. 829–953.

[23] A. Van Wijngaarden, G. W. F. Drake, and P. S. Farago, Phys. Rev. Lett. **33**, 4 (1994).

[24] G. W. F. Drake, S. P. Goldman, and A. Van Wijngaarden, Phys. Rev. A **20**, 1299 (1997).

[25] S. P. Goldman and G. W. F. Drake, Phys. Rev. A **24**, 183 (1981).

[26] W. Perrie, Ph.D. thesis, University of Stirling, 1985, and [62,63].

[27] D. Bohm and Y. Aharonov, Phys. Rev. **108**, 1070 (1957).

[28] M. Bacal *et al.*, Nucl. Instr. Meth. **114**, 407 (1974).

[29] M. Bacal and W. Reichelt, Sci. Instr. **20**, 769 (1974).

[30] J. F. Clauser, M. A. Horne, A. Shimony, and R. A. Holt, Phys. Rev. Lett. **23**, 880 (1969).

[31] R. A. Holt, Ph.D. thesis, Harvard University, Cambridge, 1973.

[32] A. J. Duncan and H. Kleinpoppen, in *Quantum mechanics Versus Local Realism - the Einstein-Podolsky-Rosen Paradox*, edited by F. Selleri (Plenum Press, New York, 1988), p. 175.

[33] J. F. Bell, Physics **1**, 195 (1964).

[34] S. J. Freedman, Ph.D. thesis, University of California, Berkeley, 1972.

[35] E. S. Fry and R. C. Thompson, Phys. Rev. Lett. **37**, 465 (1976).

[36] S. Aspect, J. Dalibard, and G. Roger, Phys. Rev. Lett. **49**, 1804 (1982).

[37] V. Paramananda and D. K. Butt, J. Phys. G **13**, 449 (1987).

[38] J. Brendel, E. Mohler, and W. Martienssen, Europhysics Lett. **20**, 575 (1992).

[39] H. Kleinpoppen, A. J. Duncan, H.-J. Beyer, and Z. A. Sheikh, Phys. Script. T **72**, 7 (1997).

[40] J. R. Torgerson, D. Branning, C. H. Monken, and L. Mandol, Phys. Rev. A **51**, 4400 (1995).

[41] T. Haji-Hassan and et al., in *Book of Abstracts*, Fifteenth Int. Conf. Physics of Electronic and Atomic Collisions (1987), p. 74.

[42] W. Tittel, J. Brendel, H. Zbinden, and N. Gisin, Phys. Rev. Lett. **81**, 3563 (1998), and [52].

[43] G. Weihs *et al.*, Phys. Rev. Lett. **81**, 5039 (1998).

[44] M. Born and E. Wolf, *Principles of Optics* (Pergamon Press, Edinburgh, Scotland, 1965 and 1987).

[45] Z. A. Sheikh, Ph.D. thesis, University of Stirling, Stirling, 1993.

[46] A. J. Duncan, Z. A. Sheikh, H. J. Beyer, and H. H. Kleinpoppen, J. Phys. B: At. Mol. Opt. Phys. **30**, 1347 (1997).

[47] J. D. Franson, Phys. Rev. Lett. **62**, 2205 (1989).

[48] Z. Y. Ou, X. Y. Zou, L. J. Wang, and L. Mandel, Phys. Rev. Lett. **65**, 321 (1990).

[49] P. G. Kwiat *et al.*, Phys. Rev. A **41**, 2910 (1990).

[50] A. M. Steinberg, P. G. Kwiat, and R. Y. Chiao, Phys. Rev. A **45**, 6659 (1992).

[51] H. Huang and J. Eberly, J. Mod. Opt. **40**, 915 (1993).

[52] U. Rathe and M. Scully, Lett. Math. Phys. **34**, 297 (1995), julian Schwinger Memorial Issue.

[53] U. Rathe, M. Scully, and S. Yellin, *Fund. Problems in Quantum Theory* (Annals N.Y., 1995).

[54] H. Hanbury-Brown and R. Twiss, Phil. Mag. **45**, 633 (1954), see also [64].

[55] Z. Y. Ou and L. Mandel, Phys. Rev. Lett. **61**, 50 (1998).

[56] Y. Shih and C. Alley, Phys. Rev. Lett. **61**, 2921 (1998), see also [65].

[57] G. Rarity and P. Tapster, Phys. Rev. Lett. **64**, 2495 (1990).

[58] P. Kwait, A. Steinberg, and R. Chiao, Phys. Rev. A **45**, 7729 (1992).

[59] T. Herzog, P. Kwait, H. Weinfurter, and A. Zeilinger, Phys. Rev. Lett. **75**, 3034 (1995).

[60] H. Gould and R. Marrus, Phys. Rev. **28**, 2001 (1983).

[61] D. M. Greenberger, M. Horne, and A. Zeilinger, in *Bell's Theorem, Quantum Theory and Conceptions of the Universe*, edited by Kafatos (Kluwer Academic, 1989), p. 73, see also [66].

[62] W. Perrie, A. J. Duncan, H. J. Beyer, and H. Kleinpoppen, Phys. Rev. Lett. **54**, 1970 (1985).

[63] A. J. Duncan, W. Perrie, H. J. Beyer, and H. Kleinpoppen, *Fundamental Processes in Atomic Physics* (Plenum Press, New York, 1985), pp. 555–572.

[64] G. Bayin, *Lect. on Quantum Mechanics* (Benjamin Press, 1955).

[65] M. Rubin, D. Klyshoko, Y. Shih, and A. Sergienk, to be published.

[66] D. M. Greenberger, M. Horne, A. Shimony, and A. Zeilinger, Am. J. Phys. **58**, 1131 (1990).

Chapter 11

QUANTUM–STATE TRANSMISSION VIA QUANTUM TELEPORTATION

Dik Bouwmeester
Centre for Quantum Computation, Clarendon Laboratory, University of Oxford, Parks Road, OX1 3PU Oxford, United Kingdom

Jian-Wei Pan
Institut für Experimentalphysik, Universität Wien, Austria

Harald Weinfurter
Sektion Physik, Ludwig-Maximilians-Universität München, Schellingstrasse 4/III, D-80799 München, Germany

Anton Zeilinger
Institut für Experimentalphysik, Universität Wien, Austria

1. INTRODUCTION

As recently as a decade ago, the issue of quantum entanglement was mainly considered to be of a philosophical nature, though a very relevant one in our attempts to understand fundamental laws of physics [1–4]. In the last few years, very much to the surprise of most of the early researchers in the field, the basic concepts of superposition and quantum entanglement have turned out to be key ingredients in novel quantum communication and quantum computation schemes [5]. Here we address an experimental implementations, based on entangled photons, of one of these novel schemes, namely "quantum teleportation".

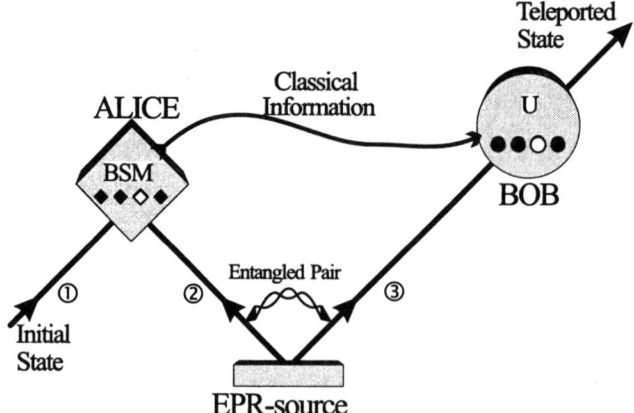

Figure 11.1 Principle of quantum teleportation: Alice has a quantum system, particle 1, in an initial state which she wants to teleport to Bob. Alice and Bob also share an ancillary entangled pair of particles 2 and 3 emitted by an Einstein-Podolsky-Rosen(EPR) source. Alice then performs a joint Bell state measurement (BSM) on the initial particle and one of the ancillaries, projecting them also onto an entangled state. After she has sent the result of her measurement as classical information to Bob, he can perform a unitary transformation (U) on the other ancillary particle resulting in it being in the state of the original particle. In the case of quantum teleportation of a qubit, Alice makes a projection measurement onto four orthogonal entangled states (the Bell states) that form a complete basis. Sending the outcome of her measurement, i.e. two bits of classical information, to Bob will enable Bob to reconstruct the initial qubit.

2. QUANTUM TELEPORTATION PROTOCOL

In this section we will review the quantum teleportation scheme as proposed by Bennett, Brassard, Crépeau, Jozsa, Peres, and Wootters [6]. The scheme is illustrated in the Fig. 11.1.

The idea is that Alice has particle 1 in a certain quantum state, the qubit

$$|\Psi\rangle_1 = \alpha|0\rangle_1 + \beta|1\rangle_1, \qquad (11.1)$$

where $|0\rangle$ and $|1\rangle$ represent two orthogonal states with complex amplitudes α and β satisfying $|\alpha|^2 + |\beta|^2 = 1$. She wishes to transfer this quantum state to Bob but suppose she cannot deliver the particle directly to him. According to the projection postulate of quantum mechanics we know that any quantum measurement performed by Alice on her particle will destroy the quantum state at hand without revealing all the necessary information for Bob to reconstruct the quantum state. So how can she provide Bob with the quantum state? The answer is to use an ancillary pair of entangled particles 2 and 3 (EPR pair), where particle 2 is given to Alice and particle 3 is given to Bob. Let us consider the case in which the entangled pair of particles 2 and 3 shared by Alice and

Bob is in the state

$$|\Psi^-\rangle_{23} = \frac{1}{\sqrt{2}}(|0\rangle_2|1\rangle_3 - |1\rangle_2|0\rangle_3). \qquad (11.2)$$

The important property of this entangled state is that as soon as a measurement on one of the particles projects it onto a certain state, which can be any normalised linear superposition of $|0\rangle$ and $|1\rangle$, the other particle has to be in the orthogonal state. The specific phase relation between the two terms on the right hand side of (11.2) (here the phase difference is π, which results in the minus sign) implies that the statement of orthogonality is independent of the basis chosen for the polarisation measurement.

Although initially particles 1 and 2 are not entangled, their joint polarisation state can always be expressed as a superposition of the four maximally entangled Bell states, given by

$$|\Psi^+\rangle_{12} = (|0\rangle_1|1\rangle_2 + |1\rangle_1|0\rangle_2)/\sqrt{2} \qquad (11.3)$$
$$|\Psi^-\rangle_{12} = (|0\rangle_1|1\rangle_2 - |1\rangle_1|0\rangle_2)/\sqrt{2} \qquad (11.4)$$
$$|\Phi^+\rangle_{12} = (|0\rangle_1|0\rangle_2 + |1\rangle_1|1\rangle_2)/\sqrt{2} \qquad (11.5)$$
$$|\Phi^-\rangle_{12} = (|0\rangle_1|0\rangle_2 - |1\rangle_1|1\rangle_2)/\sqrt{2}, \qquad (11.6)$$

since these states form a complete orthogonal basis. Therefore, the total state of the 3 particles can be written as:

$$|\Psi\rangle_{123} = |\Psi\rangle_1 \otimes |\Psi\rangle_{23} = \frac{1}{2} [\ |\Psi^-\rangle_{12}(-\alpha|0\rangle_3 - \beta|1\rangle_3)$$
$$+ \ |\Psi^+\rangle_{12}(-\alpha|0\rangle_3 + \beta|1\rangle_3)$$
$$+ \ |\Phi^-\rangle_{12}(\alpha|1\rangle_3 + \beta|0\rangle_3)$$
$$+ \ |\Phi^+\rangle_{12}(\alpha|1\rangle_3 - \beta|0\rangle_3)\ ]. \quad (11.7)$$

Alice now performs a Bell state measurement (BSM) on particles 1 and 2, that is, she projects her two particles onto one of the four Bell states. As a result of the measurement Bob's particle will be found in a state that is directly related to the initial state. For example, if the result of Alice's Bell state measurement is $|\Phi^-\rangle_{12}$ then particle 3 in the hands of Bob is in the state $\alpha|1\rangle_3 + \beta|0\rangle_3$. All that Alice has to do is to inform Bob via a classical communication channel on her measurement result and Bob can perform the appropriate unitary transformation (U) on particle 3 in order to obtain the initial state of particle 1. This completes the teleportation protocol.

Note that, during the teleportation procedure, the values of α and β remain unknown. By her Bell state measurement Alice does not obtain any information whatsoever about the teleported state. All that is achieved by the Bell state measurement is a transfer of the quantum state. Note also that during the Bell state measurement particle 1 loses its initial quantum state because it becomes entangled with particle 2. Therefore the state $|\Psi\rangle_1$ is destroyed on Alice's side during teleportation, thus obeying the no-cloning theorem of quantum mechanics [7]. Furthermore, the initial state of particle 1 can be completely unknown not only to Alice but to anyone. It could even be quantum mechanically completely undefined at the time the Bell state measurement takes place. This is the case when, as already remarked by Bennett et al. [6], particle 1 itself is a member of an entangled pair and therefore has no well-defined properties on its own. This ultimately leads to entanglement swapping, a process by which two particle that never directly interacted with oneanother become entangled [8–10].

3. EXPERIMENTAL QUANTUM TELEPORTATION

In this section an experimental demonstration of quantum teleportation of qubits, encoded in the polarisation state of single photons, will be given [11]. During teleportation, an initial photon which carries the polarisation that is to be transferred and one of a pair of entangled photons are subjected to a measurement such that the second photon of the entangled pair acquires the polarisation of the initial photon. Figure 11.2 is a schematic drawing of the experimental setup. As explained in the previous section, an experimental realisation of quantum teleportation necessitates both creation and measurement of entangled states, indicated in Fig. 11.2 by the Einstein–Podolski–Rosen (EPR) source and the Bell-state measurement (BSM) respectively. The following two subsections address the production of polarisation entangled photons and the Bell-state analyser.

3.1 POLARISATION ENTANGLED PHOTONS

Polarisation entangled photons can be created by the process of parametric down-conversion inside a nonlinear optical crystal. The nonlinear process provides the possibility that one pump photon is converted into a pair of correlated photons. Conservation of energy and momentum and the specific crystal properties places constrains on the directions, polarisations and frequencies of the created photons. In the case of the, so-called, non-collinear type-II phase matching conditions [12] the correlated photons are emitted along cones, which do not have a common axis, as is illustrated in Fig. 11.3 and Fig. 11.4. One of the cones is ordinarily polarised (along the polarisation direction of the pump photons) the other one extraordinarily (perpendicular to the polarisation of the pump photons). These cones will in general intersect along two directions. Since the two photons of each pair must always have orthogonal polarisations (type-II phase matching), we will find that along the two directions of intersection the emitted light is unpolarised, because we cannot distinguish whether a

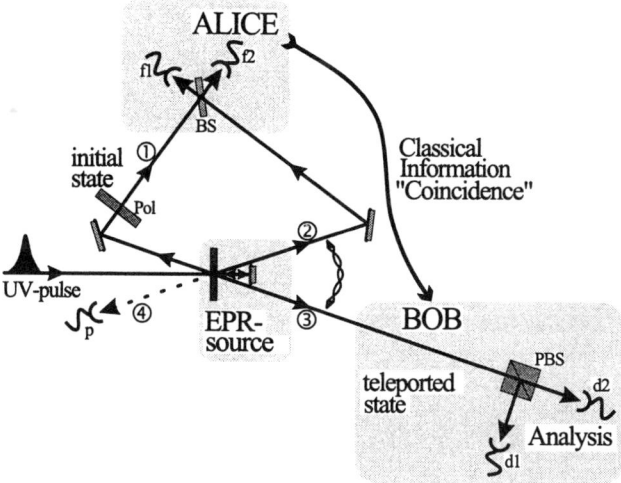

Figure 11.2 Schematic drawing of the experimental setup for quantum teleportation of a qubit. A pulse of ultraviolet (UV) light passing through a nonlinear crystal creates the ancillary pair of entangled photons 2 and 3. After retroflection during its second passage through the crystal, the ultraviolet pulse can create another pair of photons, one of which will be prepared in the initial state of photon 1 to be teleported, the other one serving as a trigger indicating that a photon to be teleported is underway. Alice then looks for coincidences after a beamsplitter (BS) where the initial photon and one of the ancillaries are superposed. Bob, after receiving the classical information that Alice obtained a coincidence count in detectors f1 and f2 identifying the $|\Psi^-\rangle_{12}$ Bell-state, knows that his photon 3 is in the initial state of photon 1 which he then can check using polarisation analysis with the polarising beamsplitter (PBS) and the detectors d1 and d2. The detector P provides the information that photon 1 is underway.

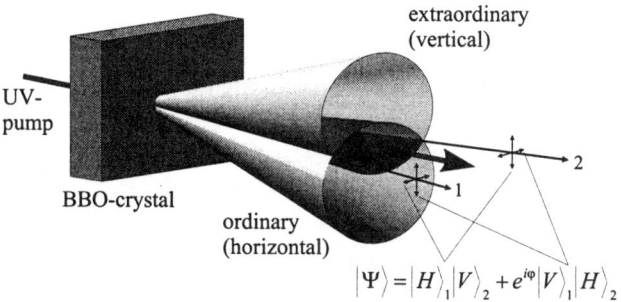

Figure 11.3 Non-collinear type-II down-conversion can produce two tilted cones of light of a certain wavelength. At same time other wavelengths are emitted, but in order to observe polarisation entanglement only we cut out a certain wavelength using narrow-band optical filters.

certain photon belongs to one or the other cone. This is not yet exactly true, because, due to birefringence in the crystal, the ordinary and extraordinary photons will propagate at different velocities and so we could at least in principle

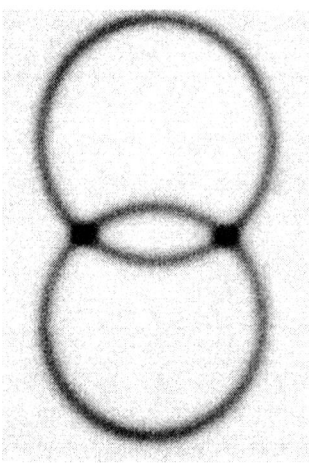

Figure 11.4 Type-II down-conversion light as seen through a narrow-band filter. The two rings are the ordinary and extraordinary cones of light rays. Along the intersecting directions we observe unpolarised light.

distinguish the two cases by the order of their detection times. It is however possible to compensate for that "walkoff" by inserting identical crystals of half the thickness rotated by 90° in each of the two beams. This procedure completely erases any such information and we have a true polarisation-entangled state which can be described by

$$|\Psi\rangle = \frac{1}{\sqrt{2}} \left[|V\rangle_1 |H\rangle_2 + e^{i\varphi} |H\rangle_1 |V\rangle_2 \right]. \quad (11.8)$$

Furthermore, we can use the birefringence of the compensator crystals to fine-tune the phase φ between the two components of the entangled state to π so that the resulting state is:

$$|\Psi^-\rangle = \frac{1}{\sqrt{2}} \left[|H\rangle_1 |V\rangle_2 - |V\rangle_1 |H\rangle_2 \right]. \quad (11.9)$$

3.2 BELL-STATE ANALYSER

Formally speaking, a Bell-state analysis is not a problem. All one has to do is project any incoming state onto the Bell state basis, (11.3)–(11.6), and one will find out by repeating this experiment with which probability the original state can be found in one of the Bell states. The Bell states depend, of course, on the type of entanglement that is present. In the case of polarisation entanglement between two photons, no complete Bell-state measurement has been achieved so far; Only the projection onto two Bell states has been realised, leaving the remaining two states degenerate in their detection [13]. However, in order to demonstrate quantum teleportation, it is sufficient to uniquely identify one of the Bell-states.

The partial Bell-state analysis of polarisation entanglement exploits the statistics of two qubits at a beamsplitter. The basic principle of that Bell-state analyser rests on the observation that of the four Bell states (11.3)–(11.6) only one state is antisymmetric under exchange of the two particles. This is the $|\Psi^-\rangle_{12}$ state (11.4), which clearly changes sign upon exchange of labels 1 and 2. The other three states are symmetric. We thus observe that the qubit obeys fermionic symmetry in the case of $|\Psi^-\rangle_{12}$ and bosonic symmetry in case of the other three states. Thus far we have not taken into account that the photons carrying the qubits are bosons. In fact, the four Bell states could equally well be those of fermions [14]. This is because the states written in (11.3)–(11.6) are not the complete states of the particles but describe only the internal (two-level) state of the particles. The total state can be obtained by adding the spatial state of the particles which could also be symmetric or antisymmetric. Then, in the case of bosons (e.g. photons), the spatial part of the wave function has to be antisymmetric for the $|\Psi^-\rangle_{12}$ state and symmetric for the other three, while for fermions this has to be just the reverse.

Consider now the case of two photons incident symmetrically onto a beamsplitter, i.e., one entering from each input mode $|a\rangle$ and $|b\rangle$, the possible external (spatial) states are

$$|\Psi_A\rangle_{12} = \frac{1}{\sqrt{2}}(|a\rangle_1|b\rangle_2 - |b\rangle_1|a\rangle_2) \tag{11.10}$$

$$|\Psi_S\rangle_{12} = \frac{1}{\sqrt{2}}(|a\rangle_1|b\rangle_2 + |b\rangle_1|a\rangle_2), \tag{11.11}$$

where $|\Psi_A\rangle_{12}$ and $|\Psi_S\rangle_{12}$ are antisymmetric and symmetric, respectively. Because of the requirement of symmetry, the total two-photon states are

$$|\Psi^+\rangle|\Psi_S\rangle, \quad |\Psi^-\rangle|\Psi_A\rangle, \quad |\Phi^+\rangle|\Psi_S\rangle, \text{ and } |\Phi^-\rangle|\Psi_S\rangle. \tag{11.12}$$

We note that only the state antisymmetric in external variables is also antisymmetric in internal variables. It is this state which also emerges from the beamsplitter in an external antisymmetric state. This can easily be found by assuming that the beamsplitter does not influence the internal state and by applying the beamsplitter operator (Hadamard transformation) on the external (spatial) state. Using

$$H|a\rangle = \frac{1}{\sqrt{2}}(|c\rangle + |d\rangle) \tag{11.13}$$

$$H|b\rangle = \frac{1}{\sqrt{2}}(|c\rangle - |d\rangle) \tag{11.14}$$

it can now easily be seen that

$$H|\Psi_A\rangle_{12} = \frac{1}{\sqrt{2}}(|c\rangle_1|d\rangle_2 - |d\rangle_1|c\rangle_2) = |\Psi_A\rangle_{12}. \quad (11.15)$$

Therefore the spatially antisymmetric state is an eigenstate of the beamsplitter operator [15, 16]. In contrast, in all three cases of the symmetric external state $|\Psi_S\rangle$, the two photons emerge together in one of the two outputs of the beamsplitter. It is therefore evident that the state $|\Psi^-\rangle$ can be clearly discriminated from all the other states. It is the only one of the four Bell states which leads to coincidences between detectors placed on each side after a beamsplitter [17–19]. How can we then identify the other three states? It turns out that distinction between $|\Psi^+\rangle$ on the one hand and $|\Phi^+\rangle$ and $|\Phi^-\rangle$ on the other hand can be based on the fact that only in $|\Psi^+\rangle$ do the two photons have different polarisation while in the other two they have the same polarisation. Thus performing polarisation measurements and observing the photons on the same side of the beamsplitter distinguishes the state $|\Psi^+\rangle$ from the states $|\Phi^+\rangle$ and $|\Phi^-\rangle$.

3.3 EXPERIMENTAL SETUP

The experimental realisation of the quantum teleportation of a qubit presented in this section is restricted to use the $|\Psi^-\rangle_{12}$ Bell-state projection only. The unitary transformation that Bob has to perform when Alice measures photon 1 and 2 in $|\Psi^-\rangle_{12}$ is simply the identity transformation, i.e. Bob should detect a photon in the same state as photon 1.

To avoid photons 1 and 2, which are created independently, being distinguished by their arrival times at the detectors, which would eliminate the possibility of performing the Bell-state measurement, the following technique is used. Photon 2, together with its entangled partner photon 3, is produced by pulsed parametric down-conversion. The pump pulse, generated by a frequency-doubled mode-locked titanium-sapphire laser, is 200 fs long. The pulse is reflected back through the crystal (see Fig. 11.2) to create a second pair of photons, photons 1 and 4. Photon 4 is used as a trigger to indicate the presence of photon 1. Photons 1 and 2 are now located within 200 fs long pulses, which can be tuned by a variable delay such that maximal spatial overlap of the photons at the detectors is obtained. However, this does not yet guarantee indistinguishability upon detection since the entangled down-converted photons typically have a coherence length corresponding to about a 50 fs long wavepacket, which is shorter than the pulses from the pump laser. Therefore, coincidence detection of photons 1 and 2 with their partners 3 and 4 with a time resolution better than 50 fs could identify which photons were created together. To achieve indistinguishability upon detection, the photon wavepackets should be stretched to a length substantially longer than that of the pump pulse. In the experiment this was done by placing 4 nm narrow interference filters in front of the detectors. These filter out photon wavepackets with a time duration of the order of 500 fs, which yields a maximum indistinguishability of photons 1 and 2 of about 85% [20].

All the important experimental components of the teleportation setup have now been discussed. This brings us to the question of how to prove experimentally that an unknown quantum state can be teleported with the above setup? For this, one has to show that teleportation works for a set of known non-orthogonal states. The test for non-orthogonal states is necessary to demonstrate the crucial role of quantum entanglement in the teleportation scheme. [1]

3.4   EXPERIMENTAL PREDICTIONS

In the first experiment photon 1, which has encoded the initial qubit, is prepared with a linear polarisation at 45°. Teleportation should work as soon as photons 1 and 2 are detected in the $|\Psi^-\rangle_{12}$ state. This implies that if a coincidence between detectors f1 and f2 (Fig. 11.2) is recorded, i.e. photons 1 and 2 are projected onto the $|\Psi^-_{12}\rangle$ state, then photon 3 should be polarised at 45° (to within an irrelevant overall minus sign, see (11.6)). The polarisation of photon 3 is analysed by passing it through a polarising beamsplitter selecting +45° and −45° polarisation. To demonstrate teleportation, only detector d2 at the +45° output of the polarising beamsplitter should detect a photon once f1 and f2 record a coincidence detection. Detector d1 at the −45° output of the polarising beamsplitter should not detect a photon. Therefore, recording a three-fold coincidence d2f1f2 (+45° analysis) together with the absence of a three-fold coincidence d1f1f2 (−45° analysis) is a proof that the polarisation of photon 1, which represents the initial qubit, has been transferred to photon 3.

To meet the condition of indistiguishability of photons 1 and 2 (see previous subsection), the arrival time of photon 2 is varied by changing the delay between the first and second down-conversion by translating the retroflection mirror (see Fig. 11.2). Within the region of temporal overlap of photons 1 and 2 at the detectors the teleportation should occur.

Outside the region of teleportation photons 1 and 2 will each go to either f1 or to f2 independently of one another. The probability of obtaining a coincidence between f1 and f2 is therefore 50%. This is twice as high as the probability inside the region of teleportation since only the $|\Psi^-\rangle$ component of the two-photon state entering the beamsplitter will give a coincidence recording. Since photon 2 is part of an entangled state it does not have a well-defined polarisation on its own, and the joint state of photons 1 and 2 is an equal superposition of all

---

[1] The reason for this is essentially the same as the reason why non-orthogonal states are used in constructing Bell's inequality.

270 COMPLETE SCATTERING EXPERIMENTS

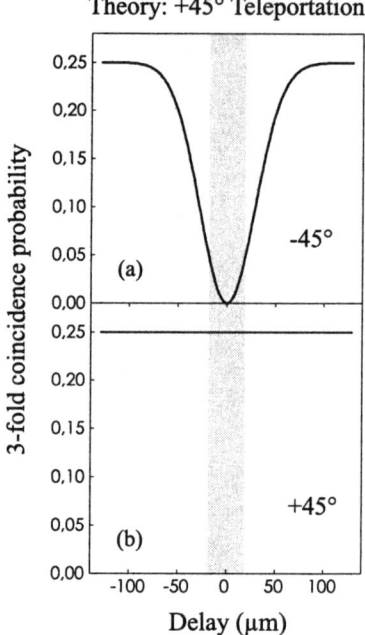

Figure 11.5 Theoretical prediction for the three-fold coincidence probability between the two Bell-state detectors (f1, f2) and one of the detectors analysing the teleported state. The signature of teleportation of a photon polarisation state at +45° is a dip to zero at zero delay in the three-fold coincidence rate with the detector analysing −45° (d1f1f2) (**a**) and a constant value for the detector analysis +45° (d2f1f2) (**b**). The shaded area indicates the region of teleportation.

four Bell states, irrespective of the state of photon 1. Photon 3 should also have no well-defined polarisation because it is entangled with photon 2. Therefore, d1 and d2 both have a 50% chance of receiving photon 3. This simple argument yields a 25% probability both for the −45° analysis (d1f1f2 coincidences) and for the +45° analysis (d2f1f2 coincidences) outside the region of teleportation.

Figure 11.5 summarises the predictions as a function of the delay. Successful teleportation of the +45° polarisation state is then characterized by a decrease to zero in the −45° analysis, see Fig. 11.5a, and by a constant value for the +45° analysis, see Fig. 11.5b. Note that the above arguments are conditional upon the detection of a trigger photon by detector p (see Fig. 11.2).

3.5   EXPERIMENTAL RESULTS

The experimental results for teleportation of photons polarised at +45° are shown in the first panel of Fig. 11.6. Figure 11.6a and 11.6b should be compared with the theoretical predictions shown in Fig. 11.5.

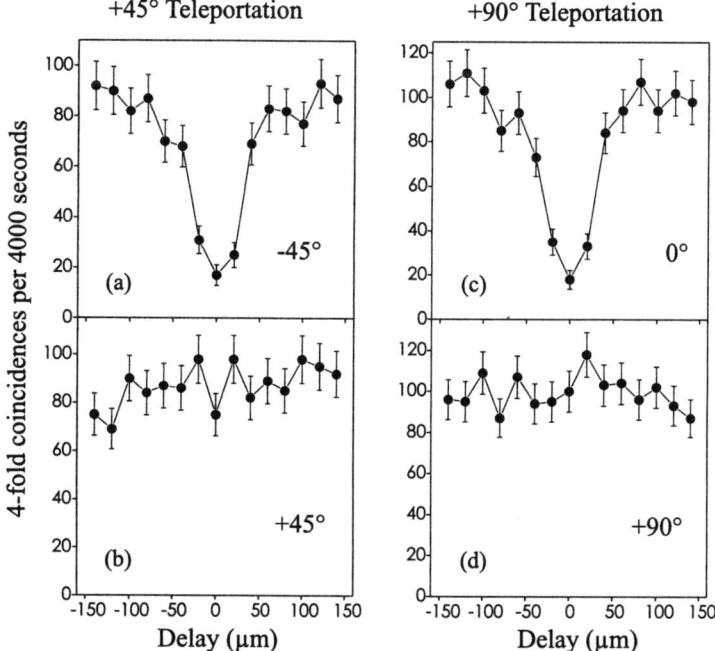

Figure 11.6 Experimental demonstration of the teleportation of qubits: Measured coincidence rates d1f1f2 (−45°) and d2f1f2 (+45°) in the case where the photon state to be teleported is polarised at +45° (**a**) and (**b**) or at +90° (**c**) and (**d**), and conditional upon the detection of the trigger photon by detector p. The four-fold coincidence rates are plotted as a function of the delay (in μm) between the arrival of photons 1 and 2 at Alice's beamsplitter (see Fig. 11.2). These data, in cunjunction with with Fig. 11.5, confirm teleportation for an arbitrary qubit state.

The strong decrease in the −45° analysis, and the constant signal for the +45° analysis, indicate that photon 3 is polarised along the direction of photon 1, consistent with the quantum teleportation protocol. Note again that a four-fold coincidence detection has been used where the fourth photon is a trigger that indicates the presence of photon 1.

To rule out any classical explanation for the experimental results, a four-fold coincidence measurement for the case of teleportation of the +90° polarisation states, that is, for a state non-orthogonal to the +45° state, has been performed. The experimental results are shown in Fig. 11.6c and 11.6d. Visibilities of 70% ± 3% are obtained for the dips in the orthogonal polarisation states.

From Fig. 11.6 one can directly obtain the measured fidelity of teleportation of a qubit encoded in the polarisation of a single-photon state. The fidelity is defined as the overlap of the input qubit with the teleported qubit and is plotted in Fig. 11.7. In the experiment, the detection of the teleported photons played the double role of filtering out the experimental runs in which there is a single

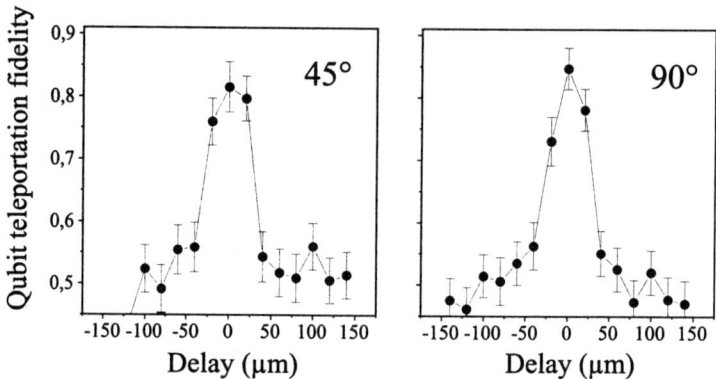

Figure 11.7 Fidelity of teleportation of a qubit encoded in the polarisation of a single-photon state: The overlap of the input qubit with the teleported qubit has been determined via a four-fold coincidence technique to be as high as 80%.

input qubit present and of measuring the fidelity of the teleportation procedure. With respect to the filtering, note that two detection events at Alice's Bell-state analyser could have been due to two pairs of photons both created during the return passage of the pump pulse. Then no photon will be observed by Bob [21], but two photons will travel towards detector p. This situation can be identified and therefore eliminated by using a detector p that can discriminate between a one-photon and a two-photon impact [22].

Whether or not such a modified detection is used, the measured fidelity will be the same [23] and is primarily determined by the degree of indistinguishability of the photons detected in Alice's Bell-state analyser. The amount of indistinguishability is directly related to the ratio of the bandwidth of the pump pulse and the interference filters. The larger this ratio the higher the fidelity but the lower the countrates.

3.6 TELEPORTATION OF ENTANGLEMENT

Instead of using the fourth photon in the experiment described above as a mere trigger to indicate that photon 1 is underway, one can explore the fact that photon 1 and 4 can also be produced in an entangled state, say in the $|\Psi^-\rangle_{14}$ state, as illustrated in Fig. 11.8.

The state of photon 1 is therefore completely undetermined and all the information is stored in joint properties of photons 1 and 4. If photon 1 is now subjected to quantum teleportation, photon 3 obtains the properties of photon 1 and therefore becomes entangled with photon 4 (see Fig. 11.8). Interestingly, photon 4 and photon 3 originate from different sources and never interacted directly with one another, yet they form an entangled pair after the quantum teleportation procedure. The experimental verification of this process of trans-

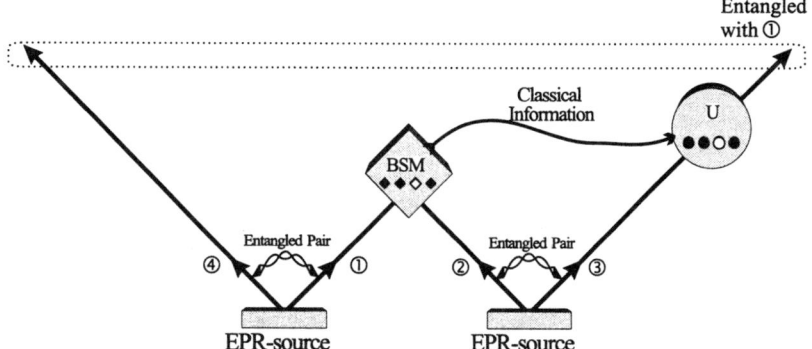

Figure 11.8 Principle of entanglement swapping: Two EPR sources produce two pairs of entangled photons, pair 1-4 and pair 2-3. Two photons, one from each pair (photons 1 and 2) are subjected to a Bell-state measurement(BSM). This results in projecting the other two outgoing photons 3 and 4 onto an entangled state.

ferring entanglement, known as entanglement swapping is presented in Ref. [9] and several possible applications can be found in Refs. [8, 10].

4. CONCLUDING REMARKS AND PROSPECTS

Pairs of polarisation entangled photons and two-photon interferometric methods have been used to transfer one qubit encoded in the polarisation state of one photon onto another one. Teleportation has also been addressed in other optical systems [24–28]. However, quantum teleportation is by no means restricted to optical experiments. In addition to pairs of entangled photons one can employ entangled atoms [29], and one can, in principle, entangle photons with atoms or phonons with ions, and so on. Then teleportation would allow the transfer of the state of, for example, fast-decohering, short-lived particles onto some more stable systems. This opens up the possibility of quantum memories, where the information of incoming photons is stored on trapped ions/atoms, carefully shielded from the environment.

Furthermore, with entanglement purification [30], a scheme for improving the quality of entanglement when it has been degraded by decoherence during storage or transmission of the particles over noisy channels, it becomes possible to send the quantum state of a particle to some place, even if the available quantum channels are of limited quality and thus sending the particle itself might destroy the fragile quantum state. If the distance over which one wants to send the quantum state through a noisy quantum channel becomes too long, the fidelity of transmission becomes too low for the application of the standard purification method. In this situation the quantum repeater method allows one to divide the quantum channel into shorter segments that are purified separately and then connected by entanglement swapping [31]. The feasibility of preserving quantum states in a hostile environment will have great advantages in the realm of quantum communication and quantum computation.

References

[1] A. Einstein, B. Podolsky, and N. Rosen, Phys. Rev. **47**, 777 (1935).

[2] N. Bohr, in *Albert Einstein: Philosopher-Scientist*, edited by P. A. Schilpp (The Library of Living Philosophers, Evanston, 1949), p. 200.

[3] J. S. Bell, Physics **1**, 195 (1964), reprinted in J.S. Bell, *Speakable and Unspeakable in Quantum Mechanics* (Cambridge U.P., Cambridge, 1987).

[4] A. Aspect, J. Dalibard, and G. Roger, Phys. Rev. Lett. **47**, 1804 (1982).

[5] A. Zeilinger, Special Issue of *Physics World* on Quantum Information **11**, 35 (1998), see also Tittle et al. page 41, Deutsch and Ekert, page 47, and DiVincenzo and Terhal, page 53.

[6] C. H. Bennett *et al.*, Phys. Rev. Lett. **70**, 1895 (1993).

[7] W. K. Wootters and W. H. Zurek, Nature **299**, 802 (1982).

[8] M. Zukowski, A. Zeilinger, M. A. Horne, and A. Ekert, Phys. Rev. Lett. **71**, 4287 (1993).

[9] J.-W. Pan, D. Bouwmeester, H. Weinfurter, and A. Zeilinger, Phys. Rev. Lett. **80**, 3891 (1998).

[10] S. Bose, V. Vedral, and P. L. Knight, Phys. Rev. A **57**, 822 (1998).

[11] D. Bouwmeester *et al.*, Nature **390**, 575 (1997).

[12] P. G. Kwiat, H. Weinfurter, T. Herzog, and A. Zeilinger, Phys. Rev. Lett. **74**, 4763 (1995).

[13] K. Mattle, H. Weinfurter, P. G. Kwiat, and A. Zeilinger, Phys. Rev. Lett. **76**, 4656 (1996).

[14] A. Zeilinger, Physica Scripta **T76**, 203 (1998).

[15] R. Loudon, in *Coherence and Quantum Optics VI*, edited by J. H. Eberly and L. Mandel (Plenum, New York, 1990), p. 703.

[16] A. Zeilinger, H. J. Bernstein, and M. A. Horne, J. Mod. Optics **41**, 2375 (1994).

[17] H. Weinfurter, Europhys. Lett. **25**, 559 (1994).

[18] S. L. Braunstein and A. Mann, Phys. Rev. A **51**, R1727 (1995).

[19] M. Michler, K. Mattle, H. Weinfurter, and A. Zeilinger, Phys. Rev. A **53**, R1209 (1996).

[20] M. Zukowski, A. Zeilinger, and H. Weinfurter, in *Annals of the New York Academy of Sciences*, edited by Greenberger and Zeilinger (Academy of Sciences, New York, 1995), Vol. 755, p. 91.

[21] Nature **394**, 840 (1998), comment by S.L. Braunstein and H.J. Kimble, and Reply by D. Bouwmeester, J-W. Pan, M. Daniell, H. Weinfurter, M. Zukowski, and A. Zeilinger.

[22] J. Kim, S. Takeuchi, Y. Yamamoto, and H. H. Hogue, Appl. Phys. Lett. **74**, 902 (1999).

[23] D. Bouwmeester, J.-W. Pan, H. Weinfurter, and A. Zeilinger, J. Mod. Opt. **47**, 279 (2000).

[24] S. Popescu, , lANL E-print quant-ph 9501020.

[25] D. Boschi *et al.*, Phys. Rev. Lett. **80**, 1121 (1998).

[26] L. Vaidman, Phys. Rev. A **49**, 1473 (1994).

[27] S. L. Braunstein and H. J. Kimble, Phys. Rev. Lett. **80**, 869 (1998).

[28] A. Furusawa *et al.*, Science **Oct23**, 706 (1998).

[29] E. Hagley *et al.*, Phys. Rev. Lett. **79**, 1 (1997).

[30] C. H. Bennett *et al.*, Phys. Rev. Lett. **76**, 722 (1996).

[31] H.-J. Briegel, W. Dür, J. I. Cirac, and P. Zoller, Phys. Rev. Lett. **81**, 5932 (1998).

Index

Adsorbed molecules, 173, 184
Aligned molecules, 171, 174
Alignment, 15, 32, 37
Amplitude, 14, 32
Angle Resolved Photoelectron-Photoion
 Coincidence technique (AR-PEPICO), 180
Angular correlation pattern, 155–165
 complementary, 155, 157–158, 160, 162, 164
Angular distribution
 of ions, 179, 184
 of photoelectrons, 168, 170, 177
  for aligned molecules, 175
  for oriented (fixed–in–space) molecules, 172, 177
  for polarized atoms, 169
Anisotropy parameter, 143
Antisymmetric states, 267
Asymptotic phase difference, 141
Auger decay
 "almost" complete experiment, 111–112, 115, 117, 123–125
 amplitude, 111–117, 121–122, 124–125
 angular distribution, 111–112, 114, 116–118, 120–122
 complete experiment, 111–113, 123–124
 relative phase, 112–117, 122, 124–125
 spin-polarization, 112, 114–115
B.C.H.S.H. inequalities, 198
Beamsplitter, 267
Bell state
 measurement, 263
Bell states, 263
Bell's
 discovery, 197
 inequalities, 189
Bell-state
 analyser, 264
 analysis, 266
 measurement, 268
Born series expansion, 130

Breit–Teller hypothesis, 243
Bremsstrahlung, 36
Calcium, 23, 25
CDAD, 169, 175
Charge clouds
 examples of, 69, 75, 83, 86
 parameterization of, 66
Chiral effects
 in unpolarized cross section, 100, 96, 98, 102, 107–108
Chiral molecules
 cross sections, 100, 95, 99
 Ingold–Prelog convention, 99
 spin asymmetries, 101–104
 steric factors, 96
Circular dichroism, 26
CO molecule, 180
Coherence analysis
 of the two–photon radiation, 223
Coherence, 7, 14
Coherent, 34
Coincidence counting, 209
Coincidence detection, 268
Coincidence, 16, 26, 32
Coincidences (photoelectron-Auger electron), 156–158, 165
Compensator crystals, 266
Complete experiments, 61, 64
Complete photoionization experiments, 167
 with atoms, 168
 with molecules, 170, 174, 181
Conflict
 with quantum mechanics, 198
Continuum mixing, 141
Convergent close coupling, 19, 132
Correlated emission spectroscopy
 of metastable hydrogen, 253
Correlation, 7, 14
Depolarization, 146
Dichroism

circular in the angular distribution (CDAD), 169
linear in the angular distribution (LDAD), 169
magnetic in the Angular Distribution (MDAD), 169
Dipole
  electric-dipole approximation, 167
  matrix element, 155–160, 162, 164–165, 168, 178
  selection rules, 168–169
Down conversion
  parametric, 216
Dynamical spin polarisation, 146
Einstein's causality, 201, 213
Einstein–Podolski–Rosen
  (EPR) source, 264
Einstein–Podolsky–Rosen Paradox, 38
Einstein–Podolsky–Rosen, 236
  (EPR) Gedankenexperiment, 190
  experiments, 223
Electron exchange, 23, 33
Electron–atom collision, 129
Electron-photon coincidence, 129
Energy distribution
  of the two–photon continuum of hydrogen–like atomic systems, 228
Entangled atoms, 273
Entangled state, 263
Entanglement purification, 273
Entanglement swapping, 273
EPR experiments, 257
Excitation, 16
Fermionic symmetry, 267
Fidelity, 271–272
Fine structure, 2
First Born approximation, 132
Five-parameter model, 156–157, 159–160, 164
Four-parameter model, 162, 164
Fourier spectral analysis, 247
Garuccio–Selleri enhancement effects, 239
Grazing incidence, 130
Guiding magnet field, 144
Hadamard transformation, 267
Hanle effect, 7
Heavy particle collision, 29
Hexapole magnet, 144
Hydrogen, 21
Hyperfine interaction, 146
Hyperfine structure, 3
Indistiguishability, 269
Inner shell ionisation, 35
Interchannel coupling, 149
Intermediate coupling, 148
Lamb shift, 2, 11, 224
LDAD, 169
Level anticrossing, 7
Level crossing, 7

Linear Magnetic Dichroism in the Angular Distribution, 144
LMDAD, 144, 147
Locality
  assumption, 201
  condition, 200
LS-approximation, 142
Magnetic linear dichroism, 27
MDAD, 169
Mercury, 24
Metal vapor, 144
Metastable hydrogen, 40
Multiple scattering (MS) method, 180
N_2 molecule, 184
NO molecule, 174
No-cloning theorem, 264
Non-collinear type-II phase matching, 264
One–channel polarizers, 210
Orientation of molecules, 100, 95
Orientation, 15, 32
Orsay experiments, 208
Outer shell excitation, 29
Oven, 144
Oxygen, 27
Parametric down-conversion, 264
Partial wave analysis, 141
Perfect scattering experiment, 14
Phase shift, 168, 178
Phase, 14, 32
Photoionization, 25, 27
  of closed shell atoms, 168
  of fixed–in–space molecules, 172–173
  of open shell atoms, 168
Polarization
  entangled photons, 264
Polarized electrons, 93, 101, 103
Polarized
  atoms, 141, 169
  thallium atoms, 141
Potassium, 22
Quadratic Stark effect, 12
Quantum mechanically complete photoionisation experiments, 141
Quantum memories, 273
Quantum
  computation, 261
  entanglement, 261
  measurement, 262
  teleportation, 261
Radial coupling, 31
Radial integrals, 143
Random phase approximation
  (RPA), 184
  relaxed, 159–164
  unrelaxed, 159, 161–164
Ratio of the matrix elements, 141

Relaxed Core Hartree-Fock (RCHF) approximation, 181
Resonance-Enchanced Multiphoton Ionization (REMPI), 174
Resonances, 5
Rolling ball, 130
Rotational coupling, 29
RPAE-method, 148
Satellite lines, 148
Scattering amplitudes
   for $S \to D$ excitation, 70
   in collision frame, 63
   in natural frame, 64
   relationship to charge cloud, 66
   spin–dependent, 77, 83
Second order distorted wave Born calculation, 132
Shape resonance, 178, 184
Sodium, 22
Spin asymmetries, 101–104
Spin polarization
   of photoelectrons, 168
Spin-orbit interaction, 168
State multipoles, 17, 171
Stereo selectivity, 99
Steric factors, 95–96
Stirling two–photon apparatus, 231
Stokes parameters, 17, 63, 171
Strontium, 23
Superelastic scattering experiments, 129
Symmetric states, 267
Teleported qubit, 271
Three–polarizer experiments, 240
Three-parameter model, 155, 158–162, 164–165
Threshold polarization, 4
Timing experiments, 218
Titanium-sapphire laser, 268
Tl-$5d$ multiplet, 141
Total asymptotic phase difference, 143
Transfer of angular momentum, 129
Two–channel
   analyzers, 211
   polarizers, 213
Two–photon correlation, 223
Two–photon decay, 38
Two–photon radiation
   from the decay of the metastable $2^2S_{1/2}$ atomic hydrogen, 223
Vector polarization, 17
Zeeman sublevels, 7

Series Publications

Below is a chronological listing of all the published volumes in the *Physics of Atoms and Molecules* series.

ELECTRON AND PHOTON INTERACTIONS WITH ATOMS
Edited by H. Kleinpoppen and M. R. C. McDowell

ATOM–MOLECULE COLLISION THEORY: A Guide for the Experimentalist
Edited by Richard B. Bernstein

COHERENCE AND CORRELATION IN ATOMIC COLLISIONS
Edited by H. Kleinpoppen and J. F. Williams

VARIATIONAL METHODS IN ELECTRON–ATOM SCATTERING THEORY
R. K. Nesbet

DENSITY MATRIX THEORY AND APPLICATIONS
Karl Blum

INNER-SHELL AND X-RAYS PHYSICS OF ATOMS AND SOLIDS
Edited by Derek J. Fabian, Hans Kleinpoppen, and Lewis M. Watson

INTRODUCTION TO THE THEORY OF LASER–ATOM INTERACTIONS
Marvin H. Mittleman

ATOMS IN ASTROPHYSICS
Edited by P. G. Burke, W. B. Eissner, D. G. Hummer, and I. C. Percival

ELECTRON–ATOM AND ELECTRON–MOLECULE COLLISIONS
Edited by Juergen Hinze

ELECTRON–MOLECULE COLLISIONS
Edited by Isao Shimamura and Kazuo Takayanagi

ISOTOPE SHIFTS IN ATOMIC SPECTRA
W. H. King

AUTOIONIZATION: Recent Developments and Applications
Edited by Aaron Temkin

ATOMIC INNER-SHELL PHYSICS
Edited by Bernd Crasemann

COLLISIONS OF ELECTRONS WITH ATOMS AND MOLECULES
G. F. Drukarev

THEORY OF MULTIPHOTON PROCESSES
Farhad H. M. Faisal

PROGRESS IN ATOMIC SPECTROSCOPY, Parts A, B, C, and D
Edited by W. Hanle, H. Kleinpoppen, and H. J. Beyer

RECENT STUDIES IN ATOMIC AND MOLECULAR PROCESSES
Edited by Arthur E. Kingston

QUANTUM MECHANICS VERSUS LOCAL REALISM: The Einstein-Podolsky-Rosen Paradox
Edited by Franco Selleri

ZERO-RANGE POTENTIALS AND THEIR APPLICATIONS IN ATOMIC PHYSICS
Yu. N. Demkov and V. N. Ostrovskii

COHERENCE IN ATOMIC COLLISION PHYSICS
Edited by H. J. Beyer, K. Blum, and R. Hippler

ELECTRON–MOLECULE SCATTERING AND PHOTOIONIZATION
Edited by P. G. Burke and J. B. West

ATOMIC SPECTRA AND COLLISIONS IN EXTERNAL FIELDS
Edited by K. T. Taylor, M. H. Nayfeh, and C. W. Clark

ATOMIC PHOTOEFFECT
M. Ya. Amusia

MOLECULAR PROCESSES IN SPACE
Edited by Tsutomu Watanabe, Isao Shimamura, Mikio Shimizu, and Yukikazu Itikawa

THE HANLE EFFECT AND LEVEL CROSSING SPECTROSCOPY
Edited by Giovanni Moruzzi and Franco Strumia

ATOMS AND LIGHT: INTERACTIONS
John N. Dodd

POLARIZATION BREMSSTRAHLUNG
Edited by V. N. Tsytovich and I. M. Ojringel

INTRODUCTION TO THE THEORY OF LASER–ATOM INTERACTIONS (Second Edition)
Marvin H. Mittleman

ELECTRON COLLISIONS WITH MOLECULES, CLUSTERS, AND SURFACES
Edited by H. Ehrhardt and L. A. Morgan

THEORY OF ELECTRON–ATOM COLLISIONS, Part 1: Potential Scattering
Philip G. Burke and Charles J. Joachain

POLARIZED ELECTRON/POLARIZED PHOTON PHYSICS
Edited by H. Kleinpoppen and W. R. Newell

INTRODUCTION TO THE THEORY OF X-RAY AND ELECTRONIC SPECTRA OF FREE ATOMS
Romas Karazija

VUV AND SOFT X-RAY PHOTOIONIZATION
Edited by Uwe Becker and David A. Shirley

DENSITY MATRIX THEORY AND APPLICATIONS (Second Edition)
Karl Blum

SELECTED TOPICS ON ELECTRON PHYSICS
Edited by D. Murray Campbell and Hans Kleinpoppen

PHOTON AND ELECTRON COLLISIONS WITH ATOMS AND MOLECULES
Edited by Philip G. Burke and Charles J. Joachain

COINCIDENCE STUDIES OF ELECTRON AND PHOTON IMPACT IONIZATION
Edited by Colm T. Whelan and H. R. J. Walters

PRACTICAL SPECTROSCOPY OF HIGH-FREQUENCY DISCHARGES
Sergei A. Kazantsev, Vyacheslav I. Khutorshchikov, Günter H. Guthöhrlein, and Laurentius Windholz

IMPACT SPECTROPOLARIMETRIC SENSING
S. A. Kazantsev, A. G. Petrashen, and N. M. Firstova

NEW DIRECTIONS IN ATOMIC PHYSICS
Edited by Colm T. Whelan, R. M. Dreizler, J. H. Macek, and H. R. J. Walters

COMPLETE SCATTERING EXPERIMENTS
Edited by Uwe Becker and Albert Crowe